资源化学丛书

非均相催化水介质清洁有机合成

朱凤霞　张　昉　李和兴　著

科学出版社

北　京

内容简介

绿色化学关乎生态平衡、人类健康和社会经济可持续发展，而水介质清洁有机合成是其重要组成部分，涉及材料、化学、化工、环境、生物医药等学科。本书着重介绍非均相催化水介质清洁有机合成的前沿动态和研究重点，包括清洁有机合成的概念及基本原理，传统催化剂、金属催化剂、负载型有机金属催化剂和负载型多功能催化剂的制备及其在水介质有机合成中的应用。本书中列举了大量已报道的最新成果，力求充分反映催化水介质清洁有机合成的研究动态，为发展绿色理论和推动绿色化工生产提供指导。

本书内容新颖，参考资料详实，可供化学、化工、材料、生物医药及环境等相关领域的研究人员参考使用，特别适合高等院校和科研院所相关专业的师生阅读参考。

图书在版编目(CIP)数据

非均相催化水介质清洁有机合成/朱凤霞，张昉，李和兴著. —北京：科学出版社，2017.6
(资源化学丛书)
ISBN 978-7-03-053483-5

Ⅰ.①非… Ⅱ.①朱… ②张… ③李… Ⅲ.①催化–应用–有机合成–研究 Ⅳ.①O621.3

中国版本图书馆 CIP 数据核字(2017)第 137863 号

责任编辑：张　析　宁　倩／责任校对：何艳萍
责任印制：张　伟／封面设计：东方人华

科学出版社 出版
北京东黄城根北街 16 号
邮政编码：100717
http://www.sciencep.com

北京九州迅驰传媒文化有限公司 印刷
科学出版社发行　各地新华书店经销
*

2017 年 6 月第　一　版　　开本：720×1000 B5
2019 年 1 月第二次印刷　　印张：16 1/4　插页：2
字数：328 000
定价：108.00 元

(如有印装质量问题，我社负责调换)

前　言

建设生态文明是关系人民福祉和民族未来的大计，是实现中华民族伟大复兴中国梦的重要步骤。要实现生态环境保护，必须树立绿色发展理念。化学工业是国民经济的支柱产业，不可或缺，但化工生产又是污染重点，因此，以源头污染控制为目标的绿色化学应运而生，由水代替有机溶剂作为反应介质进行有机合成是绿色化学的重要组成部分。

众所周知，传统的有机合成一般在有机溶剂中进行，而大量有机溶剂的挥发和排放构成严重的环境污染。水是最清洁的溶剂，但受溶解性限制，水介质有机反应难以工业推广。生命体中存在大量水介质生化反应，关键原因是酶的作用。因此，实现水介质中清洁有机合成的核心问题是开发高效催化剂。

本书隶属于"资源化学丛书"系列，是作者结合自己的研究工作，并在查阅大量文献的基础上完成。全书共分五章，第一章总体介绍了绿色化学的基本概念、清洁有机合成原理及技术，第二章至第五章按照催化剂的种类分别介绍了传统和非均相催化水介质有机合成，其中非均相催化水介质包括金属、负载型有机金属及负载型多功能催化水介质。本书可作为从事有机合成的科研人员和广大高校师生的教材及参考读物，也可供化工生产管理人员和技术人员阅读。

参与本书编写的有淮阴师范学院朱凤霞教授以及上海师范大学张昉教授和李和兴教授。感谢华南理工大学黄建林博士和淮阴师范学院周建峰教授对本书编写提供的帮助，感谢科学出版社的大力支持，同时，感谢书中所引文献的所有作者。

由于作者水平有限，书中难免存在疏漏之处，敬请同行和读者予以批评指正。

李和兴
2017 年 3 月

目　　录

前言
第一章　清洁有机合成 1
　第一节　绿色化学概论 1
　第二节　清洁有机合成基本原理 7
　第三节　清洁有机合成技术 10
　参考文献 17
第二章　传统催化水介质有机合成 20
　第一节　酸催化剂 20
　第二节　碱催化剂 44
　第三节　有机金属催化剂 60
　第四节　有机小分子不对称催化剂 105
　参考文献 128
第三章　金属催化水介质有机合成 151
　第一节　非负载型金属催化剂 151
　第二节　负载型金属催化剂 158
　第三节　非晶态合金催化剂 194
　参考文献 200
第四章　负载型有机金属催化水介质有机合成 214
　第一节　有序介孔有机金属催化剂 214
　第二节　负载型磁性有机金属催化剂 221
　第三节　负载型手性催化剂 224
　参考文献 236
第五章　负载型多功能催化水介质有机合成 241
　第一节　负载型双有机金属催化剂 242
　第二节　负载型酸-碱催化剂 243
　第三节　负载型金属-酸双功能催化剂 249
　参考文献 252

彩图

第一章 清洁有机合成

化学在给人类社会发展做出重大贡献的同时，也使人类面临严重的环境污染和资源危机。随着人口增加，资源的消耗也随之增加，淡水、矿产及人均耕地等资源逐渐减少，从而使资源与人口的矛盾越来越明显；环境的严重污染以及生态环境的破坏日益加剧，这不仅影响生态平衡，而且严重危害人类健康和社会经济的可持续发展，因此受到各国政府高度重视。2006 年我国颁布的《国家中长期科学和技术发展规划纲要(2006—2020 年)》中指出，国家发展战略之一是发展循环经济、节能减排、清洁生产和生态文明建设。化学工业是国家经济发展的主要支柱产业之一，与其他产业也密切相关，化学在给人类创造丰富的物质文明的同时，也给环境带来了严重污染。当前全球环境问题，如臭氧层破坏、大气污染、气候变暖、海洋污染、淡水资源流失和污染、生物多样性减少、固体废物污染、土地退化和沙漠化、森林锐减、酸雨蔓延均直接或间接与化学工业污染有关，化学工业正面临着可持续发展的严重挑战。随着化学工业的迅速发展，仅通过末端治理已无法阻止化学污染，因此，以污染源头控制为目标的绿色化学是化学工业未来发展的必由之路[1,2]。

第一节 绿色化学概论

绿色化学的提出源于人们对于化学工业对环境影响认识的加深，进而关注环境问题。1984 年美国国家环境保护局(U.S. Environmental Protection Agency，EPA)率先提出"废物最小化"，1990 年美国联邦政府颁布了《污染防止法案》[3]，将污染防止确立为美国的国策。"污染防止"就是不再产生废物，不再有废物处理的问题。绿色化学正是实现污染防止的基础和重要工具。1995 年美国前副总统 Gore 宣布了美国的环境技术战略目标，即到 2020 年，将废弃物减少 40%~50%，生产中消耗的原材料减少 20%~25%。1996 年美国设立了"总统绿色化学挑战奖"，这是对学校或工业界已经或将要通过绿色化学显著提高人类健康和改善环境的先驱者的国家级奖励，也是世界上第一个由国家政府出台的对绿色化学实施的奖励。美国政府的这些行为对绿色化学的蓬勃发展起到很大的促进作用。此外，1993 年日本又实施了"新阳光计划"，在环境技术的开发与研究领域，确定了环境无害制造技术、减少环境污染技术和 CO_2 固定与利用技术等绿色化学的内容。总之，绿色化学的研究已成为全球各界重要的研究方向。

一、绿色化学的含义

1991年美国化学会(the American Chemical Society, ACS)提出绿色化学(green chemistry)概念，这也成为美国国家环境保护局的口号。所谓绿色化学，又称"环境无害化学"(environmentally benign chemistry)、"环境友好化学"(environmentally friendly chemistry)和"清洁化学"(clean chemistry)，是近二十多年来全球新兴起的一门交叉学科。绿色化学主要提倡利用与化学有关的原理及技术，从源头阻止一些对人类的生活环境和健康产生不利影响的药品、催化剂和有机溶剂以及产生的副产品，尽量使参加反应的所有原子全部转化到目标产物中，不产生无关产物。绿色化学的主要目标是将现有的化学和化工生产技术路线从"先污染、后治理"改变为"从源头上根除污染"[4]；其最终目的是不再使用和产生影响人类健康及生态环境的物质，不需再处理大量的废弃物。绿色化学从杜绝污染和减少资源的浪费两部分来重新看待和调整目前的整套化学工业体系。从科学的角度来看，绿色化学是在传统化学理念基础上的思维创新；从环保的角度来看，绿色化学是从根源上杜绝污染；从经济发展的角度来看，它具有资源利用合理化、生产成本节约化的特点，从而保证了经济的可持续发展。因此，绿色化学追求人类与自然和谐，其对实现人类社会和生态环境的可持续发展具有重要的战略意义[5, 6]。

提出绿色化学的目的就是探寻化工生产过程和工艺的最优化、绿色化和无害化，实现效益最大化，尤其是要利用安全原料，提高利用率，减少浪费，使生产过程实现无污染，不排放有害物质。这些指标主要是通过以下几方面(五个"R"：reduction, reuse, recycling, regeneration, rejection)来达到最佳效果[7]：

Reduction（减少），主要是指在生产过程中减少"三废"(废气、废水、固体废弃物)排放；reuse（重复使用），要求在化学工业过程中使用的催化剂、载体等能重复循环使用，这也是降低成本和减少"三废"的需要；recycling（回收），即回收未反应的原料、副产物、助溶剂、催化剂、稳定剂等反应试剂；regeneration（再生），即各化工生产企业要尽可能变废为宝，它是节约能源、资源、减少污染的有效途径；rejection（拒用），是指在化学过程中拒绝使用一些无法回收、替代、再生和重复使用，有毒及污染明显的原料，这也是防止污染的最根本方法[7]。

二、绿色化学的研究内容

绿色化学与传统化学的不同之处在于前者更多地考虑可持续发展，促进人与自然协调发展，研究和开发能减少或消除有害物质的使用与生产的环境友好化学品，从源头上防止污染[8]。因此，绿色化学的研究内容主要包括以下几个方面：

(1) 原子经济性反应和零排放。利用清洁合成工艺和技术，减少废物排放，目标是"零排放"。

(2) 反应原料绿色化。提高原材料和能源的利用率，使用可再生资源。

(3) 化学反应的绿色化，包括催化剂的绿色化，溶剂的绿色化。寻找对环境无害的绿色催化剂和环境友好型的绿色溶剂。

(4) 产品的绿色化。即产品不应该对环境产生危害，设计可降解产品。

(5) 技术和工艺过程绿色化，开发防止生产事故的安全工艺。

(6) 绿色化学教育。用绿色化学变革社会生活，促进社会经济和环境的协调发展。

绿色化学作为一门前沿学科，其研究领域在不断深化。绿色化学以"原子经济性"为基本原则，以研究新的反应体系(包括新的合成方法和路线)为核心问题，寻求新的化学原料(包括生物资源)，探索新的反应条件(如超临界流体等环境友好介质)，设计和开发对社会安全、对环境友好、对人体健康的绿色产品[1, 8]。

三、绿色化学的十二条原则及附加原则

1998年，美国化学家 Anastas 和 Warner 在 *Green Chemistry: Theory and Practice* 一书中提出了一些评价化学反应过程和化工生产流程是否符合绿色化学的原则，共十二条。这十二条原则是[6, 9, 10]：

(1) 防止污染优于污染治理。防止废物的产生优于在其生成后再处理。

(2) 原子经济性。合成方法应具有"原子经济性"，即尽量使参加反应的原子都进入最终产物。

(3) 绿色化学合成。在合成中尽量不使用和不产生对人类健康和环境有毒、有害的物质。

(4) 设计安全化学品。设计具有高使用功效和低环境毒性的化学品。

(5) 采用安全的溶剂和助剂。尽量不使用溶剂等辅助物质，必须使用时应选用无毒、无害的。

(6) 合理使用和节省能源。生产过程应该在温和的温度和压力下进行，而且能耗应最低。

(7) 利用可再生资源合成化学品。尽量采用可再生的原料，特别是用生物质代替矿物燃料。

(8) 减少不必要的衍生化步骤，尽量减少副产品。

(9) 采用高选择性的催化剂。

(10) 设计可降解化学品。化学品在使用完后应能够降解成无毒、无害的物质，并且能进入自然生态循环。

(11) 进行预防污染的现场实时分析。开发实时分析技术，以便监控有毒、有害物质的生成。

(12) 使用安全工艺。选择合适的参加化学过程的物质及生产工艺，尽量减少发生意外事故的风险。

绿色化学十二条原则目前被国际化学界所公认，它不仅是绿色化学的基石，也指明了未来绿色化学的发展方向。

针对工艺技术放大、应用和实施的潜在能力，N. Winterton 提出了绿色化学十二条附加原则：

(1) 鉴定并量化副产品。

(2) 报告转化率、选择性和产率。

(3) 对工艺过程建立完全的物料平衡计算。

(4) 定量核算生产过程中催化剂和溶剂的损失。

(5) 充分研究基本的热化学规律，特别是放热规律，以保证安全。

(6) 预测热量和质量转移的极限。

(7) 与化学或化工工程人员协作。

(8) 考虑整个工艺对化学品选择性的影响。

(9) 协作开发和实施可持续发展措施。

(10) 量化并尽量减少功效的使用。

(11) 识别安全和废弃物最小化不兼容的地方。

(12) 对试验或工艺过程向环境中排放的废物要监视、呈报，并尽可能地使之最小化。

这些附加原则既是对以上绿色化学基本原则的补充，又可为深化绿色化学研究以及科学评价绿色化学提供参考依据。以下举例加以说明。

1. 化学合成的"原子经济性"

有机合成中最常见的反应原子经济性有：重排反应(包括 Beckmann 重排、Claisen 重排、Fries 重排、Wolff 重排等)、加成反应、取代反应(烷基化、芳基化、酰基化反应等)、消除反应(包括脱氢、脱卤素、脱卤化氢、脱水、脱醇、脱氨以及一些降解反应)[10]。在传统化学反应中，一直以产率的大小为标准评价一个合成过程的效率高低，往往忽视环境友好和节能。为此，1991 年美国著名有机化学家 Trost 首次提出"原子经济性"概念[10]，认为高效有机合成应最大限度地利用原料分子，使之完全结合到目标分子中，既不产生副产物或废弃物，也不产生其他环境污染。合成效率是关键，包括高选择性(化学选择性、区域选择性、顺反选择性、非对映选择性和对映选择性)和高转化率。例如，对于一般的有机合成反应 A+B$\longrightarrow$C+D，其中 D 是副产物，可能对环境有害，从原子利用的角度来看也是一种浪费。如果开发一种新的工艺,以 E 和 F 为原料,也可以全部生成产物 C(E+F⟶C),反应分子全部得到利用,则属于理想的原子经济性反应。原子经济性可以用原子利用率来衡量:

$$原子利用率 = \frac{目标产物的相对分子质量}{化学反应计量式中反应物的相对分子质量总和}$$

提高原子利用率的根本途径是研制高效催化剂、开发新型高效化学反应途径和工艺。

2. 环境因子与环境商

1992 年荷兰有机化学家 Sheldon 提出了"环境因子"的概念[11],以化工产品生产过程中产生的废物量的多少来衡量合成反应对环境造成的影响,计算公式如下:

$$环境因子(E) = \frac{废弃物的质量 (kg)}{预期产物的质量 (kg)} \tag{1-1}$$

这里的废弃物是指除了目标产物以外的任何副产物,生产过程中产生的废物越多,E 值越大,环境危害越严重。用原子经济性或环境因子考察化工过程的优劣过于简单,还应考虑废弃物的危害程度,因为同样原子利用率的不同反应,副产物不同,其对环境的危害程度可能截然不同。环境商(EQ)是以化工产品生产过程中产生的废物量的多少、化工产品的物理性质、化学性质及其在环境中的毒性行为等综合评价指标来衡量合成反应对环境造成的影响:

$$EQ = E \times Q \tag{1-2}$$

式中,E 为环境因子;Q 为根据废物在环境中的行为所给出的对环境的不友好度。环境商的相对大小可以作为化学合成和化工生产选择合成路线、生产过程和生产工艺的重要因素。

3. 源头控制污染优于污染末端治理

种类繁多、性能多样的化学品确实给社会发展作出了重要的贡献。然而化学品的生产、加工、储存、运输和废弃物处理等各个环节会或多或少产生对环境有毒有害的物质。洛杉矶光化学烟雾、伦敦烟雾、日本水俣病等一系列环境污染引发的恶性事件让人类不得不反思。因此"先污染,后治理"的末端治理成为早期减少或消除环境污染的主要方式,但是随着化学工业和社会经济的高速发展,污染物排放量剧增,末端治理的局限性日益凸显。

4. 清洁溶剂和助剂

传统有机反应中，常用有机溶剂包括一大批挥发性有机溶剂作为反应介质，是工业污染的主要原因之一。助剂的使用也非常广泛，如分离助剂，可将产物与副产物、杂质分开，通常用量大且浪费多。因此采用环境友好的溶剂和助剂对抑制环境污染极为必要。可能的方法有以下四种：①用超临界流体作为溶剂和助剂，研究较多的是超临界二氧化碳流体[9]。②用水代替有机溶剂作为反应介质和助剂[12]，因为水是最安全和最清洁的；存在的问题是污水处理。③采用固定化溶剂，抑制其挥发产生环境污染[13]。④最好的办法是采用无溶剂反应体系，不仅彻底消除了溶剂和助剂的污染，而且也有利于降低生产成本。

5. 节能

化学原料的获取、化学反应的发生、反应速率的控制、反应物的分离和纯化等各个环节均伴随着能量的产生和消耗，节约能源十分必要。化学工业中所需要的能量主要有热能、电能和光能。热能是常见的能量形式；电能普遍存在于化学反应过程中，电化学反应是一种清洁技术，代表着绿色化学的发展趋势；光能是潜力最大的一种能源，通过利用太阳能，开发清洁、廉价的光化学反应代替传统有毒有害的化学过程，是绿色化学追求的目标[6]。

除了以上提到的三种传统的能量以外，还可以用微波、超声波等新形式的能量来促进化学反应的进行。微波是一种在固态下进行快速化学转化的反应技术，而传统上这些反应是在液态下完成的。其优点为不需要为某一反应进行长时间加热，同时，在固态下反应也避免了对辅助物质的额外加热(液态下这种额外加热是必需的)。超声波则能对一些类型的转化反应(环化加成、电环化加成)起到催化作用[6]。

6. 可再生资源

所谓可再生资源，是指能重新利用的资源，或在短期内可再生，或可循环利用的自然资源。可再生资源主要有土地资源、生物资源、气候、水能资源等，也包括经使用、消耗、加工、燃烧、废弃等程序后，在一定周期内能具有自我更新、重复形成、复原的特性，并可被持续利用的一类自然资源，它是可持续发展中被推广应用的清洁能源。如以秸秆等为原料可以合成乳酸；也可将稻草、甘蔗渣等农业废弃物经过加工、处理后得到酸、醇、酮等化工产品。

7. 可降解化学品的设计

要实现绿色化学，在设计化学品时必须要注意当它们功能用尽时，待报废的产品不能对环境产生污染，即在废弃时它们能被降解为无毒无害的物质。因此可降解化学品已经成为化学领域的首选。设计这种化学品时，一般考虑在合成的过程中引入一些易水解、光解或具有其他裂解功能的功能团，这不仅在使用化学产品时不存在对环境的污染，且化学品降解后所产生的物质无毒无害。例如，与传统聚合物相比，多糖聚合物表现出很多优越性，如多糖本身是一种可再生资源，毒理数据显示其危害程度较低，此外，多糖聚合物在生态环境中能够降解，因此，多糖聚合物产品的发展越来越快。

除以上几点以外，减少或消除制备和使用过程中的事故隐患、减少不必要的衍生化步骤、研制高效催化剂等都是绿色化学的基本原则。传统化学强调化合物的功能、化学反应的效率，忽视了环境和安全问题[13]，绿色化学需要充分关注原料的可再生性及有效利用性、环境的友好和安全、能源的节约、生产的安全性等问题，在源头实现污染控制。

第二节 清洁有机合成基本原理

有机合成是利用化学方法和技术将原料制备成新产品的过程。早期的有机合成主要是合成自然界中已经存在但含量稀少的有机化合物，随着化学理论和技术的发展，根据结构和性质的规律性和实际需求，逐渐发展到合成自然界不存在的新型有机化合物，满足各方面的需求。有机反应的特点是存在主反应和副反应，传统有机合成比较重视产物产率，而不太注重副产物或废弃物的生成及其对环境的危害。清洁有机合成的核心是提高反应的"原子经济性"。实现"原子经济性"是一个漫长的过程，有机合成工作者应该自觉地用"原子经济性"的原则审视已有的有机合成方法和技术，努力使研发的新反应满足"原子经济性"原则。

一、清洁有机合成的目标

衡量一个有机合成工艺是否成功的基本原则是以最低廉的原料、最短的路线、最高的收率、最安全的生产环境来合成目标分子，完全达到这些条件的可称为理想的有机合成。

绿色有机合成是现代化工的理想目标。美国斯坦福大学 Wender 教授在 *Chemical Review* 中，对理想的有机合成作了完整定义：理想的有机合成是用简单的、安全的、环境友好的、资源有效的操作，快速、定量地把廉价易得的起始原料全部转化为天然产品或设定的目标分子，其在大方向上指出了实现绿色有机合成的主要途径[14-16]。

二、清洁有机合成原理

作为一个多学科交叉的新研究领域，绿色化学尚有许多基本科学问题需要深入研究，清洁有机合成是绿色化学的最重要组成部分。清洁有机合成就是要求有机合成工作者以绿色化学作标准，提高效率、减少污染、降低能耗[19]。

1. 原子经济性

1991 年，著名化学家 Trost 提出[10]以"原子经济性"的观念来评估化学反应的效率，即要考察有多少反应物分子进入最后的产物分子。理想的"原子经济性"反应，应该是 100%的反应物分子转化到最终产物中，而没有副产物生成，这是有机合成的重大发展。例如，Wittig 试剂与醛酮的反应是合成烯烃非常经典的有机反应，但从绿色化学的角度来看，它生成了较多的副产物，"原子经济性"很差，因此必须开发高效催化剂和新工艺。目前，Trost 教授提出的"原子经济性"观念被学术界和工业界广泛接受，他也因此在 1998 年获得了"美国总统绿色化学挑战奖"。当然，目前真正属于高"原子经济性"的有机合成反应，特别是适于工业化生产的高"原子经济性"的有机合成反应并不多。

2. 高效催化剂

有机化学反应效率包括活性和选择性，其中选择性与副产物相关，从而直接影响环境污染，主要包括化学选择性、区域选择性和立体选择性三种。化学选择性是指同一个分子内不同官能团在不同条件下发生的反应不同，区域选择性是指同一分子内不同部位的同一官能团的选择，立体选择性是指反应产生不同的立体构型。立体选择性可用立体选向百分率表征，对称合成的立体选向百分率为零，而不对称合成的选择性为 0~100%，其中 100%意味着立体专一性反应。

反应选择性不但与合成的效率直接相关，更因为产物精确的空间结构直接影响其生理活性，而且这类反应涉及反应的控制。近二十年来，选择性合成，尤其是不对称合成一直是有机合成化学的中心问题和研究热点。绿色化学的目标是开发高"原子经济性"的反应，研制高效的催化剂是最终实现"零排放"的关键。例如，外消旋体的拆分是合成单一手性分子的重要途径。然而，目标手性分子的最高产率为 50%，而其对映异构体一般被当成废弃物，不仅浪费，且可能造成环境的污染，根据绿色化学的要求，其原子经济性较差。所以，不对称催化反应对获得单一手性分子显得尤为重要，广泛应用于手性药物合成。2001 年，Knowles 博士、Noyori 教授和 Sharpless 教授因在不对称催化反应的研究中所取得的卓越成就，获得了诺贝尔化学奖，这也表明开展催化不对称反应研究的重要意义。

3. 简化反应步骤，减少污染排放，开发新的合成工艺

随着绿色化学的发展，对有机合成化学的要求越来越高，开发新工艺势在必行。例如，Roche Colorado 公司在刚开始研发抗病毒药物 Cytovene 时，他们认为所采取的全硅烷化合成路线是最行之有效的方法。然而，随着市场发展和需求量的增加，生产规模扩大，原工艺暴露出很多问题，随后该公司对原工艺进行了较大的改进，即采用以鸟嘌呤三酯为原料合成的新路线。与原工艺相比，新工艺将反应底物和中间产物的种类减少了 11 种，这大大减少了废气和固体废弃物的排放，产率也提高了 2 倍。

三、实现绿色有机合成的途径

随着化学工业的不断发展，现代绿色合成化学正向着生态友好型、原子经济性及高选择性的方向发展，因此，有机化学需要发展新型绿色合成路线和技术，实现高得率、低能耗、无污染且安全可靠[17]。

1. 运用协同合成技术

在精细化学品，如药物、农用化学品等合成中，常常会遇到分离中间体的多步骤反应，其操作繁琐且存在污染风险。发展"一锅"串联反应是减少污染和生产成本的有效途径。串联反应主要有多组分串联反应和多步串联反应，多组分串联反应一般有三种以上不同原料在同一反应器中进行反应，而每步反应都是下一步反应所必需的，而且原料分子的主体部分都融入最终产物中，这是一类高效的合成方法；而多步串联反应主要是模仿生物体内的多步连锁式反应，即在同一个反应器中，所有反应从原料到目标产物多个步骤连续进行，反应过程中中间体无需从反应体系中分离出，避免一些废弃物的产生。

2. 采用安全化学原料

绿色合成要求采用无毒无害原料合成产品。例如，在有机合成中常使用硫酸二甲酯 [$SO_2(OCH_3)_2$] 和光气($COCl_2$)，其毒性非常强，且在生产过程中还会产生 HCl，这不仅污染环境，还腐蚀生产设备，生产成本大大增加。现在化学工作者合成了 $CO_2(OCH_3)_2$ 代替生产中使用的 $SO_2(OCH_3)_2$ 和 $COCl_2$，其不仅毒性低，且反应活性高，因而也被誉为新世纪的"绿色化工原料"。腈类化合物由于腈类分解而产生剧毒的 CN^-，但是如采取某种方法将腈类化合物的 α-位取代，腈就很难生成自由基而不产生CN^-，从而可使其毒性大大降低，且不影响其反应能力。

3. 开发新型合成技术

近几十年来，化学工作者主要致力于研究开发新的合成技术方法，使用无毒、污染、安全的有机试剂及溶剂。例如，离子液体和CO_2超临界流体中的反应，可大大减少有机溶剂的使用，从而降低了环境污染，同时提高反应的选择性；又如，改变合成的环境，引入超声、微波、电等手段使化学反应速率更快，选择性更好，原子经济性更好，也促进了绿色化学的发展。

第三节 清洁有机合成技术

有机合成在自然科学的发展过程中起到了非常大的推动作用，对人类社会的生产和生活也具有重大的意义。如药物、洗涤剂、保鲜剂、人造纤维、化肥、杀虫剂、染料及具有各种性能的现代材料等，都是通过有机合成获得的。由此可以看出，如今有关民生、不同科学研究的领域都依赖有机合成。长期以来，为了提高有机物的溶解度，确保物料能够混合均匀，能量稳定交换，有机反应都是在有机溶剂中进行。然而，有机溶剂的挥发性、毒性以及不可回收性对环境造成了重大污染，不仅影响和阻止了人类社会的可持续发展，还威胁到人类健康和生态环境，这给化学工作者带来新的挑战，即要去探索清洁合成方法和技术。美国于1996年设立了"总统绿色化学挑战奖"，用以奖励在从源头上减少化学污染方面取得成就的化学工作者。作为化学合成的重要组成部分，有机合成在绿色化学中也起着举足轻重的作用，在绿色化学理念的指导下，最终达到清洁有机合成。

开发环境友好、高效节能的绿色化学技术是当今化学工业、化学工艺及化学工程的发展趋势。催化是有机反应的核心，传统催化反应大多采用热化学方法，具有高温、高压、有毒、有害、易燃和易爆的特点，开发新技术和工艺的目的是使反应在常温、常压下安全、可靠、高效地进行。目前，清洁有机合成技术主要有以下几种：①变有毒、易燃、易挥发的有机溶剂为不易挥发、低毒甚至无毒的溶剂——绿色溶剂，包括水相体系、氟两相体系、离子液体介质、超临界乃至无溶剂体系；②采用超声或微波技术，缩短反应时间，提高反应的选择性，从而可以实现生产能源的节约；③将昂贵及有毒的催化剂或试剂固载化，以实现催化剂或试剂的回收，减少有毒有害废液的排放[18]。

一、绿色溶剂

绝大多数有机反应需要在有机溶剂中进行，全世界每年用于反应介质和化工分离的有机溶剂达数亿万吨，大多数有机溶剂如卤代烃、芳香化合物、醚等均对人体和自然界中的动植物具有危害作用，有的甚至还具有致癌作用；此外，使用的挥发性有机溶剂及有机试剂的排放是导致化学工业污染的主要原因[19]。因此开辟绿色介质中的化学反应新途径是未来绿色化工的重要发展方向之一。绿色介质主要包括超临界流体、水、离子液体[20]和全氟溶剂等。

1. 超临界流体

超临界流体是指处于临界温度和压力以上的流体，其气液界面消失，体系性质均一，此时既不是气体也不是液体，呈流体状态(也被称为物质的第四态)。超临界合成是以超临界流体为介质(有时也作为反应物)而发生的化学反应。超临界合成用清洁超临界流体替代了传统污染性溶剂作为反应介质，同时还能大大提高反应速率，增加选择性，从而减少副产物，甚至可以简化后续分离单元，既节省了资源和能源，同时又减少了废气、废物等的排放，如 Pinacol 重排和 Beckmann 重排反应在 450℃和 25 MPa 的超临界水中不加任何催化剂就可以顺利进行，并且反应速率大幅度增加，Yokata 等认为这是因超临界水具有酸性，而此类反应在弱酸 HBr 的催化下即可完成[21]。

2. 水

水为最清洁、安全和廉价的溶剂，而且水资源丰富，地球 70%以上的表面都被水所覆盖。但长期以来，受溶解性的限制，使用水作为溶剂的有机反应主要局限于简单的水解反应。以水为反应溶剂的优点是可以利用有机反应物在水介质中的"憎水性"来促进反应。因此，有人针对水介质中的有机反应，提出了"增益憎水性作用"，进而指出憎水性作用下，有机反应物和有机试剂分子为尽量减少与溶剂水分子之间的相互接触面积，限制各自的运动，从而将反应底物和有机试剂"挤"在一起，产生比在一般有机溶剂中更强的相互作用，并使反应得以在水相中进行[22]。与在有机溶剂中的有机反应相比，水相中的有机反应不仅可以提高反应的产率和选择性，还可以加快反应速率。传统的立体化学规则一般是建立在有机溶剂中，如 Cram 规则等[23, 24]，而水中的立体化学规则可能会有所不同，如 Grieco 等[25]报道了较复杂分子的水相 Diels-Alder 反应[式(1-3)]的例子，观测到比有机溶剂反应中更快的反应速率和相反的选择性，这为有机合成提供了一种新的方法。

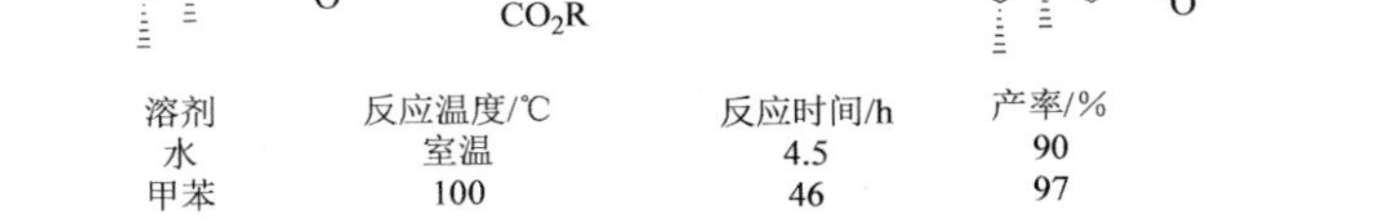

溶剂	反应温度度/℃	反应时间/h	产率/%
水	室温	4.5	90
甲苯	100	46	97

(1-3)

3. 离子液体

离子液体(ionic liquid)指在室温或接近室温下呈现液态的、完全由阴阳离子所组成的盐，也称为低温熔融盐。常见的一些离子液体即使在-96℃时仍可保持液态，与一般高温下熔融的离子型化合物有较大的区别。它一般由有机阳离子和无机阴离子组成，常见的阳离子有季膦盐离子、季铵盐离子、吡咯盐离子和咪唑盐离子等，阴离子有X^-、PF_6^-、BF_4^-等。阳离子结构对称性差，这可以降低盐晶体的晶格能，从而具有较低的熔点。离子液体有如下特点：

(1) 蒸气压低，不易挥发；

(2) 离子液体的极化能力较强，而其配位能力较弱，因此不会造成催化剂的失活；

(3) 在较大的温度范围内离子液体均能呈现非常稳定的液态，这有利于动力学控制；

(4) 对有机和无机物具有高的溶解性；

(5) 不能与一些直链烷烃、环烷烃类等有机溶剂互溶，这有利于反应产物的分离；

(6) 热稳定性好。

离子液体在大多数场合仅充当溶剂，一般是通过调节阴阳离子的结构，改变其溶解度达到溶解某一物质的目的。在一定条件下使反应发生在离子液体相中，一方面利用离子液体对金属催化剂和有机反应物有较好的溶解性，另一方面利用离子液体与某种有机溶剂的难溶性，可通过简单的萃取就能实现产物的分离，同时又能将催化剂进行回收并重复使用[26]。

近年来，室温离子液体在有机合成领域内的应用备受关注，有望发展成为实用绿色化学工业的反应介质。例如，在加氢反应中，离子液体可以起到溶剂与催化剂的双重作用。又如，Wang 等[27]报道了在 [PMIm$^+$][PF$_6^-$]离子液体中，用 5%(摩尔分数)I_2 催化 2-氨基苯甲酰肼和 2-甲酰基苯甲酸反应制备喹唑啉衍生物，获得89%的得率，这种体系适用于较大范围的底物，均可得到较高得率。

4. 全氟溶剂

全氟溶剂，也称为氟溶剂或全氟碳，主要是指碳原子上的氢原子全部被氟原子取代的烷烃、醚和胺，它是一种新兴的绿色反应溶剂。常见的氟溶剂主要有全氟己烷、全氟甲基环己烷、全氟环己烷、全氟庚烷、全氟甲苯和全氟三烷基胺、全氟二烷基醚等。氟溶剂的密度比普通有机溶剂大，其沸点范围广，无色无毒，具有非常高的热稳定性，此外，其折射率、表面张力和介电常数都较低。全氟溶剂也是一种较好的气体溶剂，它能溶解大量的 O_2、H_2、N_2 和 CO_2 等[28]。全氟溶剂分子中氟原子的高电负性及范德华半径与氢原子相近，所以 C—F 键具有高度稳定性。在较低的温度下，由于全氟溶剂与大多数有机溶剂的混溶性低，溶液分成两相，即有机相和氟相；当升高温度时，普通有机溶剂在全氟溶剂中的溶解度急剧上升，为有机反应提供了良好的均相反应条件。反应结束后，温度一旦降低，体系又会回复到含催化剂的氟相和含产物的有机相[29]。例如，Mikami 等[30]以 $Sc[C(SO_2C_8F_{17})_3]_3$ 和 $Sc[N(SO_2C_8F_{17})_3]_3$ 为催化剂，在 1,2-二氯乙烷与全氟环己烷的两相体系中，将 2,3-二甲基-1,3-丁二烯与甲基乙烯酮在 35℃下反应 8 h，反应结束后经过相分离即可获得理想的乙酰基环己烯产物[式(1-4)]，催化剂回收率也能达 99.9%，连续使用 4 次，产物的得率均能保持在 94%~95%。

$$\text{（反应式）} \quad \xrightarrow{Sc[C(SO_2C_8F_{17})_3]_3 或 Sc[N(SO_2C_8F_{17})_3]_3}_{35℃,8h} \quad \text{（产物）} \quad (1\text{-}4)$$

二、绿色催化剂

催化剂在化学合成中扮演着十分重要的角色，每一种新催化剂的诞生及催化工艺的成功研制，都能引起化学工业的重大变化。因而绿色催化也就成为绿色化学研究的一个重要方向。催化剂不仅可以加快反应速率，还可以提高转化率和选择性。此外，绿色催化剂还应该具备清洁无毒和可循环使用的特点。

1. 固体酸催化剂

固体酸是指能使碱性指示剂改变颜色或者能化学吸附碱性物质的固体，固体超强酸是应用较广的固体酸之一，其主要是指比 100%硫酸的酸度还高的固体酸，可分为三种类型：①金属氧化物负载硫酸根型的固体超强酸，此类型催化剂具有催化活性高、不腐蚀设备、耐高温及可重复使用的特点；②强 Lewis 酸负载型固体超强酸，此类催化剂存在活性组分溶脱和其中卤离子对设备有较强的腐蚀作用等问题，并不适合于在较高温度条件下使用；③其他类型固体超强酸，如杂多酸、分子筛型及高分子树脂型。固体超强酸的合成方法主要有沉淀浸渍法、溶胶-凝胶法和固相合成法，固体超强酸通常应用于烷基化反应、酯化反应、酰基化反应、低聚反应、缩醛和缩酮反应及异构化反应等。

分子筛不仅具有高热稳定性和水热稳定性，还具有酸性，因此也是一种优良的固体酸催化剂，通常应用于石油加工中酸催化反应、氧化反应、烷基化反应、羟烷基化反应、酰基化反应、N-烷基化和 O-烷基化反应、芳香烃硝化反应、氨化反应及碳化反应等，其优点是无毒无害，且对设备无腐蚀[31, 32]。

2. 固体碱催化剂

固体碱是指能使酸性指示剂改变颜色或者能化学吸附酸性物质的固体，主要包括有机固体碱、有机/无机复合固体碱和无机固体碱三大类。有机固体碱主要是指端基为叔胺或叔膦基团的碱性树脂类物质；有机/无机复合固体碱主要是指负载有机胺或季铵碱的分子筛；无机固体碱包括金属氧化物、金属氢氧化物、水滑石类、碱性离子交换树脂和负载型固体碱等，具有制备简单、碱强度分布范围宽、热稳定性好等优点，已成为固体碱催化剂的首选。固体碱催化剂具有活性高、选择性高、反应条件温和、产物易于分离、可循环使用等诸多优点，在精细化学品合成方面发挥着越来越重要的作用，广泛应用于氧化反应、氨化反应、氢化反应、还原反应、加成反应、烷基化反应、聚合反应、脱氢反应和醇醛缩合反应等[32]。

3. 离子液体催化剂

如前所述，离子液体是指有机阳离子和无机阴离子构成的、在室温或近室温下呈液体的盐类化合物，也称为室温熔融盐或室温离子液体，既可作为反应介质，又可作为反应催化剂，通常应用于 Friedel-Crafts 反应、氧化反应、酯化反应、还原反应及羰基合成等[33]。

4. 生物催化剂

生物催化剂是指从生物体，主要是微生物细胞中提取出的游离酶或经固定化技术加工后的酶，它是游离或固定化的酶或活细胞的总称。与化学催化剂相比，生物催化剂具有如下特点：①催化效率高；②反应条件温和；③生物催化剂活性具有可调节性；④具有较高的立体选择性和专一性；⑤酶可被生物降解，反应结束后没有废弃物，对环境无污染；⑥具有相容性，即一个反应器中可以发生多个生物催化的反应；⑦催化反应类型多等。

生物催化剂应用在有机合成中的主要反应有：①还原反应，如 C=O 和 C=C 等的还原；②氧化反应，如 C—H 键、醇及不饱和键的氧化；③酮醇缩合反应；④脱羧反应；⑤连锁反应，如 Diels-Alder 反应、亲电和亲核反应、重排反应；⑥水解反应，如酰胺和对称酯的水解；⑦过渡金属与酶共同催化光学转化的反应。尽管生物催化具有很多化学催化所不能比拟的优势，但其在实际应用还存在诸多问题：①影响酶催化反应的因素多；②酶容易失活；③产物分离困难；④酶稳定性差；⑤酶催化成本高；⑥存在对不需要的生物质及对底物的浪费[34]。

5. 膜催化剂

膜催化是近年来多相催化领域中出现的一种新技术，该技术是将催化材料制成膜反应器或将催化剂置于膜反应器中进行操作。反应物和产物可选择性地穿过膜，从而有效地调节某一反应物或产物在反应器中的区域浓度，摆脱化学反应受热力学平衡态的限制，实现高选择性和转化率。膜催化的显著优点有：①催化活性高。这主要是因为膜具有较大的比表面积，使单位表面积的原子或分子的占有率高，有利于活性中心与反应分子充分接触，展现出优异的催化性能。②选择性高。高选择性一方面归因于膜具有较多的微孔；另一方面其孔径分布广，可以通过不同方法对孔径、孔体积以及空隙率进行有效的控制，从而提高催化剂的选择性。尤其是生物膜催化剂，其选择性可达到100%。③稳定性高。膜催化剂耐高温、耐化学腐蚀，同时机械强度高，有利于延长催化寿命。膜催化在加氢、脱氢、氧化、酯化、生化等许多反应中得到应用，如芳烃加氢、丙烷脱氢环化、丙烯氧化制丙烯醛等反应[35]。

三、绿色合成方法

近年来，磁、声、光、电在有机合成中备受关注，可显著缩短反应时间，提高反应产率，减少能耗与污染，为实现绿色化学开辟了新途径。

1. 微波

微波加热具有快速、均质与选择性高的特点，已被广泛应用于各种材料的合成、加工中。早在1986年，化学家Gedye和Smith成功地将微波用于化学有机合成，这也引发了微波合成化学的热潮[36]。目前，微波已成功应用于烷基化、酯化、皂化、烯烃加成、磺化、氧化、还原、环合以及碳离子负缩合等反应中[37,38]。学术界对微波加速化学反应的机理仍存有一定的争议，其争论的焦点是是否有一种所谓非热活化效应。假设微波能像光那样激活分子，那么微波的量子能至少应接近或超过分子键能，或可产生能量积累。但微波量子能远低于化学键能，仅与布朗运动的能量处于同一数量级，而且能量积累将被快速的分子弛豫所释放，无法达到产生激发态所需的量值[39]。因此，现在较普遍的一种看法是，微波主要是使极性分子在电磁场中快速旋转，从而使单位量的分子吸收跃过反应能垒所需的能量或分子快速旋转碰撞的概率增加，从而加速化学反应的发生。此外，微波能选择性地加热非均相催化反应中的负载型催化剂的活性点[40]，导致反应效率大幅度提高。

与常规加热过程不同的是，微波是直接作用于目标分子的体相加热，不存在从界面到体相的传热过程，温度场均匀分布，热效率高，有利于提高反应速率。例如，环己烷基烯醇的Claisen重排反应，在微波辐射作用下反应10 min，即能获得得率为87%的理想产物，而在常规加热条件下，需要反应48 h才能获得与之相当的产物，微波反应效率可提高288倍[41]。另外，微波反应还具有操作简单和选择性好等特点,这些特点对于原料昂贵、生产批量小的高附加值精细化学品合成具有特殊意义。

2. 超声

超声波是指振动频率大于 20000 Hz 的声波。当超声波在介质中传播时,介质可以发生物理和化学变化,从而产生一系列超声效应,如温热效应、理化效应等。超声波在有机反应中,特别是在非均相的有机反应中表现出非常明显的加速效应,与常规方法相比,不仅反应时间能缩短数十倍甚至数百倍,反应产率和选择性也会有所提高。此外,实验操作也比较简单,如避免搅拌操作,有些常规条件下不能发生的反应,在超声作用下可以完成。正由于超声合成具有时间短、产率和选择性高、操作简单等优点,现已成为有机精细化工合成领域中一种普遍的反应方法。尽管目前超声化学反应的机理尚不完全清晰,但超声空化是推进化学反应主动力的观点已得到了多数人的认同[42]。例如,在超声波促进下,2′-羟基查尔酮环氧化反应在较短的时间(30 min)内给出高得率(96%)的产物,而采用常规加热,产物得率仅为 22%[43]。这也证明超声确实能改变反应的进程,提高反应的选择性,增强化学反应的速率和产率,降低能耗和减少废物的排放。

3. 激光

激光用于温和条件下的化学反应,同时,当催化剂失去活性时,可以用激光照射使之恢复活性。当用高强度短脉冲激光照射时,可生成瞬态活化中心,加速反应,并减少其他副反应的发生。另外,激光容易激发产生自由基,可诱导自由基反应。

Bonn 等[44]考察了 Ru(0001)在吸附一氧化碳和氧原子时形成二氧化碳的反应,常规加热无法使反应进行,但在红外光激发下,可顺利生成二氧化碳,表明激光诱导新的氧化反应途径。

4. 电场

在包含热传递、质量传递、动量传递的许多物理和化学过程中,把流体置于强电场下,可显著提高其传递系数。若把足够强的电场(如 1010 V/m)作用于化学反应体系,可显著影响热力学平衡,甚至还会出现非平衡态热力学。另外,电场还可以使分子的化学键产生极化,降低反应活化能,从而加快化学反应速率[45, 46]。在许多反应过程中组合使用这些强化方法,可以达到良好的效果。例如,采用等离子体与电场相结合的方法,将其应用在 CH_4 氧化偶联的反应中,结果显示,CH_4 单程转化率能达到 22%;同时,C_2 的总选择性大于 90%[46]。

综上所述,各种新技术的使用能有效提高反应效率和选择性,符合绿色化学的基本要求，即不再使用和产生有毒有害物质，从根本上消除环境污染，实现社会经济的可持续发展。在保证反应高选择性和高收率的前提下，如何找到绿色环保、适用范围广、操作简单、成本低廉的合成方法，将这些技术推广至工业化，并阐述理论和机理，仍是化学工作者们急需解决的问题[46]。

四、有机电化学合成

利用电化学氧化或还原方法合成有机化合物的方法称为有机电化学合成，简称有机电合成，是绿色合成技术的重要组成部分。有机电化学聚合反应是在含有聚合物单体、溶剂、电解质、引发剂等的溶液中通入阳极电流而发生的聚合反应，其中工作电极可以使用石墨、金属、金属氧化物和半导体材料等。当前，电化学正全面介入能源开发、环境保护和生命科学的领域。太阳能是未来最可靠的能源之一，利用太阳能，电化学将在两个途径上极有发展前途，一是通过光电化学反应设计太阳能电池板，二是通过光催化分解水产氢或光电转化再电解水产氢，提供可靠的清洁能源。同样地，在生命科学研究方面，电化学过程已成为洞察人体信息的一种重要手段。利用电化学传感器，对人工脏器或病态脏器进行人工调节，这一工作使电化学工程与医学工程相结合，促进人体医学和生命科学的发展。

与传统化学合成相比，电化学合成具有许多优点：①很多有机电化学合成中采用"电子"为反应试剂时，反应体系中不需再加入还原剂或氧化剂，因此，反应体系中所含有的反应物种相对较少，这样产物不仅比较容易分离和纯化、产品纯度高，且对环境污染也较小，有的反应甚至对环境完全无危害；②在有机电化学合成中，可以选择不同的电极调节溶剂和电极电位等，从而控制反应的方向，不仅可以提高反应的选择性，还可以提高产品收率；③有机电化学合成一般在常温、常压下进行，无需特殊的加热和加压设备，工艺流程短，设备投资简单，噪声和热污染少；④有机电化学合成过程的电流、电压等参数采集和控制方便，容易实现自动化。

有机电化学合成有望在以下方面有所突破[47-49]：①新型化合物的合成或者极难合成化合物的合成；②步骤复杂或产率低的化学合成；③要消耗大量氧化剂或还原剂，且这些试剂在大批生产时有困难或者价格昂贵的化学合成；④"三废"(废水、废气、固体废弃物)污染严重或处理费用高的化学合成；⑤反应选择性极差的化学合成；⑥市场需求小，但附加值特别高的有机产品。

参 考 文 献

[1] 麻生明. 科学发展报告. 北京: 科学出版社, 1999: 101.
[2] 闵恩泽, 傅军. 绿色化学的进展. 化学通报, 1999, 1: 10-15.
[3] 胡纯, 杨明. 基于应用型人才培养的绿色化学教学的探索与实践. 广东化工, 2012, 39: 255-256.
[4] 彭实, 刘鸿. 中学化学实验绿色化研究. 北京:人民教育出版社, 2010: 1-28.

[5] 张继红. 绿色化学. 芜湖: 安徽师范大学出版社, 2012.
[6] 张龙, 贡长生, 代斌. 绿色化学. 武汉: 华中科技大学出版社, 2014.
[7] 杨水金, 侯倩, 蔡干喜, 徐明波. 中学化学教学中绿色化学理念的渗透. 湖北师范学院学报(自然科学版), 2013, 3: 3.
[8] 吴棣华. 几种值得注意的有机原料清洁工艺技术. 化学进展, 1998, 10: 131-136.
[9] Anastas P T, Williamson T C. Green chemistry: designing chemistry for the environment// Green Chemistry. Washington, DC: American Chemical Society, 1996, 626: 1-17.
[10] Trost B M. The atom economy—a search for synthetic efficiency. Science, 1991, 254: 1471-1477.
[11] Sheldon R A. Organic synthesis: past, present and future chemistry and industry. Chemsitry and Industry, 1992, 7: 903-906.
[12] Li C J, Chan T H. Introduction//Comprehensive organic reactions in aqueous media. New York: John Wiley & Sons, Inc.2006: 1-13.
[13] 周淑晶, 冯艳茹, 李淑贤. 绿色化学. 北京: 化学工业出版社, 2014.
[14] Wender P A. Introduction: frontiers in organic synthesis. Chemical Reviews, 1996, 96: 1-2.
[15] 周云. 浅议合成化学中的原子经济性. 化学教育, 2003, 10: 3-39.
[16] 陆ががが. 绿色化学与有机合成及有机合成中的原子经济性. 化学进展，1998, 10: 123-127.
[17] 袁学玲, 卢鹏祥. 现代有机合成方法和技术的最新进展. 河南化工, 2010, 27: 5.
[18] Guillier F, Orain D, Bradley M. Linkers and cleavage strategies in solid-phase organic synthesis and combinatorial chemistry. Chemical Reviews, 2000, 100: 2091-2158.
[19] Anastas P T, Kirchhoff M M. Origins, current status, and future challenges of green chemistry. Accounts of Chemical Research, 2002, 35: 686-694.
[20] Dupont J, de Souza R F, Suarez P A Z. Ionic liquid (molten salt) phase organometallic catalysis. Chemical Reviews, 2002, 102: 3667-3692.
[21] 张敬畅, 吴向阳, 曹维良. 超临界流体技术在催化反应和金属有机反应中的应用. 化学通报, 2001, 6: 214-216.
[22] Blokzijl W, Blandamer M J, Engberts J B F N. Diels-Alder reactions in aqueous solutions. enforced hydrophobic interactions between diene and dienophile. Journal of the American Chemical Society, 1991, 113: 4241-4246.
[23] Kauffmann T. In search of new organometallic reagents for organic synthesis. Berlin: Springer Berlin Heidelberg, 1980: 109-147.
[24] Cram D J, Elhafez F A A. Studies in stereochemistry. X.the rule of "steric control of asymmetric induction" in the syntheses of acyclic systems. Journal of the American Chemical Society, 1952, 74: 5828-5835.
[25] Grieco P A, Yoshida K, He Z M. Sodium 6-methoxy-(E,E)-3,5-hexadienoate: A useful diene in the aqueous Diels-Alder reaction. Tetrahedron Letters, 1984, 25：5715-5718.
[26] Wu J X, Beck B, Ren R X. Catalytic Rosenmund-von Braun reaction in halide-based ionic liquids. Tetrahedron Letters, 2002, 43: 387-389.
[27] Jin R Z, Zhang W T, Zhou Y J, Wang X S. Iodine-catalyzed synthesis of 5H-phthalazino[1,2-b]quinazoline and isoindolo[2,1-a]quinazoline derivatives via a chemoselective reaction of 2-aminobenzohydrazide and 2-formylbenzoic acid in ionic liquids. Tetrahedron Letters, 2016, 57: 2515-2519.
[28] Cavazzini M, Montanari F, Pozzi G, Quici S. Perfluorocarbon-soluble catalysts and reagents and the application of FBS (fluorous biphase system) to organic synthesis. Journal of Fluorine Chemistry, 1999, 94: 183-193.
[29] Horváth I T, Rábai J. Facile catalyst separation without water: Fluorous biphase hydroformylation of olefins. Science, 1994, 266: 72-75.
[30] Mikami K, Mikami Y, Matsuzawa H, Matsumoto Y, Nishikido J, Yamamoto F, Nakajima H. Lanthanide catalysts with tris(perfluorooctanesulfonyl)methide and bis(perfluorooctanesulfonyl)amide ponytails: Recyclable Lewis acid catalysts in fluorous phases or as solids. Tetrahedron, 2002, 58: 4015-4021.
[31] 于世涛, 王大全. 固体酸与精细化工. 北京: 化学工业出版社, 2006.
[32] 田部浩三, 小野嘉夫. 新固体酸和碱及其催化作用. 北京: 化学工业出版社, 1992.

[33] 王均凤, 张锁江, 陈慧萍, 李闲, 张密林. 离子液体的性质及其在催化反应中的应用. 过程工程学报, 2003, 3: 177-185.
[34] 梁辰宇. 浅谈生物催化在精细有机化工中的应用. 化工管理, 2015, 20: 145-145.
[35] 杜长海, 吴树新, 高伟民, 秦永宁. 膜催化技术的研究进展. 长春工业大学学报, 2003, 24: 19-23.
[36] Gedye R, Smith F, Westaway K, Ali H, Baldisera L, Laberge L, Rousell J. The use of microwave ovens for rapid organic synthesis. Tetrahedron Letters, 1986, 27: 279-282.
[37] Caddick S. Microwave assisted organic reactions. Tetrahedron, 1995, 51: 10403-10432.
[38] Varma R S. Solvent-free accelerated organic syntheses using microwaves. Pure and Applied Chemistry, 2001, 73: 193-198.
[39] Berlan J. Microwaves in chemistry: Another way of heating reaction mixtures. Radiation Physics and Chemistry, 1995, 45: 581-589.
[40] Hájek M. Microwave activation of homogeneous and heterogeneous catalytic reactions. Collection of Czechoslovak Chemical Communications, 1997, 62: 347-354.
[41] Srikrishna A, Nagaraju S. Acceleration of orthoester Claisen rearrangement by a commercial microwave oven. Journal of the Chemical Society, Perkin Transactions, 1992, 1: 311-312.
[42] Pugin B. Qualitative characterization of ultrasound reactors for heterogeneous sonochemistry. Ultrasonics, 1987, 25: 49-55.
[43] Lahyani A, Trabelsi M. Ultrasonic-assisted synthesis of flavones by oxidative cyclization of 2′-hydroxychalcones using iodine monochloride. Ultrasonics Sonochemistry, 2016, 31: 626-630.
[44] Bonn M, Funk S, Hess C, Denzler D N, Stampfl C, Scheffler M, Wolf M, Ertl, G. Phonon-versus electron-mediated desorption and oxidation of CO on Ru(0001). Science, 1999, 285: 1042-1045.
[45] Pacchioni G, Lomas J R, Illas F. Electric field effects in heterogeneous catalysis. Journal of Molecular Catalysis A: Chemical, 1997, 119: 263-273.
[46] Wan B W, Xu G H. Catalysis conversion methane into C_2 hydrocarbons via electric field enhanced plasma. Chinese Chemical Letters, 2003, 14: 1236-1238.
[47] Polat K, Aksu M L, Pekel A T. Cathodic reduction of 4-nitroso-N, N-dimethylaniline at various electrode materials. Journal of Applied Electrochemistry, 2000, 30: 733-736.
[48] 赵崇涛, 朱则善, 王清萍, 翁家宝. 间接电解氧化法合成丙酮酸乙酯. 精细化工, 2000, 17: 61-63.
[49] Bondon A, Porhiel E, Pebay C, Thouin L, Leroy J, Moinet C. Tetraarylporphyrin synthesis by electrochemical oxidation of porphyrinogens. Electrochimica Acta, 2001, 46: 1899-1903.

第二章 传统催化水介质有机合成

最初水介质中有机合成反应的研究主要集中在均相催化剂，这是由于均相催化剂以分子或离子独立起作用，活性中心均一，具有高活性和高选择性。同时，均相催化剂易于开展反应动力学和机理研究。由于均相催化剂可溶于反应介质，分子间的扩散不受影响，且在均相催化剂中，每个活性位都能发挥催化活性，而在多相催化剂中，表面的活性位才能发挥催化作用，这就决定了均相催化剂比多相催化剂的催化活性高得多。另外，均相催化剂还具有反应条件温和、副反应少、易于控制等优点，因而降低了能量损耗[1,2]。此外，均相催化剂中的配位基可以选择和调节，使金属原子周围有特定的电子与空间环境，从而优化了反应途径，提高了选择性。最早的均相催化水介质有机反应是酸碱催化，近年来有机金属以及小分子催化的水相化学反应取得了很大突破。

第一节 酸 催 化 剂

一些 Lewis 酸如 $AlCl_3$，在 Friedel-Crafts 反应、磺化反应、芳烃氯化、硝化反应、异构化等反应中起着重要作用，H_2SO_4 等质子酸和杂多酸等在酯化、缩合反应中也有广泛应用。

一、Lewis 酸催化剂

很多常用的 Lewis 酸遇水容易分解，实现 Lewis 酸催化的水介质反应必须要寻找在水中稳定的 Lewis 酸。1992 年，Bosnich 等[3]用 $Ti(Cp^*)·H_2O$ 催化水介质 Diels-Alder 反应，这指引了 Lewis 酸催化水介质清洁有机反应的新方向。

1. 缩合反应

缩合反应是形成 C—C 键的重要反应之一，被广泛应用于天然产物和具有生物活性化合物的合成中。早在 1988 年，Lubineau 和 Meyer[4]研究水介质中烯醇硅醚与羰基化合物的交叉-羟醛缩合反应，即 Mukaiyama 羟醛缩合[式(2-1)]时，发现在反应中加入催化量的镧系元素三氟甲基磺酸盐时，可大大提高反应速率和目标产物得率。在三氟甲基磺酸镧系元素盐中，$Yb(OTf)_3$、$Ge(OTf)_3$ 和 $Lu(OTf)_3$ 等能更有效地催化羟醛缩合反应，显示较好的催化活性，但非对映选择性不高。随后，又发展了多种 Lewis

酸用于催化该类反应,显著改善了催化效率。研究表明,Lewis 酸在水介质有机化学反应中的催化活性与其水解速率和替换内层水配体的交换速率有关[5]。

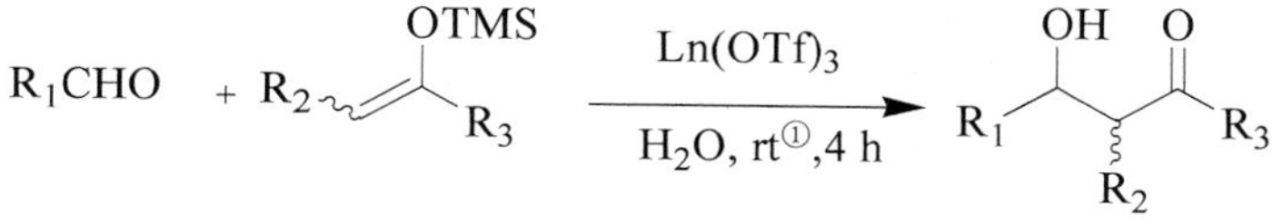

$$R_1CHO + R_2 \underset{R_3}{\overset{OTMS}{=}} \xrightarrow[H_2O, rt^{①}, 4h]{Ln(OTf)_3} R_1 \underset{R_2}{\overset{OH}{\underset{}{\wedge}}} \overset{O}{\underset{}{\vee}} R_3 \quad (2-1)$$

1993 年,Kobayashi 等[6]发现 $Ln(OTf)_3$ 可以催化水介质中的烯醇硅醚与醛的醛醇缩合反应。他们发现在此体系中加入一定量的十二烷基硫酸钠(SDS)表面活性剂时,能在水中形成胶束,这有利于稳定烯醇硅醚[7],从而加快反应速率,提高产物的得率。他们也设计并合成了具有表面活性的有机钪催化剂 $Sc(O_3SC_{12}H_{25})_3$(STDS),并考察了其在水介质醛醇缩合反应中的催化性能[8, 9]。同时,他们也报道了 $InCl_3$ 催化剂,但与 $Sc(OTf)_3$ 相比,其催化活性相对较低[10]。

Li 等[11]报道杯[6]芳烃磺酸盐(1)催化水介质中不稳定的烯醇硅醚参与的醛醇缩合反应,而且产物得率高。研究表明,在反应体系中加入催化量的二苯基硼酸,烯醇硅醚可转化为烯醇硼烷,再在水介质中与醛发生醛醇缩合反应,产物的非对映选择性达到 94:6[12]。比较突出的进展是不对称醛醇缩合反应可以在水介质中进行,Kobayashi 等用 $Pb(OTf)_2$/手性冠醚配体[13]和 $Ln(OTf)_3$/手性双吡啶基-18-冠-6[14]等催化水、乙醇介质中的醛醇缩合反应,产物 ee 值②达 79%~87%。Fan 等[15]报道了手性树枝状双噁唑啉配体与 $Cu(OTf)_2$ 催化水介质醛醇缩合反应,但产物的 ee 值仅为 61%。

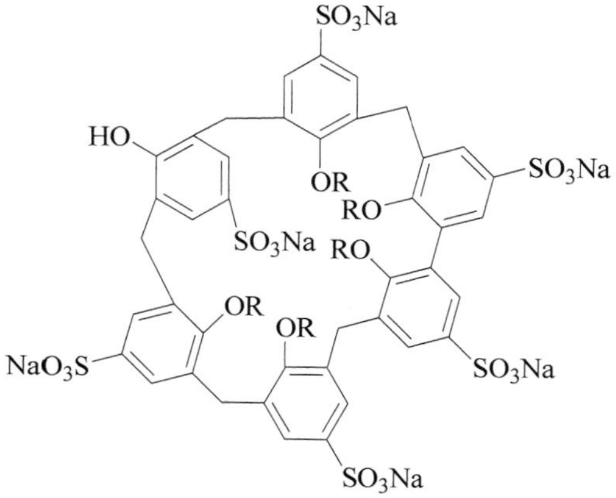

1

---

① rt 表示室温,下同。

② ee, enantionmeric excesses,表示某一对映体的过量值,代表对映选择性。ee = $\frac{[R]-[S]}{[R]+[S]} \times 100\%$。

Li等[16]发现，1,1'-联苯二萘酚的磺酸盐衍生物(2)可作为手性阴离子表面活性剂催化水介质不对称 Mukaiyama-Aldol 反应，得到羟醛缩合加成物，非对映和对映选择性中等到良好。作为 Lewis 酸催化剂，$Ga(OTf)_3$ 和 $Cu(OTf)_2$ 比 $Sc(OTf)_3$ 活性更高，如使用四丁基氟化铵活化剂、烯醇三甲基硅醚与甲醛在水介质中也能顺利实现不对称缩合反应。

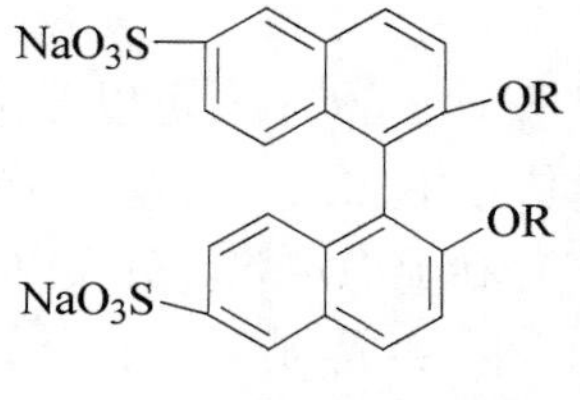

2

Tao等[17]考察了不同碱金属和过渡金属氯化物 Lewis 酸($LiCl$，$ZnCl_2$ 和 $SnCl_2$)与含有双亲性的脯氨酰胺酚（3）共同催化不对称的 Aldol 缩合反应[式(2-2)]。在最佳反应条件(10 mol%①脯氨酰胺、10 mol% MCl_2 或 20 mol% $LiCl$)下，反应在室温下就能给出 anti-产物，ee 值达 99%以上。

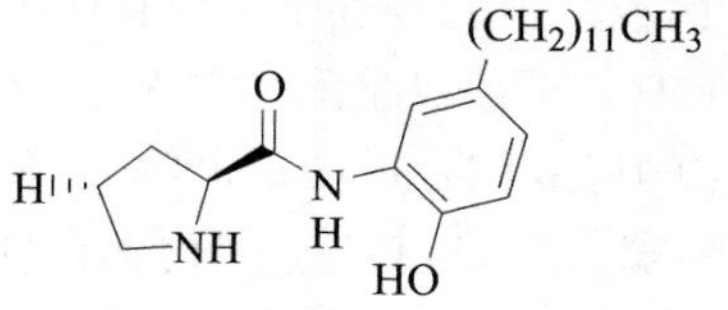

3

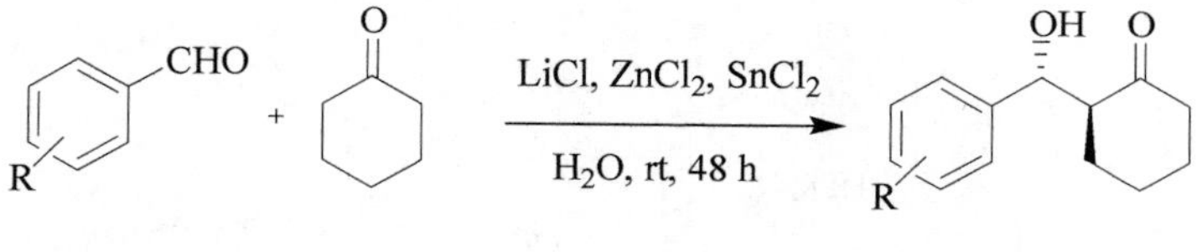

$$(2\text{-}2)$$

2. Mannich 反应

含有活泼氢的化合物与甲醛和氨或二级胺缩合生成 β-氨基(羰基)的反应称为 Mannich 反应。Loh 和 Wei 报道了以 $InCl_3$ 为催化水介质，醛、胺和烯醇硅醚或烯酮硅缩醛在水中的"一锅"反应[18]，以高产率得到了 β-氨基羰基化合物[式(2-3)]。

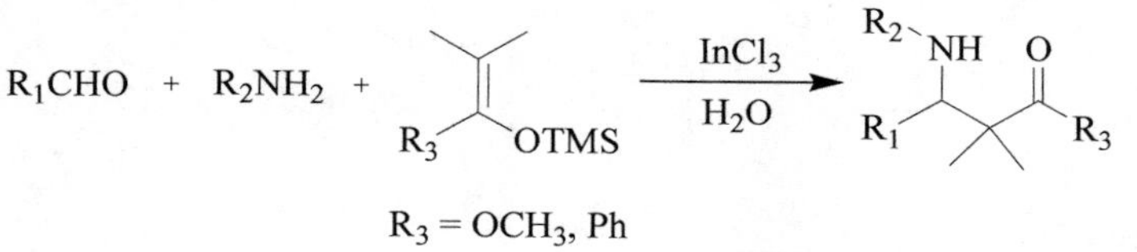

$$R_1CHO + R_2NH_2 + \underset{OTMS}{\overset{R_3}{\diagup}} \xrightarrow[H_2O]{InCl_3} \underset{R_1}{\overset{R_2-NH}{\diagup}} \overset{O}{\underset{R_3}{\diagup}} \quad (2\text{-}3)$$

$R_3 = OCH_3$, Ph

---

① mol%表示摩尔分数，下同。

Kobayashi 等[9]报道了 Lewis 酸-表面活性剂结合(LASC)的催化剂 $Cu(DS)_2$ 催化不同醛(脂肪醛、杂环醛)参与的 Mannich 反应[式(2-4)]。在室温下给出相应产物的得率为 72%~95%。研究也发现，其他 Lewis 酸如 Sc(Ⅲ)和 Cu(Ⅱ)与阴离子表面活性剂，如十二烷基苯磺酸盐对羟醛缩合反应也具有明显的催化效果[式(2-5)]，只要有底物存在，即使底物是固体，此类型催化剂在水中都能迅速形成稳定的胶体分散体。他们发现在中性盐，如 NaOTf 中催化亚胺与硅基烯醇化合物的 Mannich 反应时，在水中形成悬浮的反应介质[19]。动力学行为研究也表明，Mannich 反应加成物的存在加速了悬浮液的形成。

$$R_1CHO + \text{ArNH}_2 + R_2R_3C=C(OTMS)R_2 \xrightarrow[\text{H}_2\text{O, rt, 5h}]{5 \text{ mol\% LASC}} \text{产物} \quad (2-4)$$

$$R_1CHO + NH_3 + R_2R_3C=CHCH_2Bpin \xrightarrow[\text{表面活性剂}]{Sc(Ⅲ)\text{或}Cu(Ⅱ)} \text{产物} \quad (2-5)$$

Ishimaru 和 Kojima 报道了水中不对称 Mannich 类型反应。在 $Zn(OTf)_2$ 催化下，手性醛亚胺和 2-硅氧基丁二烯可以顺利地进行立体选择性的 Mannich 类型反应 [式(2-6)]，给出 74%~90% de 值①的目标产物[20]。

$$\text{手性亚胺} + \text{TMSO-二烯} \xrightarrow{Zn(OTf)_2, H_2O} \text{产物} \quad (2-6)$$

3. Michael 加成反应

1887 年 Michael 研究发现芳基或杂芳基的碳亲核试剂与羰基或 α, β-不饱和羰基化合物的加成反应，这是有机合成中增长碳链的常用方法之一。早在 20 世纪 70 年代，Wiechert 等[21]及 Hajos 和 Parrish[22]分别独立报道了 2-甲基环戊烷-1,3-二酮与乙烯基酮在水中进行的 Michael 加成反应，生成相应的共轭加成产物，反应体系中不需加入碱催化剂。与在甲醇溶剂中进行的 Michael 加成反应相比，以水作

---

① de 值表示立体选择性的高低。

溶剂可以大大提高反应产物的得率和纯度。与在 2-甲基环己烷-1,3-二酮与乙烯基酮的 Michael 加成反应中，具有类似的现象，反应最后得到光学纯的 Wieland-Miescher 酮[23]。但是反应需要更苛刻的反应条件。Lavallee 和 Deslongchamps 报道[24]，在以无机酸而非碱作催化剂的作用下，抗坏血酸与环己烯酮的 Michael 加成反应能在水中高效地发生[式(2-7)]。Keller 和 Feringa 将这一结果用于 13-α-甲基-14-α-羟基甾族化合物的合成，他们在反应体系中加入少量的 Yb(OTf)$_3$，有利于加快 Michael 加成反应速率，在室温下反应 5 d 就能给出较高得率的产物(>98%)[25][式(2-8)]。

$$\text{环己烯酮} + \text{抗坏血酸} \xrightarrow{H_2O} \text{产物1} + \text{产物2} \quad (2\text{-}7)$$

$$R_1\text{COCH(R)COOR}_2 + R_3\text{CH=CHCOR} \xrightarrow[H_2O, rt]{Yb(OTf)_3} \text{产物} \quad (2\text{-}8)$$

Kobayashi 等[26]首次报道了 Lewis 酸 AgOTf 高效催化的水中不对称 Michael 加成反应[式(2-9)]。在低温下反应 18 h，相应产物的得率达 74%，ee 值为 38%。

$$\text{茚酮-CO}_2\text{CH}_3 \xrightarrow[\substack{0.15 \text{ eq Yb(OTf)}_3 \\ 0.1 \text{ eq (R)-BINAP} \\ H_2O, 0℃, 18 \text{ h}}]{3.0 \text{ eq MVK}, 0.15 \text{ eq}^①} \text{产物} \quad (2\text{-}9)$$

Firouzabadi 等[27]用十二烷基苯磺酸铝([Al(DS)$_3$]·3H$_2$O)催化水中甲基乙烯基甲酮和吲哚的加成[式(2-10)]。研究发现，如果体系中不加[Al(DS)$_3$]·3H$_2$O，反应进行得很慢，反应 48 h，产物的得率不到 20%，当加入十二烷基苯磺酸钠（SDBS）后，反应速率虽然有所提高，但产物得率仍然没有明显的提高。由此可以推测，[Al(DS)$_3$]·3H$_2$O 是该反应非常有效的催化剂，与 InCl$_3$、Sc(DS)$_3$、CeCl$_3$·7H$_2$O-NaI 相比，该催化剂显示出优越的催化性能。

$$\text{吲哚} + R_2\text{C=CR}_2\text{EWG} \xrightarrow[H_2O, rt]{(2.5\sim10 \text{ mol\%})Al(DS)_3} \text{产物} \quad (2\text{-}10)$$

---

① eq 表示当量。

Kobayashi 等[28]和 Vaccaro 等[29]两个课题组同时报道了一种用于水中硫醇与 α,β-不饱和羰基化合物加成反应[式(2-11)]的催化剂——Sc(OTf)$_3$-手性联吡啶配合物(4)。该催化剂在室温下即可催化该反应，并给出高得率和高对映选择性 (ee 值 52%~97%) 的硫化物产物，且无需添加任何有机溶剂。

$$\text{R}\overset{}{=}\overset{O}{\underset{}{\text{C}}}\text{R}_1 + \text{R}_2\text{SH} \xrightarrow[\text{H}_2\text{O, pH= 6~7, 30℃}]{4\text{-Sc(OTf)}_3} \text{R}\overset{SR_2}{\underset{}{\text{C}}}\overset{O}{\underset{}{\text{C}}}\text{R}_1 \quad (2\text{-}11)$$

Wennerberg 等[30]考察了将 Ln(OTf)$_3$/α-氨基酸催化体系应用于中性水中 α,β-不饱和羰基化合物与活泼亚甲基化合物的加成反应 [式(2-12)]。作者考察了不同的 Lewis 酸和不同氨基酸对反应活性的影响，研究结果证实，Yb(OTf)$_3$/D-丙氨酸具有最高的催化效率，其在 60℃催化乙酰丙酮与环己烯酮反应，反应 24 h 后就能给出 92%的得率和 52% ee 值的产物。

$$(2\text{-}12)$$

Veisi 等[31]首次报道将 FeCl$_3$·6H$_2$O 与表面活性剂 SDS 加入反应体系中，原位形成 Lewis 酸–表面活性剂结合的催化剂 Fe(DS)$_3$，其在吲哚与 α,β-不饱和羰基化合物的 Michael 反应[式(2-13)] 以及二(吲哚)甲烷的合成中均显示了较高的活性，这个催化体系也适用于在室温下合成 1,1,3-三吲哚基化合物，产物得率较高。

$$(2\text{-}13)$$

4. Friedel-Crafts 反应

Friedel-Crafts 反应是芳香化合物最经典的亲电取代反应类型之一。传统的 Friedel-Crafts 常用 Lewis 酸，如 AlCl$_3$、ZnCl$_2$ 作催化剂，该催化剂的缺点是在水中非常不稳定，所以该反应一般在有机溶剂中进行。Zhuang 和 Jørgensen 报道在

1 mol/L NaH_2PO_4-Na_2HPO_4 水溶液中，吲哚与乙醛酸乙酯能顺利地进行 Friedel-Crafts 反应[式(2-14)]，得到吲哚衍生物产物，而反应体系中不必采用任何金属催化剂[32]。

$$R_1 \text{—indole—H} + H_5C_2O_2C\text{—CHO} \xrightarrow{NaH_2PO_4-Na_2HPO_4} R_1\text{—indole—CH(OH)CO}_2C_2H_5 \quad (2\text{-}14)$$

Kobayashi 等[33]研究发现，$Bi(OTf)_3$ 和 $BiCl_3$ 对活化及非活化的苯衍生物的 Friedel-Crafts 乙酰基化反应都是比较有效的催化剂。$Ga(OTf)_3$ 溶于水，比以前报道的 $M(OTf)_3$ (M = Al, Ga, Ln, Sc)催化剂更能有效地催化该类反应[34]。$Sc(DS)_3$ 也能有效地催化水中吲哚与缺电子的烯烃进行共轭加成反应 [式(2-15)]。采用催化量的 $LnBr_3$，吲哚与硝基烯烃的 1,4-共轭加成反应能在水介质中进行。$Ln(OTf)_3$ 能催化水中芳香化合物与三氟丙酮酸酯进行 Friedel-Crafts 反应，生成多种 α-羟基酯 [式(2-16)]。

$$\text{indole} + \text{MVK} \xrightarrow[30\ °C,\ 1\ h]{2.5\ mol\%\ Sc(DS)_3,\ H_2O} \text{3-substituted indole} \quad (2\text{-}15)$$

$$\text{ArX(R)} + F_3C\text{-CO-CO}_2CH_3 \xrightarrow[H_2O,\ rt]{Ln(OTf)_3} \text{Ar-C(CF}_3\text{)(OH)CO}_2CH_3 \quad (2\text{-}16)$$

5. Diels-Alder 反应

Diels-Alder 反应是一个涉及 1,3-二烯与含有不饱和键化合物的反应，该反应可以用于合成六元环的化合物。早在 1980 年 Breslow 和 Rideout 首次报道了纯水中环戊二烯与丁烯酮的反应[35]，这也开创了现代水相有机合成反应的先河。他们研究发现，Diels-Alder 反应在水中的反应速率比在有机溶剂如异辛烷中要快 700 多倍，而且具有更高的立体选择性。对于水中的 Diels-Alder 反应，二烯羧酸的钠盐、二烯醇醚、带有磷酸盐的二烯等都可以与不同种类的亲二烯体反应得到理想的加成产物。Engberts 等[36-38]详细研究了 Co^{2+}，Ni^{2+}，Cu^{2+} 和 Zn^{2+} 等 Lewis 酸催化剂对水相 Diels-Alder 反应速率和内式-外式选择性的影响。研究也表明，亲二烯体的双齿特性对反应是否能够发生具有重要影响。

含杂原子的反应底物参与的 Diels-Alder 反应是合成含氮或含氧六元杂环化合

物的重要方法。1985 年，Grieco 和 Larsen[39]首次报道了水中杂 Diels-Alder 反应。1987 年他们报道了 N-甲基马来酰亚胺与 N-苄基-2-氮杂双环[2.2.1]-5-庚烯在 50℃的氯化铵水溶液中的加成反应[式(2-17)][40]。

$$\text{[反应式]} \quad \xrightarrow[\text{H}_2\text{O, 50℃, 2 h}]{\text{HCl}} \quad \text{[产物]} \quad (2\text{-}17)$$

后来，Wang 等[41]采用稀土 Lewis 酸催化水中亚胺的 Diels-Alder 反应 [式(2-18)]，他们研究几种不同的 $Ln(OTf)_3$ 的催化性能，结果表明，当 $Nd(OTf)_3$ 用量为 40%时，加成产物的得率可达到 72%。随后，他们又采用 AgOTf 催化这类反应，令人兴奋的是该反应在纯水中就能顺利进行。重要的是，该催化剂还可以直接催化醛、胺与 Danishefsky 二烯的加成反应[式(2-19)]，同样也能给出理想的产物，当向此反应中加入一定量的非离子表面活性剂 Triton-100，可以大大提高产物的得率。

$$R_1CHO + R_2NH_2 \cdot HCl + R_3\text{-CH=CH}_2 \xrightarrow[\text{H}_2\text{O}]{Ln(OTf)_3} \quad \text{[产物]} \quad (2\text{-}18)$$

$$R_1CHO + R_2NH_2 + \underset{\text{TMSO}}{\overset{\text{OCH}_3}{\text{[二烯]}}} \xrightarrow[\text{H}_2\text{O, rt, 2~3h}]{\text{AgOTf}} \quad \text{[产物]} \quad (2\text{-}19)$$

1998 年 Engberts 和 Otto[38]首次报道了用手性 Lewis 酸催化剂通过催化水中环戊二烯与 3-苯基-1-(2-吡啶基)-2-丙烯-1-酮的 Diels-Alder 反应，不对称形成 C—C 键的反应。所用的手性 Lewis 催化剂是 10% $Cu(NO_3)_2$/手性氨基酸，当 N-甲基色氨酸的用量为 17%时，反应产物的对映选择性可达 76%。研究也发现，水溶剂对手性催化剂与底物所形成的络合物有一定的稳定作用。

含氧的亲双烯体也可在水中发生氧杂- Diels-Alder 反应。例如，乙醛酸和环戊二烯或环己二烯在酸性水中反应得到加成产物，产物再经重排可得到 α-羟基-γ-内酯[式(2-20)][42, 43]。

$$\text{[环烯]}_n + \text{HCOCO}_2\text{H} \xrightarrow{\text{H}_2\text{O, pH}=1} \text{[产物1]} + \text{[产物2]} \quad (2\text{-}20)$$

$n = 1$, 1.5h, 40℃, 83%(73/27)
$n = 2$, 48 h, 90℃, 85%(60/40)

Li 和 Zhang[44]在研究酸性树脂催化水相合成四氢吡喃衍生物方法的基础上，采用 LnCl$_3$ 催化水中苯胺与 2,3-二氢呋喃或二氢吡喃的 Diels-Alder 反应，得到各种四氢喹啉的衍生物[式(2-21)]。用这种方法，也能以 2-羟基四氢呋喃或 2-羟基四氢吡喃为原料合成四氢喹啉的衍生物。

$$\text{(2-21)}$$

6. 其他反应

亚胺的烯丙基化作用是形成有用的烯丙胺的重要方法，1998 年 Kobayashi[45]报道了醛、胺和三丁基烯丙基锡的三组分反应[式(2-22)]，该反应采用少量的 Sc(OTf)$_3$ 为催化剂，水为溶剂。Rao 等[46]用 β-环糊精催化醛进行烯丙基化，从而得到胺；Zhao 和 Li[47]以醛、胺及三丁基烯丙基锡为原料，以廉价的羧酸为催化剂，在温和条件下合成烯丙胺。在有机溶剂如 DMSO 中检测不到产物，在 THF 中，产物的得率为 87%，在水中产物得率约为 97%，由此可见，水可以促进此反应的发生。2009 年他们又报道了用磷钼酸催化该反应，并获得了较高的收率[48]。

$$R_1CHO + R_2NH_2 + \diagup\!\!\!\diagdown SnBu_3 \xrightarrow[H_2O, \text{rt}, 24\text{ h}]{Sc(OTf)_3} \quad \text{(2-22)}$$

Kobayashi 等[9]报道了在室温下，用 Sc(DS)$_3$ 催化水中醛与烯丙基锡的烯丙基反应，生成 82%得率的高烯醇化合物。Manabe 和 Kobayashi 用 Sc(DBS)$_3$ 催化水中醛与胺及亚磷酸酯的 Kabachnik-Fields 反应，生成 α-氨基膦酸酯[式(2-23)][49]。

$$R_1CHO + R_2NH_2 + P(OC_2H_5)_3 \xrightarrow[H_2O, 30^\circ C]{10\text{ mol\% Sc(DS)}_3} \quad \text{(2-23)}$$

Wu 等[50]报道在超声作用下，以 Lewis 酸与表面活性剂结合的催化剂催化 2-炔基苯甲醛与苯胺亲核试剂(炔、硝基甲烷、二乙基膦酸酯)反应，分别给出中等得率的 1,2-二氢异喹啉衍生物[式(2-24)~式(2-26)]。

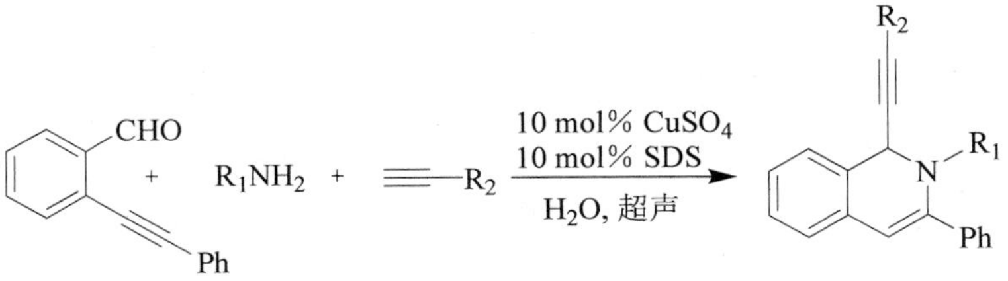

[式(2-24)]

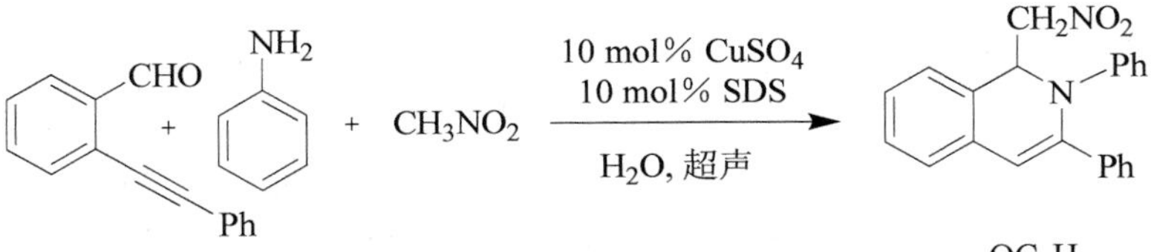

[式(2-25)]

[式(2-26)]

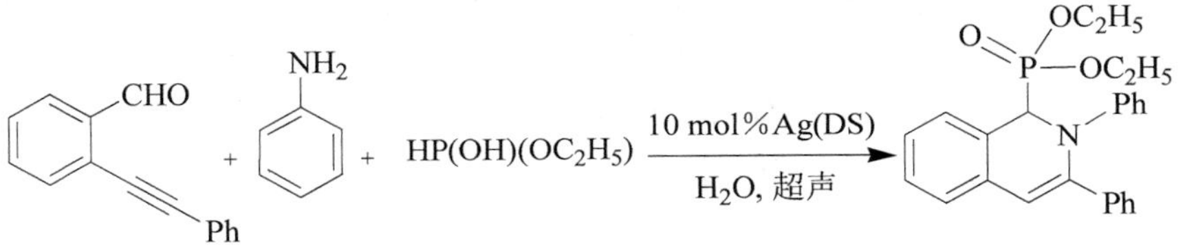

Liu 等[51]报道用 InBr$_3$/SDBS 催化体系催化水中仲芳香胺与烯糖的反应[式(2-27)]。该反应主要是通过 4π顺旋电子环化过程，给出高得率和高非对映选择性的 4-氨基环戊烯酮及其衍生物[式(2-27)]。

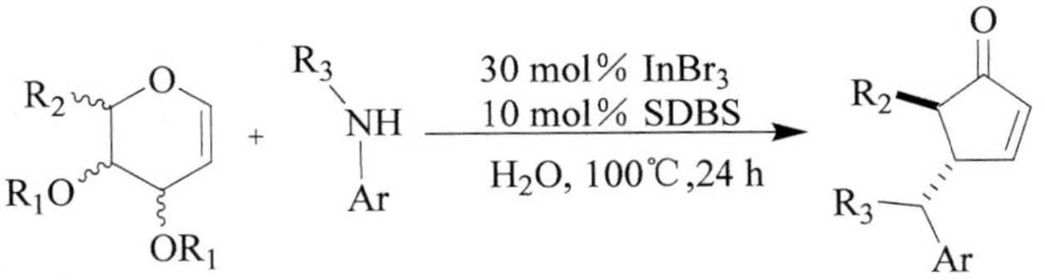

[式(2-27)]

Vaccaro 等[52]报道 α, β-环氧羧酸被溴离子或碘离子开环，生成碘代或溴代的羟基羧酸。InCl$_3$ 在 pH = 1.5 时，各种 α, β-环氧羧酸都能给出得率为 88%~95%的反式碘代羟基羧酸[式(2-28)]。随后，Fringuelli 等[53]用 InCl$_3$ 催化 1,2-环氧化物的硫解反应[式(2-29)]。

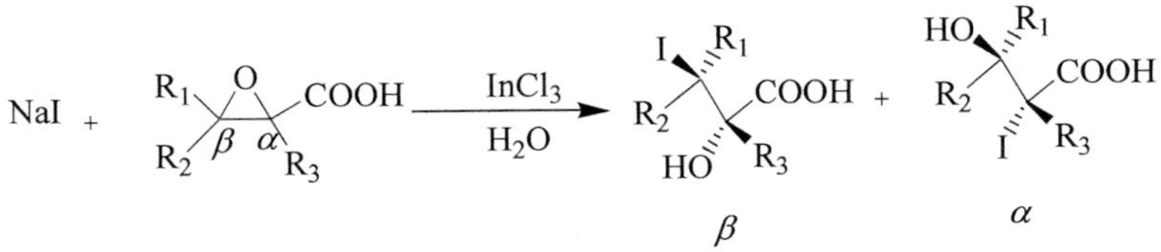

[式(2-28)]

[式(2-29)]

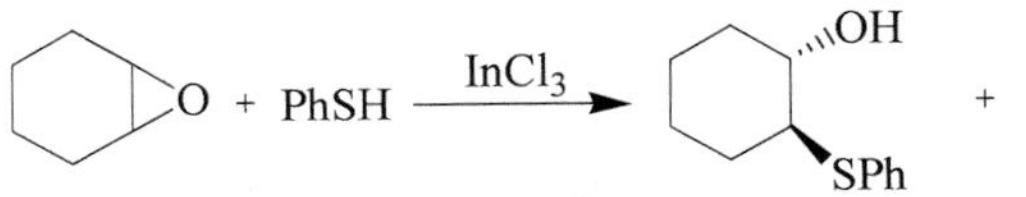

Procopio 等[54]报道 Er(OTf)$_3$ 催化水中脂肪胺或芳香胺参与的环氧化物开环反应,生成 β-氨基醇[式(2-30)],此方法的优点是反应条件温和,成本低,毒性小。

$$\underset{O}{\overset{RR_1}{\triangle}} + R_2R_3NH_2 \xrightarrow[H_2O]{Er(OTf)_3} \underset{OH}{\overset{NR_2R_3}{R-CH-CH-R_1}} + \underset{NR_2R_3}{\overset{OH}{R-CH-CH-R_1}} \quad (2\text{-}30)$$

末端炔烃对亚胺的加成反应生成的炔丙基胺是一种非常有用的合成中间体,其本身也显示了宽泛的生物活性,因此在有机合成中具有重大意义。2002 年,Li 和 Wei[55]首次报道了高效催化炔烃与各种亚胺的加成反应,通过在水中温和条件下,AuBr$_3$ 对 C—H 键的活化作用生成炔丙基胺 [式(2-31)]。但是这个方法仅限于芳香胺与醛反应生成的亚胺。后来,他们又发展用 4,5-Cu(OTf)络合物实现了末端炔烃对亚胺的对映选择性加成反应,给出相应的产物(+)-炔丙胺在水中的催化活性(产率 48%~86%)和对映选择性(78%~91%)[56]。

$$R_1CHO + R_2\text{—}\equiv\text{—} + R_3NHR_3 \xrightarrow{AuBr_3} R_2\text{—}\equiv\text{—}\underset{R_1}{\overset{R_3NR_3}{CH}} \quad (2\text{-}31)$$

偶氮化合物可与炔烃的 1,3-偶极环发生加成反应生成吡唑,这种化合物显示了抗肿瘤、抗炎症、抗病菌和抗神经的活性,Li 和 Jiang[57]报道了重氮羰基化合物与炔烃的分子间 1,3-偶极环加成-氢(烷基或芳基)转移的多米诺序列反应 [式(2-32)]。有趣的是,水在反应中不仅起到关键作用,还可以使 InCl$_3$ 催化反应。对重氮乙酸乙酯和丙炔酸酯的反应 [式(2-33)],InCl$_3$ 可再重复使用两次而保持催化活性,使用到第三次,产物的得率仍保持在 90%,而在重氮反应常用的有机溶剂如 CH$_2$Cl$_2$、苯中,仅有微量的目标产物生成。

$$(2\text{-}32)$$

$$N_2CHCO_2C_2H_5 + \underset{CO_2CH_3}{|||} \xrightarrow[H_2O, rt]{20\ mol\%\ InCl_3} \quad (2\text{-}33)$$

Khurana 等[58]报道以水为溶剂,InCl$_3$ 为催化剂,在回流反应体系中各种醛、

1,3-二酮与富电子的氨基杂环化合物反应[式(2-34)~式(2-37)],如 6-氨基-1,3-二甲基尿嘧啶反应[式(2-34)],生成具有生物活性的嘧啶和吡唑衍生物,反应 45~60 min,给出相应的产物得率为 87%~93%。

$$RCHO + \text{(1,3-二甲基-6-氨基尿嘧啶)} + \text{(5,5-二甲基-1,3-环己二酮)} \xrightarrow{InCl_3, H_2O, 回流} \text{产物} \quad (2\text{-}34)$$

$$RCHO + \text{(1,3-二甲基-6-氨基尿嘧啶)} + \text{(1,3-环己二酮)} \xrightarrow{InCl_3, H_2O, 回流} \text{产物} \quad (2\text{-}35)$$

$$RCHO + \text{(二氨基哌啶酮)} + \text{(5-甲基-1,3-环己二酮)} \xrightarrow{InCl_3, H_2O, 回流} \text{产物} \quad (2\text{-}36)$$

$$RCHO + \text{(1,3-二甲基-6-氨基尿嘧啶)} + \text{(1,3-茚二酮)} \xrightarrow{InCl_3, H_2O, 回流} \text{产物} \quad (2\text{-}37)$$

酯化反应是一个可逆反应,醇和羧酸的反应在水溶液中很难发生,研究发现,在水中仅有 1.2%的羧酸发生反应,即反应达到平衡。如向反应体系中加入一定量的 HCl 和 H_2SO_4 等质子酸可以明显提高酯化反应的速率,但反应很快达到平衡,反应物的转化率仅为 19%。如再加入一定量的酸,虽然酯化反应速率有所提升,但反应物的转化率没有增加。若加入十二烷基苯磺酸钠表面活性剂,反应物在水中则通过缩合反应选择性地生成醚、硫醚和酯[59, 60]。

Prins 反应是指在有酸性催化剂存在下,简单烯烃与醛的反应,反应可以用醛的水溶液,给出形成 C—C 键的反应产物。如 Aubele 等[61]采用环状不饱和缩醛作为氧杂碳氧离子的前体,在水中 Lewis 酸性胶束(铈盐的)内,环化反应能高效地进行,并以优异的立体控制得到各种乙烯基和芳基取代的四氢吡喃化合物[式(2-38)]。

$$\text{(2-38)}$$

Lin 等[62]考察了不同盐酸盐如 $CrCl_3$、$CuCl_2$、$FeCl_3$、$AlCl_3$ 等催化水中纤维素转化成乙酰丙酸(LA)的反应,催化活性不仅与反应体系的酸性强度有关,且与金属盐的种类也有关系。研究结果显示,反应时间、温度、催化剂用量及底物浓度对 LA 的得率都具有非常重要的影响。当使用 $CrCl_3$ 为催化剂,LA 的得率可达 67%。Heeres 等[63]研究了不同金属盐用于催化水中丙糖如二羟基丙糖(DHA)和乙二醛糖(GLY)转变成乳酸,其中 Au(Ⅲ)盐显示最佳的催化能力,如以 $AlCl_3$ 为催化剂,在 140℃下反应 90 min,乳酸的得率高于 90%。

最近,Zhou 等[64]报道用相对无毒、廉价的 $FeCl_3 \cdot 6H_2O$ 催化 2-氨基苯甲酰胺与 1,3-二酮经过缩合、分子内亲核加成以及 C—C 键断裂的串联反应,生成中等到良好得率的 2-取代喹啉酮衍生物 [式(2-39)、式(2-40)]。

$$\text{(2-39)}$$

$$\text{(2-40)}$$

二、Brønsted 酸催化剂

1. 缩合反应

2002 年 Jalil[65]和 Pan 等[66]两个课题组分别报道了磷钨酸在室温下催化水中含有不同取代基的邻氨基苯甲酰胺与醛的缩合反应,给出良好到优良得率(79%~97%)的产物——2,3-二羟基-4(1H)-喹啉酮化合物[式(2-41)]。在四种杂多酸中,磷钨酸在水中的催化活性高于其他有机溶剂;且反应仅需 0.10 mol%的磷钨酸,在室温下就能使反应进行完全。作者也考察了不同的醛(芳环上含有吸电子或给电子基团)与胺的缩合反应,结果显示,芳环上的取代基对产物得率没有明显的影响,如用杂环醛代替芳醛进行反应,同样也能给出较高得率的目标产物。

$$R_1 \underset{O}{\overset{NH_2}{\bigcirc}} NH_2 + R_2CHO \xrightarrow[H_2O, \text{rt}]{0.1\text{mol\%}\ H_3PW_{12}O_{40}} R_1 \underset{H}{\overset{O}{\bigcirc}} \underset{R_2}{\overset{NH}{\bigcirc}} \quad (2\text{-}41)$$

Wu 等[67]发展了用 $H_3PW_{12}O_{40}$ 在室温催化水中肼/酰肼、二胺和伯胺与不同 1,3-二羰基化合物的缩合反应,在室温下反应,分别给出较高得率的吡唑、吖庚因以及烯胺酮/烯胺酯[式(2-42)~式(2-44)]。

$$\underset{R_2}{\overset{O\ \ O}{\bigvee}}R_1 + R_3NHNH_2 \xrightarrow[H_2O, \text{rt}, 8\sim60\ \text{min}]{1\text{mol\%}\ H_3PW_{12}O_{40}} R_1\underset{R_2}{\overset{N-N}{\bigvee}}\underset{}{\overset{R_3}{\bigvee}} + \underset{R_2}{\overset{N-N}{\bigvee}}\underset{R_1}{\overset{R_3}{\bigvee}} \quad (2\text{-}42)$$

$$\underset{R_2}{\overset{O\ \ O}{\bigvee}}R_1 + H_2N\underset{}{\overset{R_3}{\bigvee}}NH_2 \xrightarrow[H_2O, \text{rt}, 5\ \text{h}]{H_3PW_{12}O_{40}} \underset{R_3}{\overset{N}{\bigvee}}\underset{R_2}{\overset{R_1}{\bigvee}} \quad (2\text{-}43)$$

$$\underset{R_2}{\overset{O\ \ O}{\bigvee}}R_1 + R_5NH_2 \xrightarrow[H_2O, \text{rt}, 1\text{h}]{1\text{mol\%}\ H_3PW_{12}O_{40}} \underset{R_2}{\overset{R_5NH\ \ O}{\bigvee}}R_1 + \underset{R_1}{\overset{R_5NH\ \ O}{\bigvee}}\underset{R_2}{\overset{}{\bigvee}} \quad (2\text{-}44)$$

Chandra 等[68]研究发现,$H_3PW_{12}O_{40}$ 是 4-香豆素和芳醛在水中进行 Knoevenagel 类型缩合反应/Michael 反应[式(2-45)]的高效催化剂,反应在 80 ℃、15 mol% $H_3PW_{12}O_{40}$ 条件下进行 20 min,给出得率为 93%的目标产物。研究也表明,除甲醇作溶剂,其他常见有机溶剂均不利于此反应的进行,$H_3PW_{12}O_{40}$ 在催化体系中也可以重复使用,但是随着使用次数的增加,催化活性也随之降低,使用到第 6 次,产物得率不到 70%,这可能是由于反应过程中磷钨酸发生了氧化反应。

$$\underset{OH}{\overset{O\ \ O}{\bigcirc}} + \text{ArCHO} \xrightarrow[H_2O, 80\ ℃]{H_3PW_{12}O_{40}} \underset{OH}{\overset{O\ \ O}{\bigcirc}}\underset{}{\overset{}{\bigvee}}\underset{HO}{\overset{O\ \ O}{\bigcirc}} \quad (2\text{-}45)$$

有报道称,Keggin 类型的 $H_4SiW_{12}O_{40}$ 是从 1,2-二胺与 1,2-二羰基化合物合成具有生物活性喹啉衍生物的高效催化剂[式(2-46)]。这个方法的优点是,催化剂用量小，产率高，且底物范围广[69]。研究也发现 $H_3PW_{12}O_{40}$ 也是丙二腈、氰乙酸乙酯与醛在水中进行 Knoevenagel 缩合反应的最佳催化剂[70]。

$$\text{R}_1\text{-C}_6\text{H}_3(\text{NH}_2)_2 + \text{R}_2\text{COCOR}_3 \xrightarrow[\text{H}_2\text{O, rt}]{\text{H}_4\text{SiW}_{12}\text{O}_{40}} \text{quinoxaline} \quad (2\text{-}46)$$

用 Preyssler 类型杂多酸 $H_{14}[NaPW_{12}O_{40}]$ 催化水或乙醇中 3-甲基-1-苯基-1H-吡唑-5(4H)-酮或巴比妥酸与丙二腈、醛的"一锅"反应，可以给出 4-芳基吡喃[2,3-c]吡唑酮和吡喃[2,3-d]嘧啶[71][式(2-47)]。

$$\text{cyclopentenone-Ph} + \text{NCCH}_2\text{CN} + \text{ArCHO} \xrightarrow[\text{H}_2\text{O, 回流}]{\text{H}_{14}[\text{NaPW}_{12}\text{O}_{40}]} \text{product} \quad (2\text{-}47)$$

Lai 等[72]采用商用手性伯二胺，(1S,2S)-1,2-二苯基乙烷-1,2-二胺(5)和(1R,2R)-1,2-二苯基乙烷-1,2-二胺(6)与 Brønsted 酸结合，催化对硝基苯甲醛与环己酮在水中的反应[式(2-48)，式(2-49)]，反应具有较高的化学选择性和立体选择性，得出 ee 值为 99%的 β-羟基酮。2009 年 Zare 等[73]报道用十烷基磺酸锆[Zr(DS)$_4$]在室温下催化水中 1,2-二胺与 1,2-二酮的缩合反应，反应在较短的时间内给出良好到优良得率的喹啉衍生物[式(2-50)]。

5

6

$$\text{O}_2\text{N-C}_6\text{H}_4\text{-CHO} + \text{cyclohexanone} \xrightarrow[\text{H}_2\text{O, rt}]{5, 20\text{ mol\% TfOH}} \text{product} \quad (2\text{-}48)$$

$$\text{O}_2\text{N-C}_6\text{H}_4\text{-CHO} + \text{cyclohexanone} \xrightarrow[\text{H}_2\text{O, rt}]{6, 20\text{ mol\% TfOH}} \text{product} \quad (2\text{-}49)$$

$$\text{o-C}_6\text{H}_4(\text{NH}_2)_2 + \text{PhCOCOPh} \xrightarrow[\text{H}_2\text{O, rt, 30 min}]{2.5\text{ mol}\% \text{ Zr(DS)}_4} \text{2,3-diphenylquinoxaline} \quad (2\text{-}50)$$

2. Friedel-Crafts 反应

Shirakawa 和 Kobayashi[74]用正癸酸催化伯胺、醛和含有不同取代基的吲哚氮杂 Friedel-Crafts 反应[式(2-51)]。所生成的 3-取代的吲哚化合物很容易转变成具有生物活性的化合物。他们提供了一种新型高效的在室温就能合成 3-取代吲哚化合物的方法。

$$\text{Indole}(N\text{-}R_2) + \text{ArNH}_2 + R_1\text{CHO} \xrightarrow[\text{H}_2\text{O}]{C_9H_{19}COOH} \text{3-substituted indole} \quad (2\text{-}51)$$

2014 年 Halimehjani 等[75]报道了一种简单、温和、环境友好的方法用于 N,N-二烷基苯胺与硝基烯 Michael 类型的 Friedel-Crafts 烷基化反应[式(2-52)]。这种方法采用以水为溶剂，较少量(0.0175 mol%)的杂多磷钨酸催化剂(HPW)，可以给出优良得率(83%~95%)的产物。水在此反应体系中与硝基烯形成氢键，有利于反应的发生，从而提高了杂多磷钨酸在均相体系中的催化性能。因此水既是溶剂，也是共催化剂。

$$\text{Ar}(R,R) + \text{ArCH=CHNO}_2 \xrightarrow[\text{H}_2\text{O, 回流}]{\text{HPW}} \text{product} \quad (2\text{-}52)$$

X=EWG 或 EDG；R=CH$_3$或 C$_2$H$_5$

3. 加成反应

由于自然界各种有机物中酯键和缩醛的结构比较普遍，因而酯化和形成缩醛的反应是自然界中比较常见的有机反应，这种酯化和形成缩醛的反应通常是通过酶催化完成的，采用化学方法在水中进行此类反应的报道较少。Kobayashi 等研究发现，十二烷基苯磺酸(DBSA)也可以催化水中酯化和形成缩醛的反应，在 DBSA 催化下，水中各种烷基羧酸和醇能顺利地转化成酯，带有较长烷基链的羧酸比较短链羧酸更容易生成相应的酯,因此,两个有不同长度烷基链的羧酸可以实现选择性的酯化反应[式(2-53)][36]。

$$RCOOH + R_1OH \xrightarrow[H_2O, 48\ h, 40\ ℃]{10\ mol\%\ DBSA} RCOOR_1 \qquad (2\text{-}53)$$

也有报道将磷钨酸用于水中吡咯与含有几种缺电子基团的烯烃 Friedel-Crafts 烷基化反应[式(2-54)][77],在室温下,目标产物得率为 74%~97%。有趣的是,在水中吲哚与甲基烯丙基酮的 Michael 加成反应中,$CeCl_3·7H_2O$,$FeCl_3·6H_2O$,$ZrCl_4$,WCl_6 和 H_3PO_4 具有较高的催化活性,并给出较高得率的产物。杂多酸如磷钨酸、磷钼酸同样可催化这个反应,并在短时间内就能给出 74%~79% 得率的产物,除了 N-甲基吲哚给出更高的产物得率,其他取代的吲哚得率比较相近。由于烯烃是 Michael 接受体,所以对于含有缺电子的烯烃如甲基烯丙基酮、查尔酮、氯代查尔酮和 β-硝基苯乙烯与吲哚在室温下均能顺利地进行反应,产物也具有较高的选择性[77]。最近,Kobayashi 等[78]用 DBSA 在 23℃催化纯水中醛与二取代芳胺和酮的加成反应[式(2-55)]。

$$\text{吡咯-NH} + R \diagup\!\!\!\diagup X \xrightarrow[H_2O,\ rt]{H_3PMo_{12}O_{40}} X\text{-吡咯-}R + X\text{-吡咯-}R\ R\text{-}X \qquad (2\text{-}54)$$

R = H, X = COCH_3
R = Ph, X = COPh 或 NO_2

$$R_1CHO + \underset{NH_2}{\overset{R_2\ R_3}{\text{Ar}}} + \underset{R_4}{\overset{O}{\|}}R_5 \xrightarrow[H_2O,\ 23℃]{DBSA} \underset{R_1\ R_4}{\overset{R_2\ R_3}{\text{Ar-NH}}}\overset{O}{\|}R_5 \qquad (2\text{-}55)$$

4. 消除反应/取代反应

水介质中醇的经典反应是酸催化的亲核取代和消除反应。有强酸如浓硫酸存在时,醇可以发生亲核取代反应生成醚和消除产物烯烃。在超临界水中,385℃和 34.5 MPa 压力下,乙醇被清洁转化成乙烯和乙醚[79]。近来,Kobayashi 等[60]报道了 Brønsted 酸结合表面活性剂的催化剂催化醇和酸反应生成酯[式(2-56),式(2-57)],以及醇和硫醚脱水生成醚[式(2-58)]、硫醚[式(2-59)]和二硫缩醛 [式(2-60)]。反应体系中加入表面活性剂 DBSA,反应在 35~80℃水中进行[60]。与经典的强酸性催化剂催化的反应相比,该反应条件比较温和,选择性好。例如,采用 10 mol% 的 DBSA,水中苄醇的取代反应可以顺利地进行,给出高产率(~90%)的对称的醚。在相同条件下,若以 TsOH 代替 DBSA,相同底物的醚化反应仅得到微量的产物,若两种醇,如伯醇和仲醇同时存在,尽管用 2eq 苄醇,仍可选择性地进行不对称醚化反应。通过调节疏水性,可以实现选择性的醚化反应。当等量的十二烷基醇与丙醇、二苯甲醇在水中进行醚化反应时,加入 DBSA 可选择性地生成二苯甲基醚,同样也能实现硫醇和醇之间的硫醚化反应[式(2-59)]。DBSA 反应体系也适用于醛和酮与 1,2-乙二硫醇的二硫缩醛化反应,生成相应的二硫缩醛[式(2-60)],提高反应温度,产物的得率反而降低。有趣的是,增加表面活性剂浓度,产物的得率随之降低,而缩短表面活性剂的烷基碳链长度,将会丧失其催化性能。光学显微镜的观察结果表明该过程形成了胶束,从而形成了疏水环境,降低了水的有效浓度,使脱水反应更易进行。

$$R_1COOH + R_2COOH + ROH \xrightarrow[H_2O, 48\text{ h}, 40℃]{10\text{mol}\% \text{ DBSA}} R_1COOR + R_2COOR \quad (2\text{-}56)$$

$$RCOOH + R_1OH \xrightarrow[H_2O, 48\text{ h}, 40℃]{10\text{mol}\% \text{ DBSA}} RCOOR_1 \quad (2\text{-}57)$$

$$R_1OH + ROH \xrightarrow[H_2O, 24\text{ h}]{10\text{ mol}\% \text{ DBSA}} R_1OR \quad (2\text{-}58)$$

$$R_1OH + RSH \xrightarrow[H_2O, 24\text{ h}, 80℃]{10\text{mol}\% \text{ DBSA}} R_1SR \quad (2\text{-}59)$$

$$\underset{R}{\overset{O}{\|}}\!\!\!-\!\!R_1 + HS\!\!-\!\!\!\frown\!\!\!-\!\!SH \xrightarrow[H_2O, 4\text{ h}, 40℃]{10\text{mol}\%\text{DBSA}} \underset{R\ R_1}{\overset{S\ \ S}{\diagdown\diagup}} \quad (2\text{-}60)$$

$H_3PW_{12}O_{40}$ 也可用于催化水中醛和酮与吲哚或取代吲哚的亲电取代反应,在室温下生成高得率的二(吲哚基)甲烷[式(2-61)],这种方法不仅对醛具有较高的化学选择性,而且 $H_3PMo_{12}O_{40}$ 和 $H_3PW_{12}O_{40}$ 无论是在水中还是在其他有机溶剂中,均具有相同的催化效果[80]。吲哚也可与 50%乙醛酸在 $H_3PW_{12}O_{40}$ 催化下发生亲电取代反应。

$$\underset{R_1}{\text{indole-}R_2} + R_3CHO \xrightarrow[H_2O, \text{rt}, 2\sim 8\text{ h}]{H_3PW_{12}O_{40}} \text{bis(indolyl)methane} \quad (2\text{-}61)$$

Kobayashi 和 Shirakawa[81]设计一种表面活性剂类型的 Brønsted 酸,DBSA 催化水中醇的亲核取代反应[式(2-62)]。该催化体系可以应用到 1-羟基糖的糖基化反应中[式(2-63)]。

$$\underset{Ar}{\overset{R_1\ R_2}{\diagdown\diagup}}\text{OH} + \text{Nu-H} \xrightarrow[\text{H}_2\text{O}]{10\text{mol}\%\ \text{DBSA}} \underset{Ar}{\overset{R_1\ R_2}{\diagdown\diagup}}\text{Nu} \qquad (2\text{-}62)$$

$$\text{(糖-OH)} + \text{Nu-H} \xrightarrow[\text{H}_2\text{O}]{10\text{mol}\%\ \text{DBSA}} \text{(糖-Nu)} \qquad (2\text{-}63)$$

5. Mannich 类型反应

1999 年 Manabe 和 Kobayashi[82]报道 Brønsted 酸-表面活性剂结合的催化剂 DBSA 催化水中醛、胺和酮的三组分 Mannich 反应[式(2-64)]。反应在室温下给出较好得率的 β-氨基酮。有趣的是在水中产物的得率接近 100%，而在有机溶剂甲醇或 CH_2Cl_2 中，产物得率不到 10%。他们也将 DBSA 用于催化水中醛、芳香胺与甲硅烷基烯醇[式(2-65)]或酮[式(2-66)]或烯丙基锡[式(2-67)]或 3-三甲硅氧基-1-甲氧基丁二烯[式(2-68)]的三组分 Mannich 类型反应。除肉桂醛外，其他醛进行的 Mannich 反应产物的得率在 78%~90%[83]。

$$\text{RCHO} + \text{R}_1\text{NH}_2 + \underset{R_2}{\overset{O}{\underset{\|}{\text{C}}}}\text{R}_3 \xrightarrow[\text{H}_2\text{O}]{\text{DBSA}} \underset{R\ R_2}{\overset{R_1\text{NH}\ O}{\diagdown\diagup}}\text{R}_3 \qquad (2\text{-}64)$$

$$\text{R}_1\text{CHO} + \text{Ar-NH}_2(R_2) + \underset{R_4}{\overset{\text{OTMS}}{\underset{R_3}{\diagdown\diagup}}}\text{R}_5 \xrightarrow[\text{H}_2\text{O, 2 h}]{10\text{mol}\%\text{DBSA}} \text{产物} \qquad (2\text{-}65)$$

$$\text{R}_1\text{CHO} + \text{Ar}(R_2,R_3)\text{NH}_2 + \underset{R_4}{\overset{O}{\|}}\text{R}_5 \xrightarrow[\text{H}_2\text{O, 23℃}]{10\text{mol}\%\ \text{DBSA}} \text{产物} \qquad (2\text{-}66)$$

$$\text{PhCHO} + \text{R-C}_6\text{H}_4\text{-NH}_2 + \text{CH}_2=\text{CHCH}_2\text{SnBu}_3 \xrightarrow[\text{H}_2\text{O, 42 h}]{10\text{mol}\%\ \text{DBSA}} \text{产物} \qquad (2\text{-}67)$$

$$PhCHO + \underset{NH_2}{\underset{|}{C_6H_4OCH_3}} + \underset{OCH_3}{\overset{OTMS}{CH_2=C-CH=CH}} \xrightarrow[H_2O,\ 1\ h,\ 0\ ℃]{10mol\%\ DBSA} \underset{Ph}{\text{产物}} \quad (2\text{-}68)$$

2002 年 Kobayashi 和 Akiyama 课题组[84-87]报道 HBF$_4$ 是水介质 Mannich 反应的一个很好的催化剂。在加入少量的 SDS 后，醛、胺和烯醇硅醚在纯水中就能顺利地进行反应[式(2-69)]。因为在此体系中 SDS 既是一种 Brønsted 酸，也是一种表面活性剂，即作为催化剂催化该反应时，就不需要再向反应体系中加入表面活性剂。他们采用 HBF$_4$ 催化硅烯醇与醛亚胺进行的 Mannich 类型的反应，可以成功地给出高得率的 β-氨基羰基化合物，这也实现了用 Brønsted 催化水中 Mannich 反应。

$$ArCHO + R_1NH_2 + \underset{R_3}{\overset{R_4}{C=C-OTMS}}_{R_2} \xrightarrow[SDS]{HBF_4} \underset{R_2\ R_3}{\overset{ArHN\ \ O}{\underset{R_1}{C-C}}} \quad (2\text{-}69)$$

Akiyama 等[87]也将此体系应用在烯酮硅缩醛与醛亚胺的 Mannich 类型的非对映选择性反应[式(2-70)]。烯酮硅缩醛可以从 α-氧代酯制得。实验结果表明，在水/异丙醇中，采用芳基酯制备的烯酮硅缩醛生成较高选择性的 anti-β-氨基-α-硅氧基酯，而在水中，采用从甲酯得到的烯酮硅缩醛，加入一定量的 SDS，则优先给出 syn-异构体的产物。

$$(2\text{-}70)$$

2005 年 Akiyama 等[88]发展一种 HCl-SDS 催化体系用于水芳香醛、芳香胺和环己酮的 Mannich 类型的三组分反应，得到反式的 β-氨基酮[式(2-71)]。在室温下，10% HCl-10 mol%催化体系可以使反应顺利进行，反应给出较高得率(87%)和立体选择性(anti∶syn = 99∶1)的目标产物，无论是得率和立体选择性均高于催化剂 CF$_3$COOH 和 HBF$_4$。

$$Ar_1CHO + Ar_2NH_2 + \underset{\text{cyclohexanone}}{\text{C}} \xrightarrow[\text{H}_2\text{O, rt, 1d}]{10\%\text{HCl-10 mol}\%\text{SDS}} \underset{Ar_1}{\overset{Ar_2HN}{\text{C}}} + \underset{Ar_1}{\overset{Ar_2HN}{\text{C}}} \quad (2\text{-}71)$$

有报道将杂多酸 $H_3PMo_{12}O_{40}$ 应用于水中醛、胺和 Mannich 类型的反应[89]，如在室温下，加入 10~20 mg 的催化剂，反应 3~18 h，得到理想产物 β-氨基酮，得率为 63%~94%。2007 年 Heravi 等[90]报道用 $H_{14}[NaP_5W_{30}O_{110}]$ 催化醛、丙二腈与 α,β-萘酚的反应，生成 2-氨基-4H-查尔酮[式(2-72)]，在回流条件下，产物得率高达 90%~93%。

$$RCHO + NCCH_2CN + \underset{X,Y=H \text{ 或 OH}}{\text{naphthol}} \xrightarrow[\text{H}_2\text{O, 回流}]{H_{14}[NaP_5W_{30}O_{110}]} \text{product} \quad (2\text{-}72)$$

R=Ph, 4-CH$_3$Ph, 4-ClPh, 3-NO$_2$Ph, 4-NO$_2$Ph

90%~93%

最近，Gao 等[91]合成了一种水溶性的 6-磺酸基水杨酸，并将其应用在水中 Mannich 类型反应。作者考察了不同的芳香醛、芳香胺与环己酮的反应活性，在室温下反应能给出较高得率和非对映选择性的 β-氨基羰基化合物，且该催化剂不需要从反应体系中分离便可以直接循环使用，使用 6 次后，产物的得率仍保持在 90%以上。他们也成功合成了不同 Brønsted 酸-表面活性剂结合催化剂(BASCs)，并将其应用在三组分 Mannich 类型的水相反应中。在 25℃下，3-(N, N-二甲基辛基铵)戊磺酸苯磺酸盐([DOPA][Tos])能在反应过程中形成微乳，显示出了最佳的催化活性，反应 4.5 h 后产物得率达 89%。这个催化过程中催化剂通过简单的分离就能重复使用，使用 9 次后，催化活性没有明显降低(得率 85%)[92]。

Mukhopadhyay 等[93]发展了硼酸-甘油催化水中芳香醛、芳香胺和环酮的 Mannich 类型的反应。研究结果显示，甘油对反应产物的得率和选择性具有非常重要的作用，即加入 1~2 滴甘油，产物得率明显提高，但对产物立体选择性没有明显影响。

最近，Javanshir 等[94]将 Wells-Dawson 类型杂多酸用于水中 2-氨基苯并噻唑、2-萘酚和醛的"一锅"反应，在 45℃超声作用下给出 Mannich 的加合物 2′-氨基苯并噻唑甲基萘酚[式(2-73)]。含有吸电子基团(NO$_2$, Cl, CN)及给电子基团(Me, OMe)的醛都能给出较高得率(62%~92%)的 Mannich 加合物。这个反应体系也可以用于 3-氨基-1,2,4-三唑的反应，但是对于 2-氨基苯并咪唑，即使提高反应温度至 90℃，延长反应时间至 24 h，产物的得率也不是很理想。

$$RCHO + \text{[2-naphthol]} + \text{[2-aminobenzothiazole]} \xrightarrow[\text{超声}, 45°C]{H_2O/HPA} \text{[product]} \quad (2\text{-}73)$$

6. Diels-Alder 反应

Danishefsky 二烯虽然具有比较高的反应活性,但在水中非常不稳定,研究发现,如在反应体系中加入阴离子表面活性剂 SDS 后,在 HBF_4 催化下,三甲硅氧基丁二烯甲醚可以顺利地和亚胺在水中反应,得到较高产率(75%~88%)的氮杂六元环化合物[式(2-74)][95]。

$$RCHO + \text{[p-anisidine]} + \text{[TMS diene]} \xrightarrow[\text{H}_2\text{O, rt, 1 h}]{HBF_4, SDS} \text{[product]} \quad (2\text{-}74)$$

$R = o\text{-}C_6H_{11}; 75\%$
$R = Ph; 88\%$
$R = p\text{-}MePh, PhCH=CH; 86\%$

7. 氧化反应

1986 年 Ai 报道了 $H_{3+n}PMo_{12-n}V_nO_{40}(n = 1\sim2)$ 及其盐是异丁酸氧化脱氢生成甲基丙烯酸的高效催化剂[96]。Saidi 和 Azizi[97]报道称,杂多酸 $H_3PW_{12}O_{40}$ 是水中环氧化物与芳胺发生开环生成 β-氨基醇的优良催化剂[式(2-75)]。苯胺、对甲氧基苯胺、对氯苯胺、对溴苯胺和对异丙基苯胺与脂肪族的环氧化物,如环氧丙基苯基醚、环氧丙基异丙基醚、1,2-环氧丁烷、1,2-环氧戊烷和烷基 2,3-环氧丙基醚等反应生成相应的氨基醇[97]。

$$\text{[epoxide]} + ArNH_2 \xrightarrow{0.35 \text{ mol}\% \ H_3PW_{12}O_{40}} \underset{OH}{\overset{NHAr}{R}} + \underset{OH}{\overset{OH}{R}} \quad (2\text{-}75)$$

有报道采用 Pd^{2+} + HPA-n, (HPA-n=$H_{3+n}PMo_{12-n}V_nO_{40}$, $n=1\sim4$)催化体系可以使水中的丙烯选择性地氧化成丙酮,动力学研究表明,Pd^{2+}+HPA-n(n=1~4)中钒原子的数目对体系的催化活性没有明显的影响[98,99]。Matveev 等[100]报道以分子氧为氧化剂,P-Mo-VHPAs 为催化剂催化水中 2,6-二甲基和 2,6-二叔丁基苯酚氧化生成 2,6-二芳基-1,4-苯基醌[式(2-76)]。

$$\text{(phenol with R groups, R=CH}_3\text{, }t\text{-Bu)} + O_2 \xrightarrow{\text{P-Mo-VHPAs}} \text{(benzoquinone with R groups)} \quad (2\text{-}76)$$

Afonso 和 Rosatella[101]用对甲基苯磺酸(PTSA)催化 H_2O_2 溶液中环己烯的双羟基化反应，得环己二醇[式(2-77)]。该催化体系对于体积较大的烯烃的氧化反应[式(2-78)]，同样具有非常好的催化效率。在环己烯的氧化反应中，该催化剂可以重复使用 8 次，但使用到第 8 次，得到的产物得率仅为 65%。

$$\text{环己烯} \xrightarrow[50\ ^\circ\text{C}]{\substack{20\ \text{mol\% PTSA} \\ 2\ \text{eq}\ H_2O_2}} \text{环己二醇} \quad (2\text{-}77)$$

$$\underset{R_2}{\overset{R_1}{\diagdown}}= \xrightarrow[50\ ^\circ\text{C}]{\substack{20\ \text{mol\% PTSA} \\ 2\ \text{eq}\ H_2O_2}} \underset{R_2}{\overset{R_1}{\diagdown}}\text{(OH)(OH)} \quad (2\text{-}78)$$

8. "一锅"反应

Li 等[102]采用 DBSA 作为 Brønsted 酸-表面活性剂结合类型的催化剂，在水介质中实现了 1,8-二氧六氢吖啶的"一锅"合成。即 5,5-二甲基-1,3-环己烷二酮与 4-甲基苯胺和 4-氯苯甲醛的 Knoevenagel 反应/共轭加成/环化/消除反应串联在一起，生成 9-(4-氯苯基)-3,4,6,7,9,10-六氢-3,3,6,6-四甲基-1,8-(2H, 5H)-吖啶二酮[式(2-79)]。

$$\text{PhNH}_2 + \text{ArCHO} + \text{(5,5-dimethylcyclohexane-1,3-dione)} \xrightarrow[H_2O,\ 回流]{\text{DBSA}} \text{(acridinedione product)} \quad (2\text{-}79)$$

也有报道称 Wells-Dawson 类型酸 $H_6P_2W_{18}O_{62}$ (5mol%)催化水中吲哚与不同靛红的亲电取代反应，在 60 ℃反应 30 min 即给出 95%得率的 3,3-二(吲哚基)-羟吲哚[式(2-80)]。在所有反应中，亲电活化反应仅在 3-位可以发生羰基化反应，而 2-位没有活性，这可能是由于吲哚中的氮稳定[103]。

$$\text{(2-80)}$$

Rashidi 等[104]报道对甲基苯磺酸催化 4-羟基香豆素与芳香乙二醛在水中反应生成双香豆素，产物的得率在 65%以上，这种方法提供了一种直接合成各种结构香豆素衍生物的方法。最近，Amrollahi 和 Kheilkordi 报道[105]用 $H_3PW_{12}O_{40}$ 催化水中二吲哚衍生物的合成方法。这种合成方法主要以醛、吲哚和活泼亚甲基化合物三种组分为原料，在超声作用下通过"一锅"法在短时间内(10min)即可给出较好得率(95%)的产物[式(2-81)]。

$$\text{(2-81)}$$

9. 其他反应

Bamoharram 等[106]报道 $H_3PMo_{12}O_{40}$ 均相催化硝基化合物的加氢反应，反应以水合肼为还原剂，在不同醛的水溶液中，得到产物喹啉衍生物。$H_3PMo_{12}O_{40}$ 还可以在室温下催化水中高烯丙醇与醛的 Prins 环化反应，生成四羟基吡喃-4-醇[式(2-82)]，得率为 80%~90%，只有环酮能得到螺环产物[式(2-83)][107]。杂多酸也可以催化水中不同芳香胺与环氧化物的开环反应，生成中等到优良得率的 β-氨基醇[式(2-84)]。

$$\text{(2-82)}$$

$$\text{(2-83)}$$

$$R_1\triangle + R_2R_3NH \xrightarrow[H_2O]{H_3PMo_{12}O_{40}} R_1\overset{OH}{\underset{}{C}}\!\!-\!\!NR_2R_3 \quad (2\text{-}84)$$

$R_1=CO_2C_2H_5$, $R_2=Ph$, 得率= 60%
$R_1=CO_2C_2H_5$, $R_2=$ 2-Naphthyl, 得率= 61%
$R_1=CN$, $R_2=Ph$, 得率= 92%

Palkovits 等[108]考察了硅钨酸在 160~200℃ H_2 氛围中催化水介质木糖醇和山梨醇脱水生成 1,4-木糖醇酐和 1,4-山梨醇酐。Ren 等[109]用 DBSA 催化反式 3-烯基吲唑衍生物的立体选择性合成。在超声作用下以杂多酸为催化剂,2-磺酰基苯甲酸酐与酚为原料,合成羟基三芳基甲烷[式(2-85)]。其中,$H_{14}[NaP_5W_{30}O_{110}]$给出最高得率的产物,且其活性高于负载性催化剂的性能。各种杂多酸在反应体系中的活性顺序为 $H_{14}[NaP_5W_{30}O_{110}]>H_3[PW_{12}O_{40}]_4>[SiW_{12}O_{40}]$。含有给电子基团的苯酚,相应产物的得率较低。

$$(2\text{-}85)$$

第二节 碱 催 化 剂

碱催化是指催化剂与反应物分子之间通过接受质子或给出电子对作用,形成活泼的负碳离子中间化合物(活化的主要方式),继而分解为产物的催化过程。相对于酸催化剂,碱催化剂的研究起步相对较晚,但其在有机合成中的应用比较广泛,如缩合反应、加成反应、取代反应、消除反应等。

一、无机碱催化剂

1. 缩合反应

(1) Henry 反应。

1997 年,Ballini 和 Bosica 用 0.025 mol/L NaOH 催化水中硝基烷和醛的缩合反应[式(2-86)],为了提高反应底物在水中的溶解度,作者加入了阳离子表面活性剂十六烷基三甲基氯化铵(CTACl)[110]。在此反应体系中,脂肪族醛与硝基烷反应时生成的 β-硝基醇,产物得率高于芳香族醛。在这样的条件下,某些官能团可以保留,且避免了副反应的发生,如逆羟醛缩合反应或2-硝基醇的脱水反应。另外,伯硝基烷反应或仲硝基烷反应活性要高于三级硝基烷,在2~6 h 就能得到较高得率的产物。

$$\underset{R}{\overset{NO_2}{\underset{R_1}{\bigwedge}}} + R_2CHO \xrightarrow[CTACl]{NaOH} \underset{R}{\overset{O_2N}{\underset{R_1}{\bigvee}}}\overset{R_2}{\underset{OH}{\bigvee}} \qquad (2\text{-}86)$$

Ben 和 Armi 报道了采用 6~10 mol/L K_2CO_3 水溶液为碱,用于 2-烷基-1, 3-二酮与甲醛水溶液的反应,给出羟醛缩合产物,然后被碱解离给出乙烯基酮[式(2-87)][111]。Wang 等[112]用聚苯乙烯负载三丁基氯化铵作为相转移试剂,以 KOH 为催化剂催化纯水中硝基烷与芳香醛的反应,在室温下,产物的得率为 64%~94%。作者也研究了这种负载型相转移试剂的使用寿命。

$$\underset{CO_2Me}{\overset{O\quad O}{\bigvee}} \xrightarrow[K_2CO_3, H_2O]{30\%\ aq.\ HCHO} \underset{O}{\overset{CO_2Me}{\bigvee}} \qquad (2\text{-}87)$$

(2) Aldol 反应。

经典的 Aldol 反应通常都是用 NaOH、KOH、EtONa 和 Na_2CO_3 等强碱来进行催化的,最常见的溶剂是 C_2H_5OH、水、CH_3CN 等,但是通常都存在选择性差、副产物多等问题,且很容易得到脱水产物 α, β-不饱和羰基化合物。化学工作者对经典的 Aldol 反应进行了改进,取得了许多令人瞩目的成绩。

Mase 等[113]报道了一种手性的含有较长疏水基团的二元胺类催化体系。该体系是一种在纯水介质中产率高(高达 99%)、选择性高(de 值可达 88%,ee 值可达 99%)的催化 Aldol 反应,得到 β-羟基酮产物。他们认为在此种催化体系中,催化剂的疏水作用可促使反应底物进入有机相,同时可以将水排斥在外,从而在有机相中可以聚集高浓度的反应物,使反应顺利进行,给出理想的目标产物。

Jayaram 等[114]报道用 NaOH 催化水相环酮与芳香醛的交叉 Aldol 缩合反应,给出高得率的 α, α'-二(取代苯亚甲基)环烷酮产物[式(2-88)]。通过对比实验发现,CTAB 对此反应具有非常重要的作用。如在反应体系不使用 CTAB,反应几乎不能发生,但向反应体系中加入 15 mmol/L CTAB,反应 8 h 后就能给出产物,得率高达 95%。然而,如果加入较多的 CTAB(20 mmol/L),反应速率不但没有加快,反而减慢,即使反应 24 h,产物的得率也只有 80%。这个反应体系也适用于水相醛与活泼亚甲基试剂的 Knoevenagel 缩合反应[式(2-89)],给出相应产物的得率为 90%~95%。

$$\text{环酮} + \text{4-R-苯甲醛} \xrightarrow[\text{H}_2\text{O, 60 ℃}]{\text{NaOH, CTAB}} \text{双苄叉环酮} \quad \text{R = H, OCH}_3\text{, CH}_3 \tag{2-88}$$

$$\text{RCHO} + \underset{\text{CN}}{\overset{R_1}{\text{CH}}} \xrightarrow[\text{rt}]{\text{CTAB, H}_2\text{O}} \underset{\text{H}\quad\text{CN}}{\overset{R\quad R_1}{\text{C=C}}} \tag{2-89}$$

Wang 等[115]首次报道用 Na_2CO_3 在室温下催化水中醛与苯乙酮或环酮(4-硝基环己酮、3-硝基环己酮)、硝基苯甲醛的 Aldol 缩合反应,给出 β-羟基酮[式(2-90)]。此催化体系的优点是:100%原子经济性,催化剂 Na_2CO_3 便宜易得,反应容易大规模生产 β-羟基酮。

$$R_1\text{COCH}_2R_2 + \text{ArCHO} \xrightarrow[\text{H}_2\text{O, rt}]{\text{Na}_2\text{CO}_3} R_1\text{CO-CR}_2\text{-CH(OH)Ar} \tag{2-90}$$

Sobhani 和 Parizi 报道[116]用硬脂酸钠催化亚磷酸酯与芳香醛/杂环醛/脂肪醛和丙二腈在水相进行 Knoevenagel-Phospha-Michael 串联反应,给出产物 β-膦酸丙二腈衍生物[式(2-91)]。不同的醛在此反应中能给出相应的产物,产物得率为 82%~91%。这方法的优点是使用了价廉、环境友好的反应介质,且使用较少的催化剂(硬脂酸钠)就可以获得较高得率的产物。

$$\text{RCHO} + \underset{\text{CN}}{\overset{\text{CN}}{\text{CH}_2}} + \text{PH(OC}_2\text{H}_5)_3 \xrightarrow[\text{80 ℃}]{\text{CH}_3(\text{CH}_2)_{16}\text{COONa}} (C_2H_5O)_2\overset{O}{\underset{}{P}}\text{-CH(R)-CH(CN)}_2 \tag{2-91}$$

Brahmachari 报道[117]一种简单、方便、实用的方法用于在室温下"一锅"合成具有潜在生物活性的香豆素-3-羧酸。这种方法主要是通过价廉易得的 K_2CO_3 或 NaN_3 为催化剂,使各种 2-羟基苯甲醛与 Meldrum 酸发生 Knoevenagel 缩合和分子内环化反应[式(2-92)]。这方法的优点有:反应条件温和,得率高,不需经过柱色谱层析就可分离,且反应规模可以放大到克级。

$$\text{X-C}_6\text{H}_3(\text{CHO})(\text{OH}) + \text{Meldrum 酸} \xrightarrow[\text{H}_2\text{O, rt, 20 h}]{\text{K}_2\text{CO}_3 \text{ 或 NaN}_3} \text{香豆素-3-羧酸} \tag{2-92}$$

Cai 和 Lu[118]报道 K_2CO_3 碱催化的有机卤化物、硫脲和苯酰氯的三组分"一锅"反应可以在 TX100 水相胶束中完成[式(2-93)]。加入表面活性剂 TX100 可以有效地减缓苯酰氯的水解速率,从而在水介质中实现苯酰氯参与的反应。与已报道的方法相比,该方法可以有效地避免具有恶臭的硫醇和有毒、易挥发的有机溶剂的使用,同时后处理十分简单、易行,这些优点大大提高了其在全合成以及大规模生产中应用的可能性。

$$RCH_2X + \underset{NH_2}{\underset{\|}{H_2N}}\overset{S}{C} + R_1\text{-}C_6H_4\text{-}COCl \xrightarrow[H_2O, rt, 20 h]{K_2CO_3, TX100} RCH_2S\text{-}CO\text{-}C_6H_4\text{-}R_1 \quad (2\text{-}93)$$

Bhutani 等[119]报道用 LiOH 催化水中取代喹啉-2,4-二醇的 C-3 选择性烷基化反应,给出产物 C-3-二烷基喹啉-2,4-二酮衍生物[式(2-94)]。这个方法简单易操作,产物得率高。

$$\text{(quinoline-2,4-diol)} \xrightarrow[RX' (X'=Br, I)]{LiOH, H_2O, 80℃} \text{(C-3-dialkylquinoline-2,4-dione)} \quad (2\text{-}94)$$

Pizzo 等[120]采用"一锅"方法,在碱性水介质中芳香醛与芳基甲基酮的缩合反应可以生成 7- 和 2′,4′-取代的黄酮醇[式(2-95)]。羟基甲氧基取代的苯乙酮与胡椒醛的缩合反应可以在 KOH 碱性水(5 mol/L)中进行。然后在同一反应中查尔酮被 35% H_2O_2 溶液氧化给出 70% 得率的黄酮醇。水相羟醛缩合反应在合成中的另一个应用就是合成胭脂红酸。

$$\text{(2′-hydroxy-4′-methylacetophenone)} \xrightarrow[2) H_2O_2]{1) ArCHO, 5 mol/L KOH} \text{(3-hydroxyflavone)} \quad (2\text{-}95)$$

Fan 等[121]报道 NaOH 催化 5-羟甲基呋喃甲醛与乙酰丙酸在水中的缩合反应,生成混合 Aldol 产物[式(2-96)]。相似的糠醛和乙酰丙酸在此条件下反应生成线型聚合物,得率为 91%。

$$\text{HMF} + \text{levulinic acid} \xrightarrow{NaOH, H_2O} \text{product 1} + \text{product 2} \quad (2\text{-}96)$$

Tamaddon 等[122]采用 $CaMg(CO_3)_2$ 催化水中 Henry、Knoevenagel 和 Michael 反应，用于形成 C—C、C—N 和 C—S 键。例如，在硝基甲烷与苯甲醛的 Henry 反应中，催化剂在使用后经过滤、乙酸乙酯洗涤干燥后对其进行元素分析，发现催化剂的化学组分基本保持不变，表明催化剂具有较好的稳定性。其在 Henry 反应中可以重复使用三次活性不降低，在 Knoevenagel 反应中使用两次也未发现活性降低。

2. 加成反应

Iwamura 等[123]报道在 KOH 水溶液中乙酰乙酸乙酯、丙二酸酯与丙烯酸酯的 Michael 加成反应给出中等到较高得率的加合物，反应中未观测到副反应的发生。叔丁基或苯基酯在此强碱溶液中相当稳定，并顺利地进行反应[式(2-97)]。产物得率为 73%~79%。

$$R\text{-CO-CH}_2\text{-CO-CH}_2\text{-OR}' + \text{CH}_2=\text{CH-CO-OR}'' \xrightarrow[\text{H}_2\text{O, rt}]{\text{KOH}} \text{产物} \quad (2\text{-}97)$$

Mashraqui 等[124]报道了在含有 CTAB 的水胶束介质中亚苄基苯乙酮和亚苄基苯丙酮与乙酰乙酸乙酯、硝基甲烷和乙酰丙酮的 Michael 加成反应[式(2-98)]。不同的硝基甲烷与烯烃的加成反应能在含有 CTAB 的 NaOH 水溶液(0.025～0.1 mol/L)中进行，反应体系中不用加入任何有机溶剂。

$$\text{PhCH=CH-CO-R} + \text{XCH}_2\text{Y} \xrightarrow[\text{rt}]{\text{CTAB, 0.1\%aqNaOH}} \text{产物} \quad (2\text{-}98)$$

Menéndez 等[125]报道在室温下，0.07 mol/L K_2CO_3 水溶液中，很多不饱和的羰基衍生物都能与 α-硝基环烷酮选择性地进行 Michael 加成反应[式(2-99)]，相应产物的得率高达 99%。

$$\text{环己酮-NO}_2 + \text{CH}_2=\text{CH-Z} \xrightarrow{0.07 \text{ mol/L K}_2\text{CO}_3} \text{产物} \quad (2\text{-}99)$$

Z = $COCH_3$, COC_2H_5, CHO, CO_2CH_3, CN, SO_2Ph

醇可以和炔或烯烃发生加成反应，该反应的活性取决于烯烃和炔烃的性质，通过亲核或亲电的途径给出目标产物，这些反应在水中常会与水合反应发生竞争。然而，已有报道，在水存在时，以 Brønsted 碱为催化剂，1,3-丁二烯与醇的醚化反应能高效、高选择性地制备烷氧基丁烯 $CH_3CH=CHCH_2OR$ 和 $CH_2=CH(OR)CH_3$，其中 R 可以是无取代的烷基、烯基、炔基、环烷基、芳基等。反应过程中还会有还原二聚体副产物的生成。2,2,2-三氟乙醇与氟代烯烃发生加成反应生成氟代醚[式(2-100)]，增加反应中质子源的量有利于 A 产物的生成，此外，以水为溶剂是该反应最有效的质子源，同时也能使反应的后处理更加简单，这也便于扩大反应的用量规模。例如，采用有机溶剂如 1,4-二氧六环，主要生成加成-消除产物 $B^{[126]}$。

$$CF_3CH_2OH + CF_2=CXY \xrightarrow{KOH} \underset{A}{CF_3CH_2OCF_2CHXY} + \underset{B}{CF_3CH_2OCF=CXY} \quad (2\text{-}100)$$

3. 取代反应

在碱性条件下，醇或酚与卤代烃的 S_N2 亲核取代反应是经典的 Willianmson 反应，该反应是合成混合醚的常用方法。例如，在三甲胺修饰的水溶性杯芳烃($n=4, 6, 8$)如杯[4]芳烃催化 NaOH 水溶液中，醇和苯酚与烷基卤化物的烷基化反应促进作用，反应的产物得率从良好到高[127]。

芳卤在常规加热的条件下很难发生亲核取代反应，一般需要高温、较长的反应时间，且芳卤的苯环必须含有吸电子基团，才能与胺或醇等亲核试剂反应。Cherng 报道[128]在邻位和对位均含有硝基时，芳基卤才能发生亲核取代反应，如 2,4-二硝基氟苯与氨基酸在微波作用下反应较短的时间，给出 N-芳基-α-氨基酸衍生物[式(2-101)]。如向体系中加入一定量的 $NaHCO_3$ 可以进一步提高氨基酸的亲核能力，在 80℃ 微波 3~40 s 给出产物得率为 91%~93%。

$$O_2N\text{-}C_6H_3(NO_2)\text{-}F + H_2N\text{-}CHR\text{-}CO_2H \xrightarrow[80℃]{H_2O, NaHCO_3} O_2N\text{-}C_6H_3(NO_2)\text{-}NH\text{-}CHR\text{-}CO_2H \quad (2\text{-}101)$$

Semple 等[129]报道将 $NaHCO_3$ 应用于 3-硝基-4-氟苯与胺的取代反应得到多取代的芳香胺。Degani 等[130]报道在四丁基溴化铵和 30% KOH 水溶液中，$(RS)_2CO(R=CH_3, C_2H_5, n\text{-}C_4H_9)$ 与 $R_1X(X=Cl, Br, I,$ 对甲苯磺酰基，甲基磺酰基)可以顺利地进行反应[式(2-102)]，给出产物的得率为 93%~100%。

$$\begin{matrix}R\text{-}S\\R\text{-}S\end{matrix}\!\!=\!O + R_1X \xrightarrow[30℃\ \text{aq. KOH}]{(n\text{-}C_4H_9)_4NBr} 2R_1SR \quad (2\text{-}102)$$

Song 和 Peng 等[131,132]发现,在微波和超声共同作用下,苯酚和氯代烷烃或卤代芳烃可以快速地给出 Williamson 醚[式(2-103)]。与单独使用微波或超声相比,这种方法不仅将反应时间大大缩短,其产率也有很大的提高。如同时使用 50 W 超声和 200 W 微波辐射,反应 60~150 s,能给出中等到良好得率的二苯基醚和甲基苄基醚。此外,该反应体系不需使用相转移催化剂。Song 等[132]将这个反应体系应用到水相 4H-吡喃酮[2,3-c]吡唑。如以哌嗪为催化剂,吡喃与水合肼反应 1 min 就能给出最高得率(100%)的产物。如只用微波加热,反应 20 min 产物得率仅 48%,而采用超声和油浴加热相结合的方式,反应 20 min,产物得率为 99%。这个方法可以用于长时间内(40~60 s)合成高得率的吡喃酮吡唑类化合物。

$$\underset{R = H, EWG, EDG}{\text{[ethyl ester pyran with CN, NH}_2\text{, Ph-R]}} \xrightarrow[\text{MW}^{①}/\text{US, 200W, 40~60 s}]{N_2H_4 \cdot H_2O, \text{哌嗪}} \text{[pyrano-pyrazole product]} \quad (2\text{-}103)$$

2004 年 Varma 和 Ju 报道[133]在微波加热的条件下,通过伯胺和仲胺(芳香,环状和非环状的胺)与卤代烃的 N-烷基化反应合成叔胺[式(2-104)]。采用微波加热不仅可以将反应时间从 12 h 缩短到 20~30 min,且可以避免副产物的生成。与无溶剂或有机溶剂乙腈和 PEG 相比,水是合成叔胺的最环境友好的溶剂。

$$\text{PhCH}_2\text{Br} + \text{PhCH}_2\text{CH}_2\text{CH}_2\text{NH}_2 \xrightarrow[\text{MW, 45~50℃, 20 min}]{\text{NaOH, H}_2\text{O}} \text{[tertiary amine]} \quad (2\text{-}104)$$

Pironti 和 Colonna 报道了采用 NaOH 催化水相过氧环己烷与巯基苯酚的反应,生成 β-羟基苯硫醚[式(2-105)]。在微波辐射下,150℃反应 5min 得到 97%的产物[134]。Tu 等[135]报道了 N-取代吖啶衍生物的 Hantzsh 的"一锅"合成方法。该方法将 1,3-环己二酮用作二羰基化合物底物,以水为溶剂,NaOAc 为催化剂,在微波辐射下 1,3-环己二酮和甲胺在敞口容器中可以顺利地进行反应,给出高得率的双甲酮[式(2-106)]。

$$\text{[cyclohexene oxide]} + \text{PhSH} \xrightarrow[\text{H}_2\text{O, MW}]{\text{NaOH}} \text{[trans-2-(phenylthio)cyclohexanol]} \quad (2\text{-}105)$$

---

① MW 表示微波,下同。

$$RCHO + 2 \text{(cyclohexane-1,3-dione)} + CH_3NH_2 \cdot HCl \xrightarrow[MW, 6\sim10\ min]{NaOAc,\ H_2O} \text{产物} \quad 85\%\sim98\% \qquad (2\text{-}106)$$

Kidwai 等[136]报道了醛、丙二腈和间苯二酚的快速三组分"一锅"反应，生成 2-氨基-4H-苯并吡喃[式(2-107)]。反应体系中使用饱和 K_2CO_3 代替有机碱如哌嗪和有机溶剂，反应温度为 95~105℃，这种方法可用于合成 2-氨基-苯并[e]苯并吡喃。

$$PhCHO + NCCH_2CN + \text{(resorcinol)} \xrightarrow[MW,\ 95\sim100℃,\ 180\ s]{K_2CO_3,\ H_2O} \text{产物} \qquad (2\text{-}107)$$

Quesada 和 Taylor 通过 Horner-Wadsworth-Emmons(HWE)烯烃化、Claisen 重排和水解的三步串联反应合成 α-酮酸。在微波辐射下以 K_2CO_3 为催化剂，醛与膦酸酯在 105℃反应 10 min，给出得率为 53%~94%的目标产物[式(2-108)][137]。他们同样发展了一种微波辅助加热条件下，通过伯胺和肼分别和二卤代化物反应合成含氮杂环化合物的方法。这种合成方法表明多步合成反应过程以及使用过渡金属催化剂[138-140]，在微波辐射下，不仅反应时间缩短至 20 min，产物得率也有所增加，副产物也减少。在 120℃下伯胺与二卤化物或对甲苯磺酸酯发生双烷基化反应，生成乙酰唑胺环烷酮如吡咯烷等化合物或六亚甲基亚胺[式(2-109)]。而 2,3-二氢-1H-异吲哚可以通过 α,α-二卤-o-二甲苯与伯胺反应得到。在相同条件下，通过肼及其衍生物的双烷基化反应可合成 4,5-二氢吡唑、吡唑烷和 1,2-二氢酚嗪[式(2-110)]。Barnard 等[141]将该反应规模从 20 mmol 放大到 1 mol，在相同的条件下反应，产物得率保持不变。

$$ArCHO + (C_2H_5O)_2P(O)CH(O\text{-allyl})CO_2C_2H_5 \xrightarrow[MW,\ 50\ W,\ 105℃,\ 10\ min]{K_2CO_3,\ H_2O} \text{产物} \qquad (2\text{-}108)$$

$$RNH_2 + X(CH_2)_nX \xrightarrow[MW,\ 120℃,\ 20\ min]{K_2CO_3,\ H_2O} R\text{-}N(CH_2)_n \qquad (2\text{-}109)$$

$n = 3\sim6$; R = alkyl, aryl, cyclic; X = Cl, Br, I, OTs

$$R_1NHNH_2 + X\text{-}CHR_2\text{-}CH_2\text{-}CHR_3\text{-}X \xrightarrow[MW,\ 120℃,\ 20\ min]{K_2CO_3,\ H_2O} \text{产物} \qquad (2\text{-}110)$$

R_1, R_2, R_3 = alkyl, aryl; X = Cl, Br, I, OTs

4. 其他反应

在水相条件下研究的 Wittig 反应，是指羰基化合物与稳定的叶立德反应生成烯烃。该反应的优势是可以在水/有机两相体系中进行，但通常也需要使用一定量的相转移催化剂。已有报道单独使用水作溶剂，采用非常弱的碱，如 K_2CO_3 或 $KHCO_3$，反应进行得比较顺利，且不需加入相转移催化剂，可以直接使用含有对碱和酸敏感的官能团的化合物，如在此条件下，β-二甲基腙乙醛可以顺利地进行 Wittig 反应[式(2-111)][142]。

$$\text{N-methylpyrazole-CHO} + (C_2H_5O)_2POCH_2CO_2C_2H_5 \xrightarrow[\text{回流, 1 h}]{\text{aq. } K_2CO_3} \text{product} \qquad (2\text{-}111)$$

Warren 和 Russell 合成了水溶性的鏻盐 7，在 NaOH 水溶液中它们与含有给电子基团或吸电子基团的苯甲醛进行 Wittig 反应[143]，给出高得率的反式二苯乙烯衍生物产物，且产物和副产物磷化氢通过过滤就可以分离。

$$\text{HO-C}_6\text{H}_4\text{-P}^+(\text{Ph})_2\text{CH}_2\text{Ph} \cdot \text{Br}^-$$

7

Lu 等[144]报道了 NH_3 催化近临界水(NCW)中肉桂醛与水的反应，生成苯甲醛[式(2-112)]。这个反应包括两步，一是肉桂醛水解，二是苯甲醛与乙醛的 Aldol 缩合。

$$\text{PhCH=CHCHO} \xrightarrow[\text{H}_2\text{O}]{NH_3} \text{PhCHO} + CH_3CHO \qquad (2\text{-}112)$$

Tu 等[145]报道用碳酸钾催化水中芳香醛与 1,2-二苯乙酮、杂-芳胺的 Mannich 反应[式(2-113)]。作者通过微波辅助加热，控制底物的空间位阻，立体选择性地合成一种新的化合物 β-氨基酮。研究发现，水做溶剂可以促进这个反应的进行，即以水为溶剂，控制反应温度 100℃，微波辐射 20 min 就能给出得率达 65%的产物，而在乙腈、乙醇等有机溶剂中，产物的得率也不到 50%。

$$\text{ArCHO} + \underset{\text{NH}_2}{\overset{X}{\text{pyridyl}}} + \underset{\text{Ph}}{\overset{O}{\text{PhC-CH}_2}} \xrightarrow[130℃, \text{MW}]{K_2CO_3, H_2O} \text{product} \qquad (2\text{-}113)$$

二、有机碱催化剂

1. 缩合反应

有报道以三苯基膦或叔胺为催化剂催化水中硝酮与炔酸酯的缩合反应[式(2-114)]。如炔丙酸甲酯与硝酮反应给出两种比例为 1∶1 的产物, 总得率为 90%。而在甲苯、苯或二氯甲烷溶剂中的得率却很低[146, 147]。有趣的是, 如用己炔羧酸甲酯与硝酮反应, 在 PPh$_3$ 催化下, 仅得到 1,3-偶极环加成产物噁唑啉[式(2-115)]。此反应在甲苯和二氯甲烷溶剂中不能发生, 在纯水中反应微弱, 若在 3 mol/L LiCl 的溶液中, 反应活性明显提高。

$$(2\text{-}114)$$

$$(2\text{-}115)$$

最近, Zhang 等[148]在室温下, 以 1,4-二氮杂双环[2.2.2]辛烷(DACBO)为碱, 催化靛红和 3,5-二甲基-4-硝基异噁唑, 在 1.5 h 内, 能定量地给出理想产物[式(2-116)]。由于 3,5-二乙基-4-硝基异噁唑与 3,5-二甲基-4-硝基异噁唑在反应活性方面存在着一定的差异, 作者研究用不同的碱催化 3,5-二乙基-4-硝基异噁唑与靛红的缩合反应。研究发现, 三乙胺(TEA)是这个反应的最佳催化剂, 在 3 h 内, 产物的得率高达 97%, 非对映立体选择性值达 97%, 当用 5-氯靛红、5-甲基靛红、4,7-二氯靛红及 N-保护(Me, Bn)靛红与 3,5-二乙基-4-硝基异噁唑反应时, 产物的非对映立体选择性值在 92%~99%, 当用 5-氟靛红作底物时, 产物的非对映立体选择性值降为 87%, 而用 5-溴靛红或 5-三氟甲氧基靛红时, 产物没有立体选择性。

$$\text{(2-116)}$$

三乙胺同样也可以催化 3,5-二甲基-4-硝基异噁唑与芳香醛或三氟甲基酮的缩合反应[式(2-117)和式(2-118)]。有趣的是,该反应在二甲基亚砜(DMSO)、N,N-二甲基甲酰胺(DMF)、CH_3CN 等有机溶剂中反应 24 h 也检测不到产物,而在水溶剂中反应 15 h,产物的得率高达 96%[149]。

$$\text{(2-117)}$$

$$\text{(2-118)}$$

Cai 和 Lu[150]报道了一种更加简单有效的在水相中利用多组分反应合成多取代 4H-吡喃化合物的方法。研究表明,在水介质中哌啶可以有效催化三组分"一锅"反应,合成多取代 4H-吡喃化合物[式(2-119)~式(2-121)],若使用表面活性剂 SDS 作为添加剂则有利于反应的进行。阴离子型表面活性剂 SDS 加入水中,可以有效地减少传质阻力,增加水与反应物的接触面积,从而促进反应顺利进行。该方法不仅避免了易挥发、有毒有机溶剂的使用,而且反应时间短,产率高,反应后处理简单,底物使用范围十分广泛,具有大规模生产的应用前景。

$$\text{(2-119)}$$

$$\text{(2-120)}$$

$$\text{(2-121)}$$

Dickerson 和 Janda[151]报道了尼古丁的代谢产物降烟碱(8)催化磷酸盐的缓冲水溶液中对硝基苯甲醛与丙酮羟醛的缩合反应[式(2-122)]。有趣的是,在很多常用有机溶剂中,如 DMSO、DMF、THF、乙腈、苯以及氯仿溶剂中,该反应比较难进行。该交叉羟醛缩合反应也可以被有机胺催化,反应在 0.01 mol/L 磷酸盐、2.7 mol/L KCl、137 mol/L NaCl 的缓冲介质(pH = 7.4)中进行,各种醛和酮,包括碳水化合物都可以作为底物进行缩合反应。这个方法可以用来合成糖的前生源[152]。也有报道在水相阳离子胶束中进行的脯氨酸催化的不同酮与硝基苯甲醛的羟醛缩合反应,在各种类型的阴离子表面活性剂,如 SDBS、SDS 及月桂酸(SLS)等存在下,都能给出比较好的产物得率。但是,即使以脯氨酸为催化剂,无论是在中性的或阳离子表面活性剂,如 CTAB 和十八烷基三甲基氯化铵(OTAC)等中,均不能促进水中缩合反应的进行[153]。

$$\text{(2-122)}$$

2. 加成反应

Kotsuki 等[154]设计了一种新的与 4-二甲氨基吡啶(DMAP)相关的有机碱催化剂,如 4-(二癸氨基)吡啶催化纯水中的 Michael 加成反应,生成 β-酮酸酯[式(2-123)],产物的得率为 73%~99%。这种合成 β-酮酸酯的方法优点是操作简单、原料易得及反应条件温和。该反应在 THF 或二氯甲烷中不反应,但在甲醇和水溶剂中反应进行地顺利,并给出非常高得率的产物,这也表明极性溶剂可以加速反应的进行,从绿色化学角度讲,作者研究了不同 β-酮酯在水相中的合成。在室温下用 10 mol%催化剂,反应即能给出 75%~99%得率的产物。

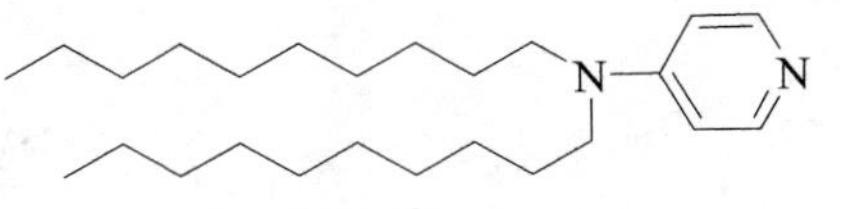

[反应式 2-123]

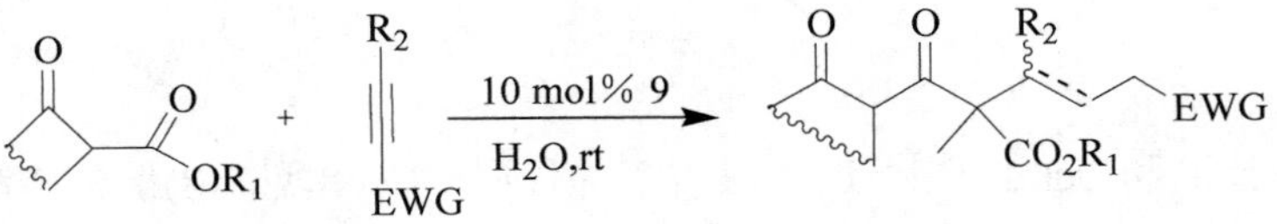

Malhotra 等[155]报道采用 1-甲基咪唑催化水中丙二腈与取代苯甲醛的缩合反应，给出非常高得率(96%~99%)的缩合产物 2-氯亚苄基丙二腈及其衍生物[式(2-124)]。反应条件温和(35~50℃)，时间短(1.0~10 min)，该方法可以用来大规模生产 2-氯亚苄基丙二腈类化合物。

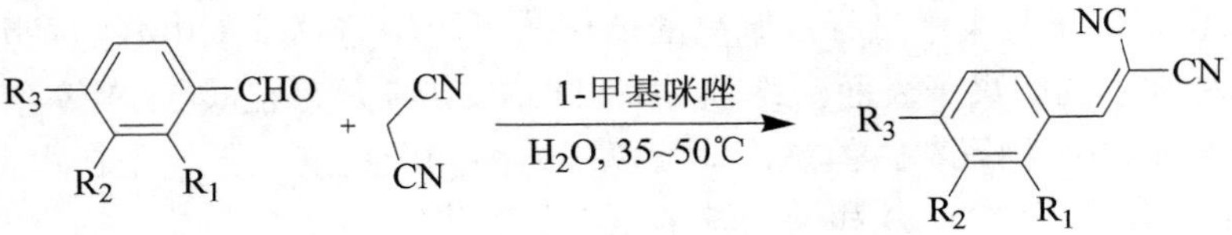

[反应式 2-124]

Zhu 等[156]报道了 TEA 催化通过 2,4,6-三异丙基-N'-(全氟苯基)苯磺酰肼经 Bamford-Stevens 反应原位产生的全氟苯基重氮甲烷与丙烯酸甲酯的 1,3-偶极环加成反应，生成产物吡唑啉[式(1-125)]。反应产物在水中的得率(99%)虽然高于 THF 中的得率(88%)，但在水中进行反应时，不溶于水的产物通过简单过滤就能实现分离，不需经过柱色谱层析操作。

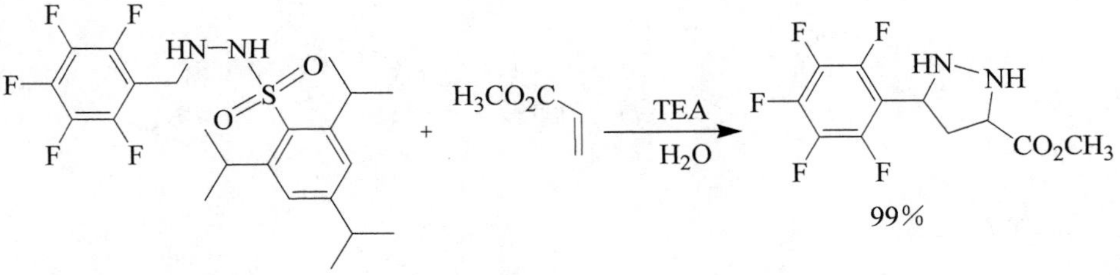

[反应式 2-125]

3. 其他反应

Luthman 等[157]为了发展一种环境友好的合成 2,5-二酮哌嗪的方法，他们以用 HCl/MeOH 脱保护的十一缩二氨酸甲酯盐为原料，水为溶剂，在微波加热或常规加热的条件下于 140℃反应 10 min 得到得率为 63%~97%的目标产物[式(2-126)]。

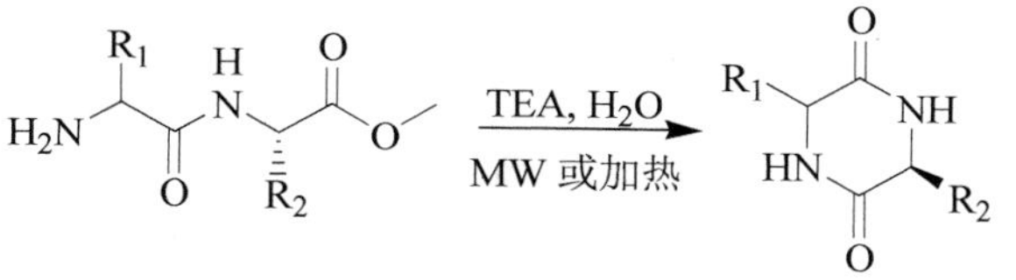

[式(2-126)反应式]

Samanta 等[158]报道了一种简单、温和、绿色的合成功能化的 1-甲氧基羰基-2-芳基/烷基-3-硝基-9H-咔唑的方法。此方法采用水为溶剂、1,4-二氮二环[2.2.2]辛烷(DABCO)为催化剂催化芳基/烷基-取代的 β-硝基烯和甲基-(3-甲酰基-4-卤-1H-吲哚-2-基)乙酸乙酯的多米诺 Michael-Henry/芳香化反应[式(2-127)],给出较高得率(82%~95%)的目标产物。

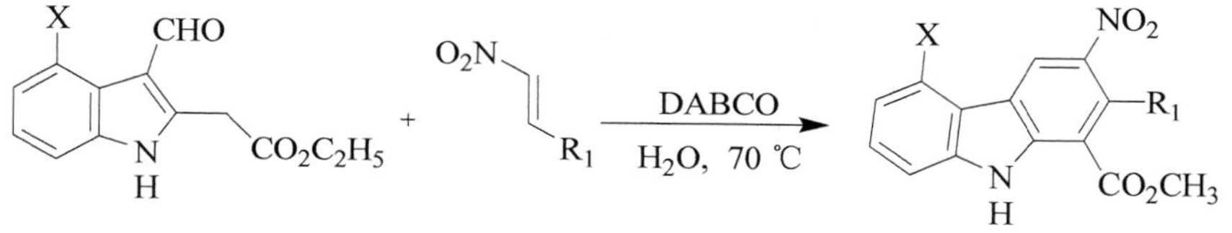

[式(2-127)反应式]

Heravi 等[159]报道了用 DABCO 催化水中醛与乙酰乙酸乙酯、肼及巴比妥酸的四组分"一锅"反应,合成具有生物活性的吡喃酮吡唑和吡喃酮嘧啶结构的三环化合物[式(2-128)]。在较短的时间(20~45 min)内给出较高得率(84%~99%)的目标化合物。

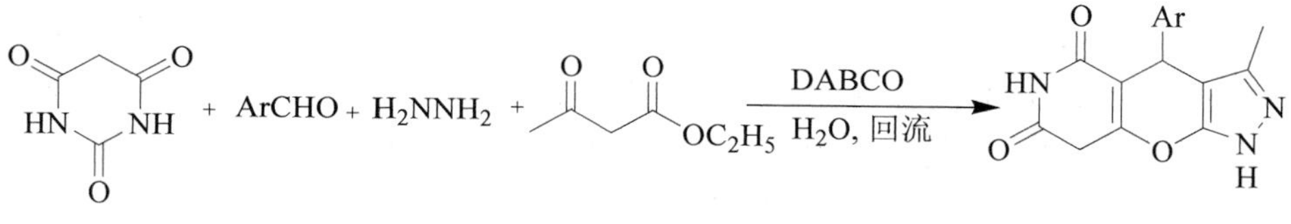

[式(2-128)反应式]

Fayazipoor 等[160]报道用 DABCO 作催化剂用于水中 5-芳基-1-(1H,3H,5H,10H)-嘧啶并[4,5-b]喹啉-2,4-二酮和硫代衍生物的合成[式(2-129)]。芳胺(含吸电子和给电子基团)、芳醛(包括杂芳醛)和巴比妥酸在微波辐射下回流 1 min,无论是吸电子芳胺还是给电子芳胺都能给出非常高得率(89%~98%)的产物。

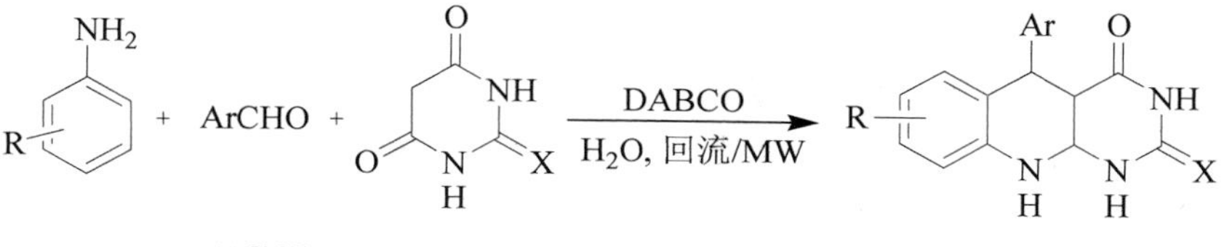

[式(2-129)反应式]

Khurana 等[161]采用 1,8-二氮杂双环[5.4.0]十一碳-7-烯(DBU)催化水中醛、活泼亚甲基化合物丙二腈/氰乙酸乙酯与活化的 C—H 酸如双甲酮[式(2-130)]、1,3-环己二酮[式(2-131)]、1,3-环戊二酮[式(2-132)]、1,3-二甲基巴比妥酸[式(2-133)]及乙酰乙酸乙酯反应,生成四氢-苯并吡喃、四氢[b]吡喃、吡喃[d]嘧啶和四氢-吡喃高到优良得率的目标产物。

$$RCHO + NC-R_1 + \text{(5,5-二甲基-1,3-环己二酮)} \xrightarrow[\text{H}_2\text{O, 回流}]{DBU} \text{产物} \quad (2\text{-}130)$$

$$RCHO + NC-R_1 + \text{(1,3-环己二酮)} \xrightarrow[\text{H}_2\text{O, 回流}]{DBU} \text{产物} \quad (2\text{-}131)$$

$$RCHO + NC-R_1 + \text{(1,3-环戊二酮)} \xrightarrow[\text{H}_2\text{O, 回流}]{DBU} \text{产物} \quad (2\text{-}132)$$

$$RCHO + NC-CH_2-CN + \text{(取代环己烯酮)} \xrightarrow[\text{H}_2\text{O, 回流}]{DBU} \text{产物} \quad (2\text{-}133)$$

Zakeri 等[162]报道了一种绿色的合成螺-4H-吡喃衍生物的方法，该方法采用 4-二甲基氨基吡啶(DMAP)为催化剂，催化水中丙二腈、乙酰乙酸乙酯和靛红发生多米诺 Knoevenagel/Michael/cyclization 串联反应[式(2-134)]，在回流条件下反应 50 min，给出理想产物的得率为 88%。实验结果表明，在极性溶剂，如水、甲醇中反应速率明显优于非极性溶剂，如 DMSO、甲苯等。

$$\text{(靛红)} + \overset{CN}{\underset{CN}{>}} + \text{(乙酰乙酸乙酯)} \xrightarrow[\text{H}_2\text{O}]{10\ mol\%\ DMAP} \text{螺环产物} \quad (2\text{-}134)$$

最近，Bharti 和 Parvin[163]报道以有机碱 10 为催化剂，催化水中醛、3-二甲基-6-氨基尿嘧啶和 2-羟基-1,4-萘醌/4-羟基香豆素的多组分"一锅"反应，生成一系列具有生物活性的三取代甲烷衍生物[式(2-135)]。底物结构中含有—OH，—NH$_2$ 和 C＝O 等基团时，在此催化体系中仍比较稳定。

10

$$R_1CHO + \text{(6-amino-1,3-dimethyluracil)} + \text{(2-hydroxy-1,4-naphthoquinone)} \xrightarrow[H_2O, 回流]{20\ mol\%\ 10} \text{产物} \quad (2\text{-}135)$$

Rimaz 和 Aali[163]报道了 DABCO 催化水中取代苯丙酮、水合肼和芳基一水合乙醛的缩合反应, 得到一系列 3,6-二芳基-4-甲基哒嗪[式(2-136)], 产物得率高到优良。此催化体系具有区域选择性高、易操作、安全等优点。

$$\text{ArCH(OH)}_2 + \text{Ar'COCH}_2CH_3 \xrightarrow[H_2O,\ 25\ ℃,\ 2\sim 4\ h]{50\ mol\%\ DABCO,\ NH_2NH_2} \text{产物} \quad (2\text{-}136)$$

Samanta 等[164]在 2014 年报道了一种无金属催化的 1-甲氧基羰基-2-芳基/烷基-9H-咔唑衍生物的一步合成方法, 该方法采用 DABCO 为催化剂催化水中芳基/烷基-β-丙烯醛与 2-(3-甲酰基-1H-吲哚-2-基)乙酸甲酯的反应[式(2-137)], 在 60℃反应 6~12 h 给出理想产物的得率为 51%~68%。随后他们将此方法应用于水中 α, β-二取代的硝基烯或 α, δ-二取代的硝基二烯和 2-(3-甲酰基-1H-吲哚-2-基)乙酸甲酯的原位芳香化反应[式(2-138)], 在室温下反应 18~24 h 后用 HCl 处理后给出目标产物, 得率为 60%~91%。这种 "一锅" 无氧化剂的合成方法还可以合成结构相近的具有生物活性的其他化合物[165]。

$$\text{(indole-CHO-CH}_2CO_2CH_3) + \text{OHC-CH=CH-Ar} \xrightarrow[2)\ 6\ mol/L\ HCl,\ 60℃,\ 48\ h]{1)\ 30\ mol\%\ DABCO,\ H_2O,\ 60℃,\ 6\sim 12\ h} \text{产物} \quad (2\text{-}137)$$

$$\text{(R-indole-CHO-CH}_2CO_2CH_3) + \text{R}_2\text{-CH=CH-CH=CH-NO}_2\text{R}_1 \xrightarrow[2)\ 6\ mol/L\ HCl,\ rt,\ 7\sim 10\ h]{1)\ 20\ mol\%\ DABCO,\ H_2O,\ rt,\ 18\sim 24\ h} \text{产物} \quad (2\text{-}138)$$

第三节 有机金属催化剂

过渡金属催化的 C—C 偶联反应因其原子经济性和高合成效率备受有机合成工作者的关注，已经成为有机合成中构建 C—C 键的强大工具，能够高选择性地、高效合成具有特殊精细骨架结构的化合物[166]。在近几十年中，涌现出了诸如 Suzuki 反应、Sonogashira 反应、Heck 反应等一系列反应，这些反应在材料、天然产物以及生物活性物质的制备中有着非常重要的地位。但是这些 C—C 偶联反应常是在传统的有机溶剂中进行的，如 THF、1,4-二氧六环、DMF、N-甲基吡咯烷酮(NMP)等。有机溶剂由于具有易挥发、易燃、有毒和难以回收等特点，是主要的环境污染源之一。目前，已经研究出来的替代有机溶剂的绿色溶剂主要有聚乙二醇、水和有水参与的有机溶剂以及超临界 CO_2、离子液体等。显然，水溶剂是一种最理想的绿色溶剂，研究前景也十分广阔，因为它具有以下几个特点[167,168]：①水是地球上最丰富的自然资源，成本低；②水分子间能形成较强的分子间相互作用——氢键；③无色无味、无毒、不燃烧、对环境无害；④有高的传热性，热效率利用率高；⑤极性高，易于与非极性溶剂分离。但是有机底物较差的水溶性使得反应在水相中难以进行。本节主要介绍有机钯、铑、钌等有机金属催化水介质有机反应。

一、有机金属钯催化剂

1. Suzuki 反应

有机金属钯金属催化卤代烃与硼酸的 Suzuki 偶联反应[式(2-139)]是形成 C—C 键的重要方法之一，被广泛应用于有机合成中。自 1990 年 Casalnuvo 和 Calabrese[169]报道水/乙腈中的 Suzuki 反应以来，有关水介质 Suzuki 反应的研究已取得了很大的进展。

$$Ar^1B(OH)_2 + Ar^2B(OH)_2 \xrightarrow[\text{碱，溶剂}]{\text{催化剂}} Ar^1\text{-}Ar^2 \qquad (2\text{-}139)$$

1) 含氮配体/钯催化体系

Kostas 等[170]发展了一种水溶性卟啉/钯配合物 11 应用于水介质中 Suzuki 反应的催化体系。催化剂用量为 0.1 mol%，对溴代芳烃在 100℃反应 1~4 h，就可得到较好收率的产物，催化剂虽然能重复使用，但是使用到第二次时，催化剂的活性明显降低，产物的得率从 100%降低到 76%。

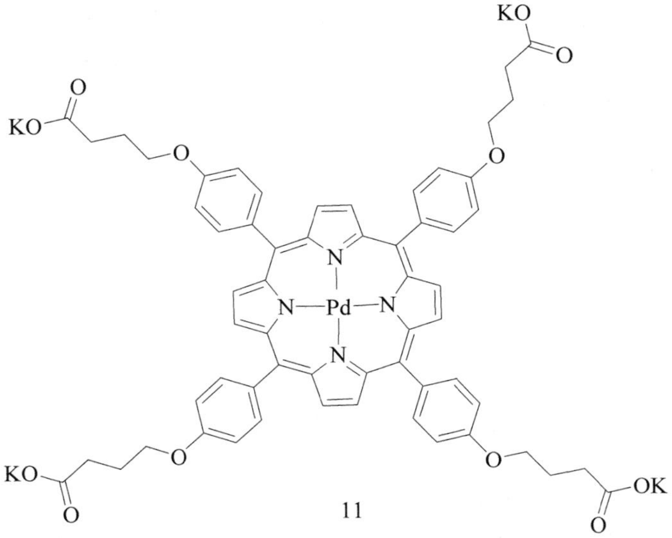

[结构式 11]

Zhang 等[171]将超声波技术应用到水介质中钯配合物 12 和 13 催化的 Suzuki 反应。在催化剂 12 用量为 0.50 mol%~1.0 mol%，以 K_3PO_4 为碱，表面活性剂存在的条件下，可高效催化碘代芳烃和溴代芳烃的偶联反应，当用催化剂 13 时，芳基氯代芳烃也可以顺利进行 Suzuki 反应。实验结果表明，超声波可明显提高反应速率。

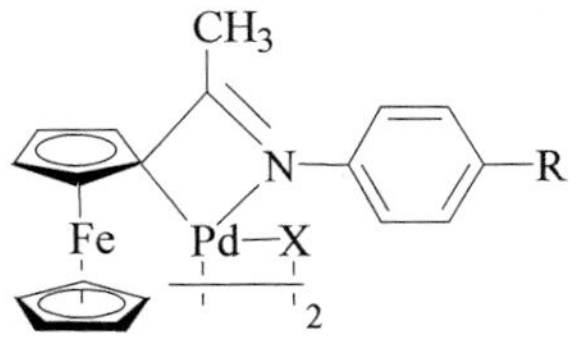

12: X = Cl, R = Cl
13: X = I, R = CH_3

Li 和 Wu[172]研究了钯配合物 14 在水介质 Suzuki 反应中的催化性能。当催化剂用量为 0.1mol%时，对含有给电子、吸电子和位阻基团的溴代芳烃均具有较好的催化性能，当催化剂负载量提高到 1mol%时，含吸电子基团的氯代芳烃，如对氯硝基苯也可以顺利进行 Suzuki 反应，产物的得率为 65%。Bai 和 Hor[173]合成了一种水溶性的钯配合物 15，并考察其在水介质中 Suzuki 催化反应的催化性能。催化剂用量为 2 mol%时，对溴苯乙酮与苯硼酸在以 Na_2CO_3 为碱，75℃条件下反应 6 h，产物的得率达 100%。

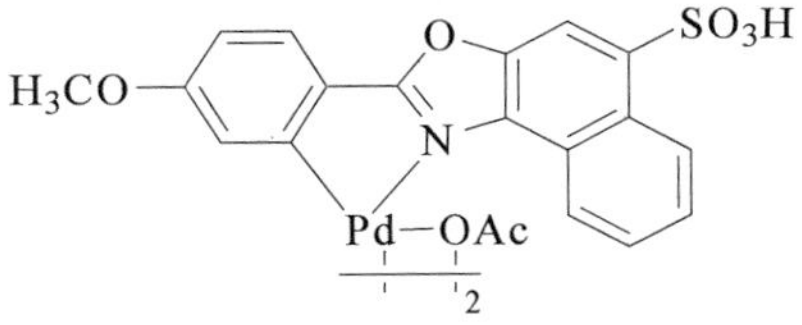

14 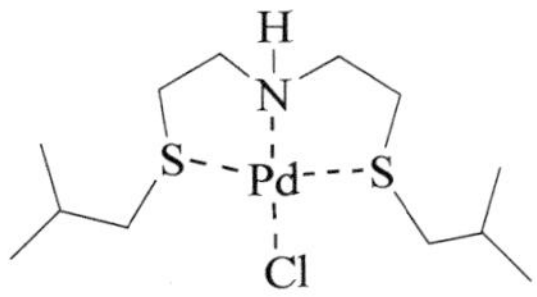   15

Zhou 等[174]将合成的一种钯配合物(16 和 17)应用在水介质中溴代芳烃的 Suzuki 反应,该催化剂在对溴苯甲醚与苯硼酸的反应中转化数(turnover number, TON)高达 9.3×10^6,且催化剂可以循环使用 5 次。

R_1 = H, SO$_3$Na; R_2 = Me. i-Pr; R_3 = H, i-Pr

Gülecemal 等[175]分别设计了 N,N-双齿钯配合物 18 和 N,O-双齿钯配合物 19,并将其应用于水介质中的 Suzuki 反应。该催化剂对溴代芳烃苯环上的取代基比较敏感,即对含有吸电子基团的溴代芳烃表现出较高的催化活性。如在对溴苯乙酮与苯硼酸的偶联反应中,TON 值高达 1.0×10^5。然而,给电子基团的溴代芳烃的反应只有在表面活性剂存在的条件下才发生。

Conelly-Espinosa 和 Morales-Morales[176]将 N, O-多齿钯配合物 20 应用在水介质中的 Suzuki 反应,同样在溴代芳烃的 Suzuki 反应中,含有催化剂的水相可循环使用 4 次,催化活性没有明显降低。Tu 等[177]发展了一种水溶性钯配合物 21,在催化水中溴代芳烃的 Suzuki 反应中,该配合物的催化剂用量非常低,仅 0.005%的用量就可以获得较高的得率。

Shao 等[178]报道了用脯氨酸合成 Pd(Ⅱ)-NHC 配合物 22,该配合物在水介质催化含有吸电子基团的碘代芳烃和溴代芳烃的 Suzuki 偶联反应中显示了优越的催化活性,研究也发现,这个反应可以放大到中等规模,对催化活性没有影响。Banik 等[179]合成了一种对空气稳定的双核钯配合物催化剂 23,该催化剂在室温下即可催化水介质中溴代芳烃与芳基硼酸的 Suzuki 偶联反应,反应无需任何添加剂;但对于氯代芳烃,需要将温度升高至 80℃,反应才能发生,也能获得较好得率的联苯芳烃化合物。

Hanhan 等[180]合成了一系列新型离子型双核金属钯配合物(24~30),并考察这些配合物在水介质 Suzuki 反应中的催化性能。研究表明,配合物的双核性质可以提高反应速率,配体骨架上的电子性质和空间结构对催化剂的活性有重要的影响。在这些配合物中 27 和 28 的催化活性明显高于其他配合物,更重要的是,这两种催化剂可以重复使用 10 次,产物的得率没有明显降低,但使用到第 14 次时,产物的得率只有 43%。

24: X = SO$_3$Na; R$_1$ = R$_2$ = R$_3$ = H

25: X = SO$_3$Na; R$_1$ = H; R$_2$ = CH$_3$; R$_3$ = H

26: X = SO$_3$Na; R$_1$ = CH$_3$; R$_2$ = H; R$_3$ = H

27: X = SO$_3$Na; R$_1$ = CH$_3$; R$_2$ = H; R$_3$ = CH$_3$

28: X = SO$_3^-$(TBA)$^+$; R$_1$ = CH$_3$; R$_2$ = H; R$_3$ = CH$_3$

Ding 等[181]用 L-脯氨酸为配体合成一种水溶性的钯配合物 31,并将其用于催化水介质 Suzuki 反应，他们主要研究了催化剂在不同卤代芳烃和芳基硼酸的偶联反应中的性能及其使用寿命，结果显示，该催化剂可以循环使用 6 次，活性没有明显降低。Schmitzer 等[182]报道了配合物 32 的合成，并用于 Suzuki 反应，反应无需加入表面活性剂，在较短时间内(2 h)就能使反应进行完全。

Dewan 等[183]报道了一种四齿螯合的钯配合物 33 的合成，并将其用于水介质含有不同取代基溴代芳烃的 Suzuki 反应，作者也研究了对甲氧基氯苯在此反应中的活性。遗憾的是，对甲氧基氯与苯硼酸的偶联反应只能在有机溶剂(异丙醇)中进行。

Xu 等[184]合成一种双齿金属钯配合物 34，并将其应用在水介质 Suzuki 反应。在间羟甲基苯硼酸与对溴甲苯的偶联反应中，当催化剂用量为 0.2 mol%，在 100℃反应 12 h 时，可以获得理想的产物，不过产率只有 36%。当向反应体系中加入四丁基溴化铵(TBAB)时，产物的得率提高到 83%。作者也研究了取代基的性质对反应的影响，研究发现，当溴代苯的苯环上含有给电子基团时，反应的得率明显降低，但是溴代苯环上吸电子基团对反应有明显的促进作用。例如，在对硝基溴苯的 Suzuki 反应中产物得率高达 94%，遗憾的是，这个催化体系对氯苯不能起催化作用，但可以催化对氯硝基苯的偶联反应，给出产物的得率为 64%。

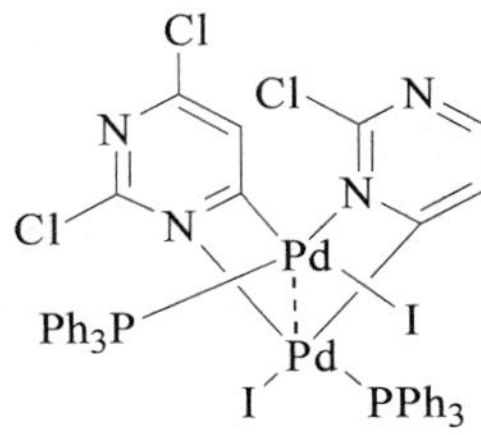

34

Planas 等[185]发展了一种 m-和 o-碳硼烷剂 NBN 螯合钯配合物 35 和 36 用于水介质 Suzuki 反应。这些催化剂用量非常少(10^{-9} mol)，并能给出非常高的催化活性。由于 B 单体较强的给电子能力，他们用一种 NBN 螯合配合物代替了 NCN 螯合配合物，因而展现了更强的反式效应。配合物 35(m-CB-L1)比 36(o-CB-L1)具有更高的催化活性和 TON 值($7.7×10^5$~$9.9×10^5$)。尽管催化剂存在潜性的手性环境，但是并没有观测到手性异构体。

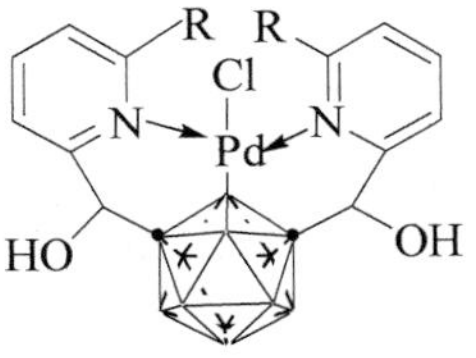

(m-CB-L1)Pd
35

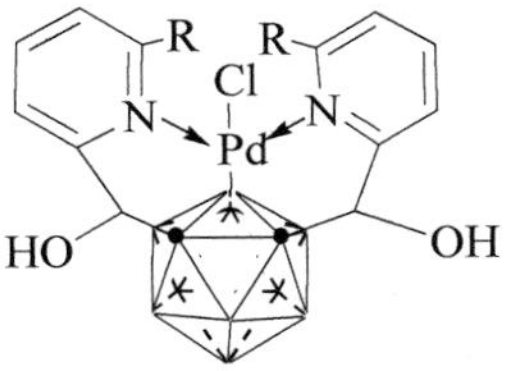

(o-CB-L1)Pd
36

Senemoglu 和 Hanhan 报道[186]微波促进不对称磺酸化的水溶性钯-吡啶亚胺配合物 37 催化的水相 Suzuki 反应，实验结果表明，这个配合物是一种有效的催化剂，TON 值达 980。最近，Lang 等[187]报道催化剂 38 在 100℃催化水中 Suzuki 交叉偶联反应中的催化性能，在以 K_3PO_4 为碱，0.05 mol%催化剂 38 在 100℃的反应中，TON 值达 1980，且催化剂 38 可以重复使用 4 次。

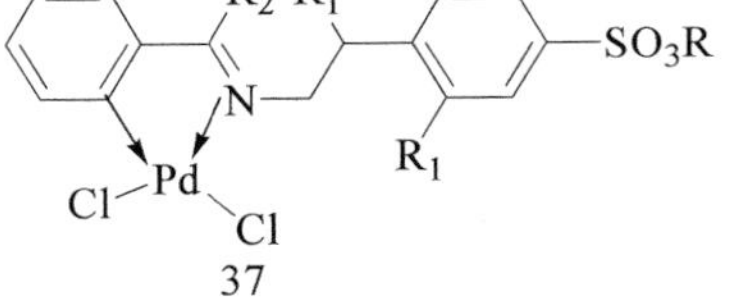

37

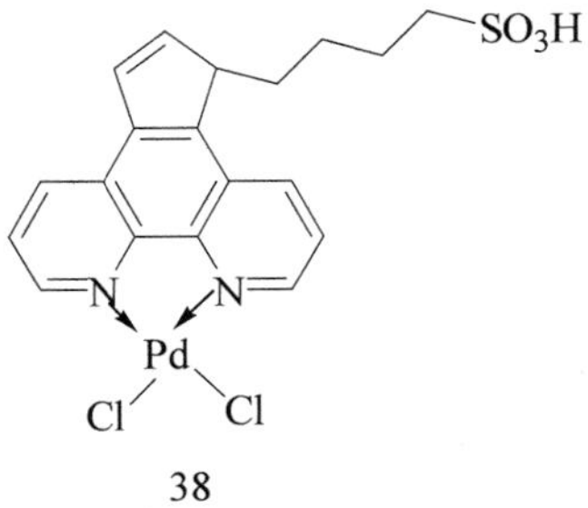

38

近来也有报道，水溶性 Pd(Ⅱ)配合物 trans-[(imidate)$_2$(PTA)$_2$][imidate = 琥珀

酰胺(suc)39，马来酰胺(mal)40，邻苯二甲酰胺(phthal)41 或邻磺酰苯甲酰胺(sacc)42]是高效的 Suzuki 反应的催化剂。在温和条件下，催化剂 41 在具有抗病毒能力的核苷 5-碘-2′-脱氧尿苷与间甲氧基苯硼酸的偶联反应[式(2-140)]中，TON 值达 96，随着反应时间的延长，较低量(0.1 mol%)的催化剂也能给出较高的得率[188]。

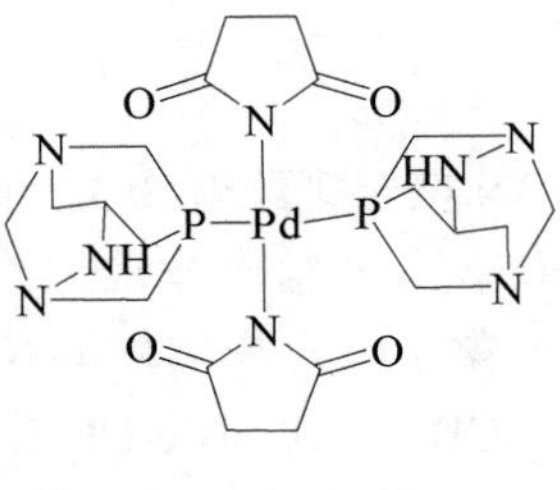

39

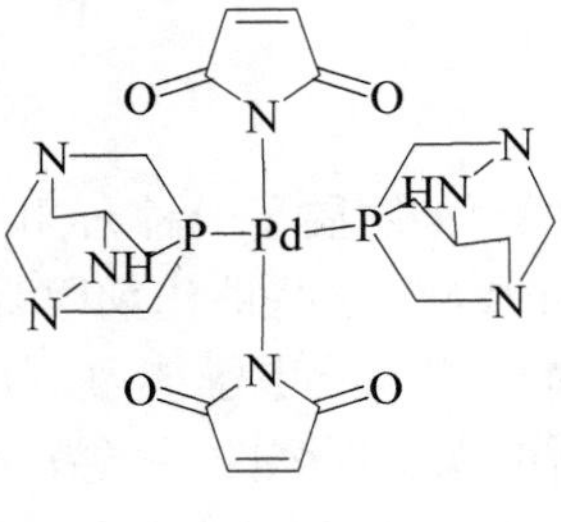

40

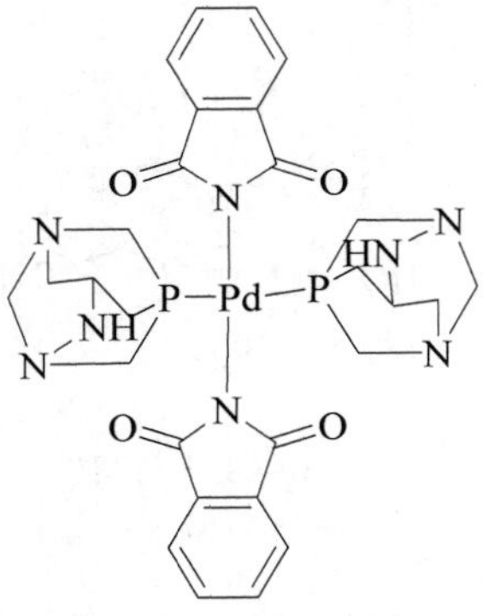

41

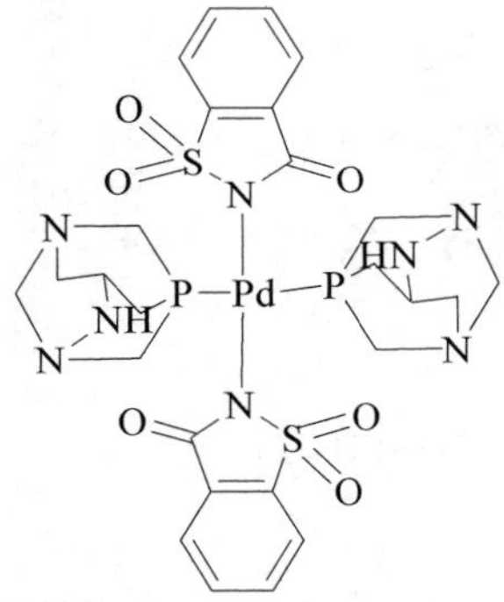

42

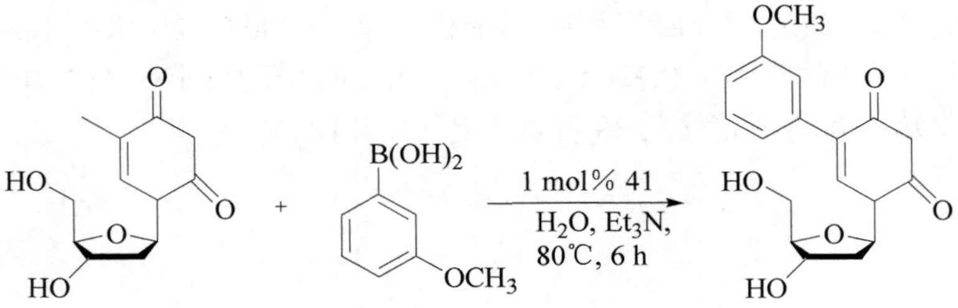

$$(2\text{-}140)$$

1 mol% 41
H$_2$O, Et$_3$N,
80℃, 6 h

Ding 等[189]报道了一种新型水溶性钯环配合物 43 催化水中 Suzuki 反应。这种催化剂(0.00005 mol%)在此反应中展示了优异的催化性能，TON 值达 9.8×10^5，且可重复使用 3 次，催化活性没有明显降低。

2) 膦配体/钯催化体系

在水相 Suzuki 反应中，与 N 配体和 NHC 配体相比，P 配体容易分解，因此其在水相有机反应中的应用受到了一定的限制。有关 P 配体/钯催化体系在 Suzuki 反应中的报道相对较少。

Ines 等[190]设计了一种三齿型钯配合物 44，并将其应用在水介质中的 Suzuki 偶联反应。在催化剂 44 用量仅为 0.01%时，100℃条件下反应 2 h，碘代芳烃和溴代芳烃均能顺利发生 Suzuki 反应，在四丁基碘化铵(TBAI)存在下，反应温度升至 150℃，反应 4 h，4-氯苯乙酮和苯硼酸也能进行 Suzuki 反应，给出产物得率为 72%。

Zhao 等[191]合成了 4 个含杂环二茂铁亚胺的钯配合物 45~48，并将其应用在水介质中的 Suzuki 反应。该催化剂对空气和水非常稳定，整个反应过程无需惰性气体保护，在空气中即可进行，对溴代芳烃表现出较好的催化活性和官能团的兼容性。

Marziale 等[192, 193]合成了一种环钯配合物 49，并将其应用于水介质中各种硼酸与溴代芳烃或碘代芳烃在室温下的 Suzuki 反应。在此反应中，环钯化合物是催化活性物种，反应结束后即可得到高纯度、高得率(99%)的偶联产物。另外，催化剂可以重复使用 4 次，活性基本保持不变。

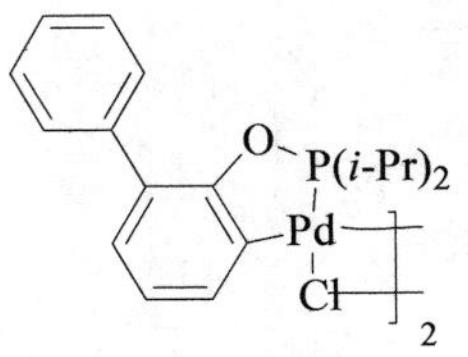

49

Qu 等[194]用 Pd(PPh$_3$)$_2$Cl$_2$ 作催化剂，Na$_2$CO$_3$ 为碱，在水介质中通过微波辐射使 6-氯嘌呤与四苯基硼化钠反应，从而合成 6-芳基嘌呤。Gholinejad 等[195]合成双核 Pt(Ⅰ)和 Pd(Ⅱ)配合物 50，并将其应用在水介质 Suzuki 反应中，与单金属 Pt(Ⅰ)配合物和 Pd(Ⅱ)配合物相比，该双核配合物催化剂显示了优越的催化性能，氯苯在此反应体系中虽然反应很慢，但也能获得中等得率的产物。

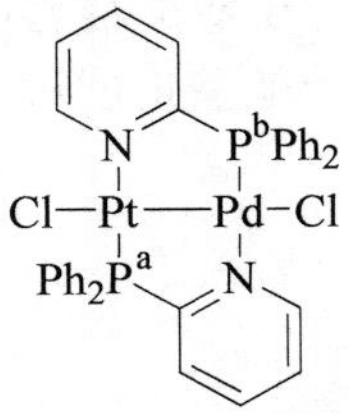

50

最近，Khazalpour 等[196]报道钯配合物 51 是水相 Suzuki 反应中可循环使用的高效催化剂。通过 P 和叶立德碳原子形成了五元螯合环的 Pd 催化剂，它在 Cs$_2$CO$_3$ 存在下可以重复使用 4 次，催化活性没有明显降低。

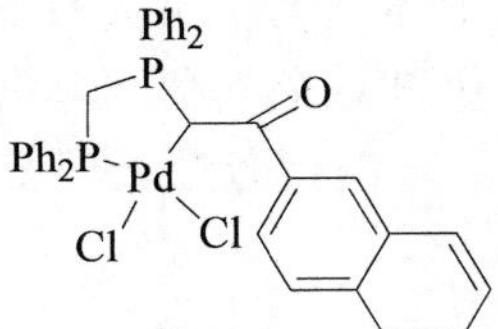

51

3) *N*-杂环卡宾配体(NHC)/钯催化体系

具有强给电子能力的 NHC 配体由于具有较强的碱性(pK_a = 20)，因此很少作为自由配体用于水相有机反应中。正是由于金属-NHC 之间强的配位作用，以 NHC 为配体的有机金属具有较好的水溶性。

Godoy 等[197]合成了一系列磺化的 *N*-杂环卡宾/钯配合物 52~55，并考察它们在水介质中溴代芳烃的 Suzuki 偶联反应中的催化活性，研究发现配合物 53 在催化偶联反应中的活性最高，如向反应体系中加入一定量的 TBAB，催化剂 53 还可以有效地催化氯代芳烃的偶联反应。

Kühn 和 Pöthig 报道了 4 种水溶性 PEPPSI 类型 Pd-NHC 配合物(56~59) 用于在空气氛围下水中的 Suzuki 反应。含有 2,6-二异丙基苯基取代基的配合物 59 在各种溴代芳烃的偶联反应中用量仅为 0.1mol%，在室温下即能表现出非常高的催化性能，这种催化剂至少可以重复使用 4 次，活性没有明显降低[198]。

Wang 等[199]合成了 pH 响应的 N-杂环二卡宾钯配合物 60，利用羧基的脱质子化提高其在水中的溶解性,从而高效地催化水中的 Suzuki 反应。最近,Nechaev 等[200]合成了六元环 NHC 钯配合物 61,催化水中溴代和氯代杂环芳烃与杂环芳烃硼酸的 Suzuki 交叉偶联反应。在此反应中,使用 0.5mol%的催化剂 61 能给出非常好的得率,TON 值达 200。Kinzhalov 等[201]报道了含有非环状二氨基卡宾配体钯配合物 62 在 80℃催化水中有机卤化物与芳基硼酸的偶联反应,给出高得率(99%)的偶联产物,在碘代芳烃及溴代芳烃的偶联反应中 TON 值分别为 9900 和 9200,对于氯代芳烃的 Suzuki 反应,提高反应温度至 100℃,并延长反应时间至 3 h,也能给出较高得率(82%)的产物。

2. Sonogashira 反应

用钯和铜结合催化的末端炔烃与有机卤化物的偶联反应,称为 Sonogashira 反应[式(2-141)],它是形成 C—C 键的重要方法之一,在有机合成中得到了广泛应用。近年来,该反应在水介质中的研究受到很多学者的关注。早在 1996 年 Vlassa 等[202]用 Pd(dba)$_2$ 催化剂催化水介质中脂肪炔与烯丙基溴的 C—C 偶联反应。如 Bhattacharya 和 Sengupta[203]用 0.5mol% Pd(PPh$_3$)$_4$ 和 1%CuI 催化水中卤代芳烃和末端炔的偶联反应,反应在 70℃快速进行,30 min 反应结束,但是该反应不能催化氯代芳烃的 Sonagashira 反应。

$$R_1\text{≡=} + R_2X \xrightarrow{Pd, CuI} R_1\text{≡=}R_2 \quad (2\text{-}141)$$

Kamali 等[204]以 Cs$_2$CO$_3$ 为碱,在十二醇硫酸钠(SLS)和 CuI 存在下,反应温度 60℃,Pd(PPh$_3$)$_4$ 可催化溴代 3-(2-丙炔基)噻唑盐与不同取代的碘代芳烃的偶联[式(2-142)]。实验结果表明,在强碱如 KOH,或有机弱碱如哌啶或 TEA 等碱中,目标产物的得率不足 39%,若用 Cs$_2$CO$_3$ 或 K$_2$CO$_3$ 为碱,产物的得率可达 78%以上,尤其是在 Cs$_2$CO$_3$ 的水溶液中,产物的得率高达 90%。

Luo 和 Lo[205]用咖啡因为配体，快速合成了钯配合物 63，并将其用于水相对硝基溴苯与苯乙炔的 Sonogashira 反应中。在 CuI、KOH 存在下，90 ℃反应 24 h，产物得率不到 5%，但若向体系中加入非离子表面活性剂 Brij30，产物的转化率可提高至 96%。Ranu 等[206]报道了在四氢吡咯存在下，$Pd(PPh_3)_4$/CuI 使不同取代的邻硝基碘苯与苯乙炔进行 Sonogashira 偶联和炔的水合反应[式(2-143)]。Park 等[207]分别研究了 $Pd(PPh_3)_2Cl_2$/dppb 和 $Pd(TPPMS)_2Cl_2$/TPPMS 两个催化体系催化含有十八烷基氯化铵的水溶液中不同芳基溴与炔丙酸的偶联反应，生成对称的二芳基炔[式(2-144)]。

$$\text{63}$$

$$(2\text{-}143)$$

$$(2\text{-}144)$$

Lipshutz 等[208]在室温无需 Cu 的条件下，用 $PdCl_2(CH_3CN)_2$ 与配体 X-Phos 64 催化体系催化溴苯与芳炔的偶联反应，当然，在这反应中需要加入 3%的非离子两亲相转移催化剂促进该反应。Bakherad 等[209]合成一种水溶性的钯配合物 65，其无需 CuI 的存在即可催化含有不同取代基的碘苯与苯乙炔的偶联反应。当然，该反应的顺利进行也需要一定量的表面活性剂 SLS 和 Cs_2CO_3 碱的存在。

$$\text{64} \qquad \text{65}$$

Liu 等[210]研制了两种两亲的钯配合物 66 和 67，这两种配合物直接催化水中碘苯和苯乙炔的 Sonogashira 反应，无需 CuI，给出非常高得率(97%和100%)的 1,2-二苯乙炔，而在乙腈介质中，产物的得率仅为 40%，这可归因于催化剂具有亲水和亲油特性。

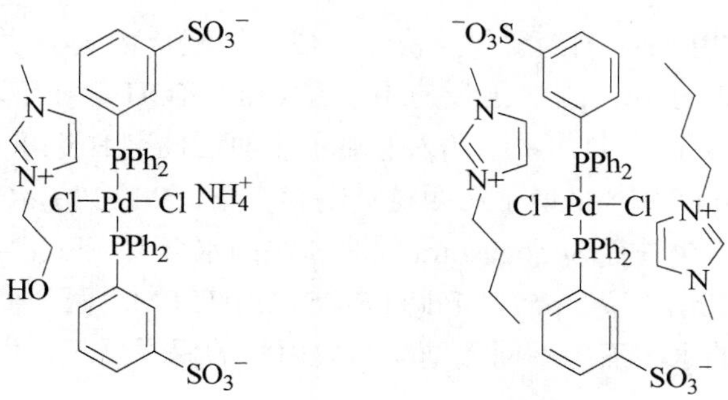

66 67

Xia 等[211]设计了一种水溶性的 Pd-NHC 配合物 68 的合成,并考察其在无膦配体存在下水相碘苯与芳炔的偶联反应。这种水溶性的催化剂可以通过萃取反应回收溶液;实验也证实,该催化剂可以重复使用 4 次,但使用 4 次后反应活性有所降低。

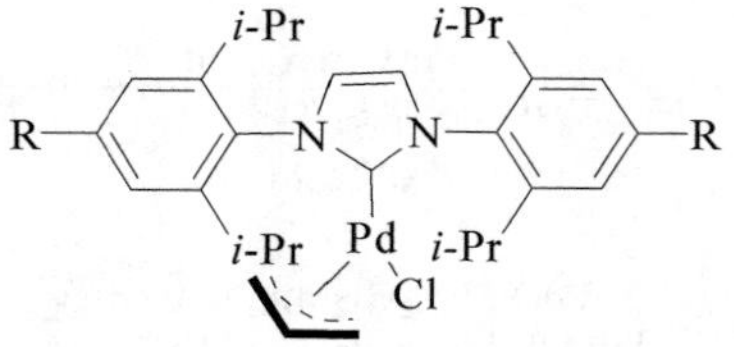

68

Sen 等[212]报道以吡啶甲醛甲基苯基腙(69)为配体,乙基二异丙基胺为碱,$Pd(C_6H_5CN)Cl_2$ 为催化剂,在 80 ℃水中催化末端芳炔与卤代 8-羟基喹啉的 Songashira 反应[式(2-145)],合成[3,2-h]喹啉衍生物,产物的得率为 83%~92%。

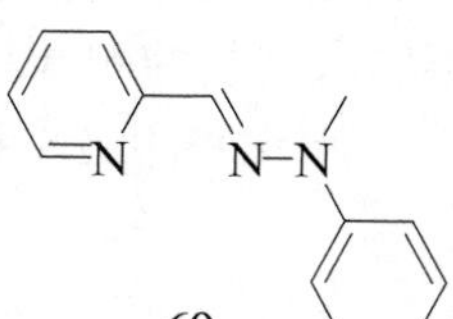

69

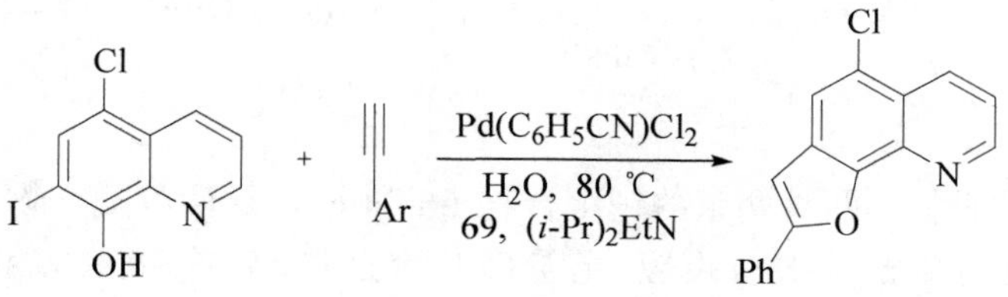

$$\text{(2-145)}$$

Pd(C_6H_5CN)Cl_2, H_2O, 80 ℃, 69, (i-Pr)_2EtN

Li 等[213]报道称$[Pd(\eta^3-C_3H_5)Cl]_2$是水介质无铜 Sonogashira 偶联反应的高效催化剂。反应以 N,N,N',N'-四(二苯基膦甲基)吡啶-2,6-二胺(70)为配体,K_3PO_4 为碱,

[Pd(η³-C₃H₅)Cl]₂ 为催化剂前驱体,用量为 1 ppm(10^{-6}),卤代芳烃及杂环芳烃卤化物均能成功地进行炔基化反应,TON 值高达 860000。

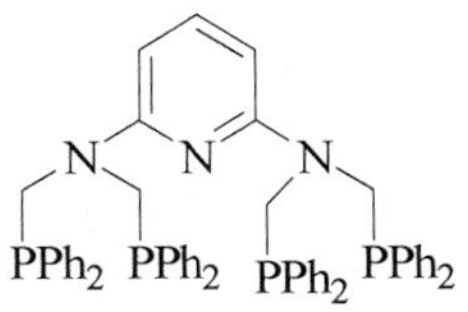

70

Joó 等[214]合成三种磺化的四氢席夫碱类型的钯配合物 71~73,这种钯配合物可以在空气存在下的水中保存数月不会发生明显的分解。它们在水中卤代芳烃与末端炔的 Sonogashira 偶联反应中展示了优越的催化性能,反应 4~24 h,产物得率为 76%~98%,转化频率(TOF 值)为 2790 h^{-1},重复使用 4 次,产物得率仍保持在 93%~98%。

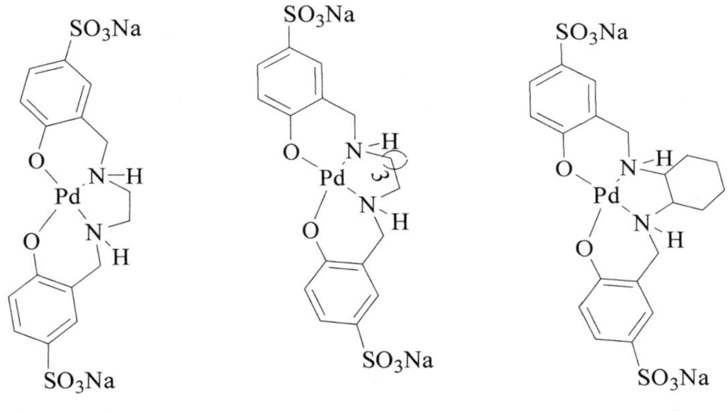

71　　　　**72**　　　　**73**

Košmrlj 等[215]报道以离子型钯配合物 74 催化水中溴代烃与末端炔的偶联反应,反应在 K_2CO_3 的碱性水溶液中 100℃反应 1 h 给出产物的得率为 58%~97%,如提高反应温度,延长反应时间,得率也将随之增加。

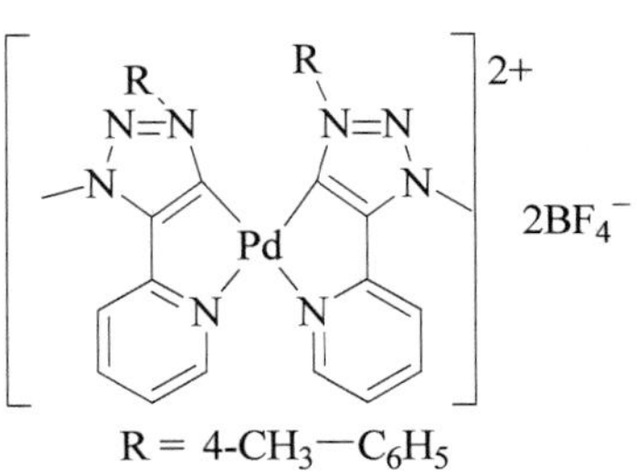

R = 4-CH₃—C₆H₅

74

3. 其他反应

在催化量的钯催化下，卤代烃与烯烃反应生成卤原子被烯烃取代的产物，通常称为 Heck 反应。简单的烯烃分子和分子内 Heck 反应都能在水介质中进行。

Kobayashi 和 Manabe[216]用 Pd(PPh$_3$)$_4$ 催化剂在含有 1-金刚羧酸中催化烯丙醇与碳亲核试剂的偶联反应[式(2-146)]。当 1-金刚羧酸用量为 10 mol%时，70℃下产物的得率达到最佳。

$$R_1\diagup\diagdown\overset{OH}{\underset{NuH}{\diagdown}}R_2 + \text{(1-金刚羧酸)} \xrightarrow[H_2O]{Pd(PPh_3)_4} R_1\diagup\diagdown\overset{Nu}{\diagdown}R_2 \quad (2\text{-}146)$$

2004 年有报道[217] 称 Pd(PPh$_3$)$_2$Cl$_2$/CuI 可以催化酰氯与末端炔的高效直接偶联反应，再结合使用催化量的十二烷基磺酸钠为表面活性剂和碳酸钾为碱，反应能在水中进行，生成得率很高(98%)的炔酮[式(2-147)]。当单独使用 Cu(Ⅰ)或 Pd(Ⅱ)作催化剂时反应几乎不能发生。不使用表面活性剂时产物的得率只有 9%。

$$R\overset{O}{-}Cl + \equiv\!-\!R_1 \xrightarrow[SDS, H_2O]{Pd(PPh_3)_2Cl_2/CuI} R\overset{O}{-}\!\equiv\!-R_1 \quad (2\text{-}147)$$

Wolf 和 Lerebours[218]设计了一种用于水介质中氯代或溴代芳烃与芳基硅氧烷的偶联反应[式(2-148)]的钯催化剂 75。该催化剂不需要任何添加剂和有机溶剂即可使不同底物的卤代芳烃与芳基硅氧烷反应顺利进行，并给出高达 99%得率的产物。

$$\text{Ar}^{Cl,Br}_X + \text{Ar}^{Si(OMe)_3}_Y \xrightarrow[\text{NaOH, H}_2\text{O}]{75} \text{biaryl}_{X,Y} \quad (2\text{-}148)$$

Lerebours 和 Wolf[219]报道用催化剂 75 催化水中芳基硅氧烷与 α,β-不饱和酮的共轭加成[式(2-149)和式(2-150)]。在微波辐射 4 h，仅用 5 mol%的钯催化剂 75，就能获得得率为 83%~96%的 β-取代的酮醛、酯、腈和硝基烷烃等产物[式(2-151)]。

$$\text{ArSi(OMe)}_3 + \text{环戊烯酮} \xrightarrow[\substack{10\text{ mol\% Cu(ACN)}_4\text{BF}_4 \\ \text{H}_2\text{O, MW 4 h}}]{5\text{ mol\% 75}} \text{产物} \quad (2\text{-}149)$$

$$\text{ArSi(OMe)}_3 + \text{环己烯酮} \xrightarrow[\substack{10\text{ mol\% Cu(ACN)}_4\text{BF}_4 \\ \text{H}_2\text{O, MW 4 h}}]{5\text{ mol\% 75}} \text{产物} \quad (2\text{-}150)$$

$$\text{ArSi(OMe)}_3 + R\text{-CH=CH-}R_2 \xrightarrow[\substack{10\text{ mol\% Cu(ACN)}_4\text{BF}_4 \\ \text{H}_2\text{O, MW 4 h}}]{5\text{ mol\% 75}} \text{产物} \quad (2\text{-}151)$$

$R_2 =$ CN, NO_2, CO_2H, $CO_2C(CH_3)_3$

Lipshutz 等[220]报道了不用有机锌并在室温下用钯配合物 76 催化碘化物或溴化物与溴代芳烃的交叉偶联反应[式(2-152)]。作者也考察了不同溴代芳烃在此催化体系中的反应活性，结果表明，即使溴苯的苯环间位有大体积的基团，反应也能给出 74%的产物得率。

76

$$RX + \text{Ar-Br} \xrightarrow[\substack{\text{Zn-二胺} \\ 2\% \text{ PTS/H}_2\text{O, rt}}]{2\text{ mol\% 76}} \text{Ar-R} \quad (2\text{-}152)$$

X = I, Br

Lipshutz 和 Nishikate[221]报道双离子[Pd(MeCN)$_4$](BF$_4$)$_2$ 在室温下催化非酸性水中芳酰胺与丙烯酸酯的 Fujiwara-Moritani 反应(芳烃的 C—H 键活化),成功地获得富电子的肉桂酸酯[式(2-153)]。在 1,4-苯醌、AgNO$_3$ 2 wt%(质量分数)PTS(聚氧乙基-生育酚基-癸二酸酯)/水中,10 mol%[Pd(MeCN)$_4$](BF$_4$)$_2$ 催化此反应,除位阻较大的棕榈酰胺给出理想产物的得率较低(43%)外,其他反应的得率为 72%~96%。

$$\text{ArNHAc} + \text{CH}_2=\text{CHCO}_2\text{R} \xrightarrow[\text{PTS/H}_2\text{O, rt}]{[Pd(MeCN)_4](BF_4)_2} \text{ArCH=CHCO}_2\text{R} \quad (2\text{-}153)$$

Shao 等[222]将钯配合物 77 催化剂应用在水介质中的 Mizoroki-Heck 反应[式(2-154)]。对于含有不同取代基团(吸电子和给电子)的底物的偶联反应,该催化剂在 KOtBu 强碱性环境中都具有较好的催化性能。

$$\text{PhBr} + \text{CH}_2=\text{CHCO}_2\text{H} \xrightarrow[\text{碱, H}_2\text{O, 24 h}]{1.0 \text{ mol\% 77}} \text{PhCH=CHCO}_2\text{H} \quad (2\text{-}154)$$

Yang 等[223]报道在 1-金刚羧酸和水中,Pd(acac)$_2$ 可以催化烯丙醇的 C—O 键活化与芳胺的反应[式(2-155)]。金刚酸的加入可以使烯丙醇中 C—O 键更易断开,从而提高反应的速率和产物的得率。实验结果也证实,在回流条件下,反应 30 min 即能给出较高得率的目标产物,而在位阻较大的芳胺的反应中产物的得率相对较低。

$$\text{ArNHR}_1 + \text{HOCH}_2\text{CH=CHPh} \xrightarrow[\text{1-AdCO}_2\text{H, H}_2\text{O}]{Pd(acac)_2, PPh_3} \text{R}_1\text{N(Ar)CH}_2\text{CH=CHPh} + \text{ArN(CH}_2\text{CH=CHPh)}_2 \quad (2\text{-}155)$$

Shao 等[224]合成了一种咪唑-单原子螯合的钯配合物 NHC-Pd(Ⅱ)-lm 78,该配合物可以催化水中芳基硼酸与苯甲酸酐的交叉偶联反应[式(2-156)]。考察不同芳基硼酸的取代基和空间位阻对反应活性的影响,结果显示,无论苯环含有吸电子取代基还是给电子取代基,该反应均能在室温下顺利进行,且对于位阻较大的 3,5-

二甲基苯硼酸与苯甲酸酐，反应也能很好地进行，并给出了 86%的得率。随后 Lu 等[225]合成了钯配合物 79，同样将其用于水介质芳香酸酐与芳基硼酸的交叉偶联反应。与 Shao 等工作不同的是，他们将反应底物扩大到不同芳香羧酸的酸酐与芳基硼酸的偶联，且反应时间也大大缩短。他们用催化剂 79 催化水中烯丙醇与芳基硼酸进行芳基-芳基偶联反应[式(2-157)和(2-158)]。在最佳条件下，所有反应都能顺利进行，并给出中等到良好得率的目标产物[226]。

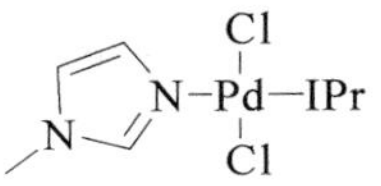

78

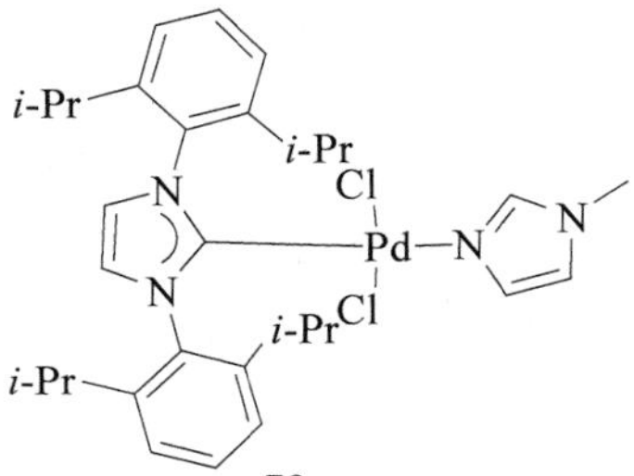

79

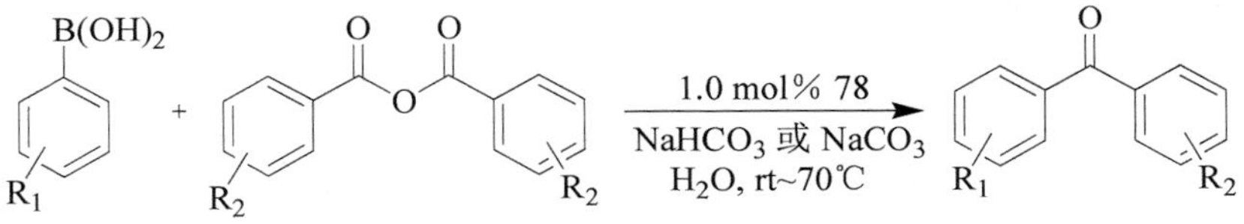

(2-156)

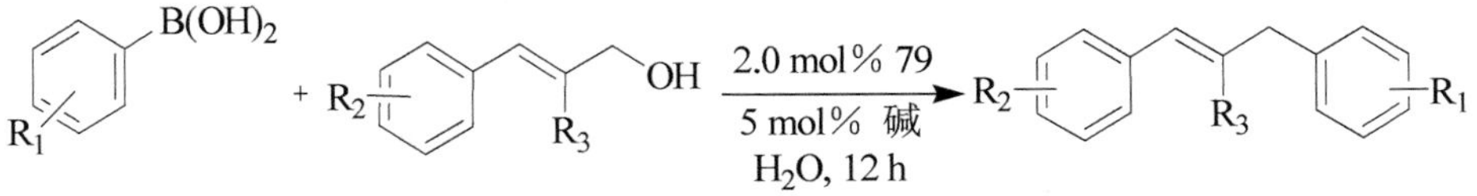

(2-157)

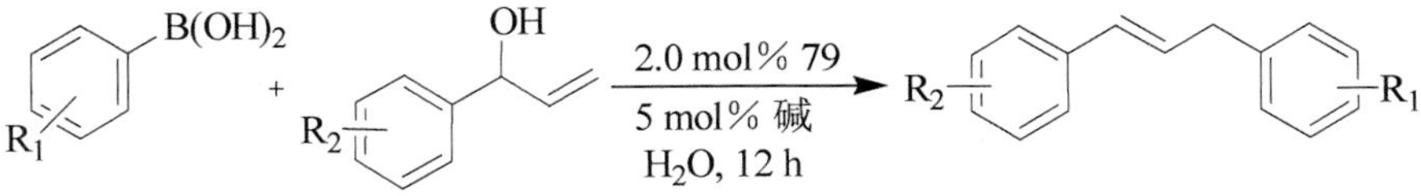

(2-158)

二、有机金属钌催化剂

有机金属钌不仅是构建 C—C 键的一种重要催化剂，也是还原反应、烯丙醇异构化反应等非常有用的催化剂，本节主要介绍了金属钌配合物催化剂在水介质还原反应、异构化反应和环丙化反应等中的应用。

1. 还原反应

1) 醛和酮的加氢反应

在水介质中，很多有机金属钌催化剂可以催化醛和酮的加氢反应。例如，1989

年 Joó 和 Bényei[227]用甲酸钠作为氢源，加入水溶性的 RuCl(TPPMS)$_2$ 作催化剂，水介质中不同种类的芳香醛和 α,β-不饱和醛均可被还原成相应的饱和醇。他们又用 RuHCl(TPPMS)$_3$ 催化酮酸的加氢反应生成羟基羧酸[式(2-159)]。

$$\text{CH}_3\text{CH}_2\text{CH}_2\text{COCOOH} \xrightarrow[\text{H}_2,\ 60\ ℃]{\text{RuHCl(TPPMS)}_3} \text{CH}_3\text{CH}_2\text{CH}_2\text{CH(OH)COOH} \qquad (2\text{-}159)$$

1994 年 Daigle 等[228]报道了水溶性的 cis-RuCl$_2$(PTA)$_4$ (PTA = 1,3,5-三氮杂-7-磷杂金刚烷)催化醛的加氢反应。采用手性钌离子催化剂同样可以催化水介质中酮的不对称加氢反应。而在酮加氢反应的体系中加入表面活性剂，能明显提高催化剂的催化活性，且催化剂还能重复使用。Georg 等[229]合成了 8 种不同对映异构体钌配合物(80~87)，并考察它们在水介质苯乙酮加氢反应中的应用，结果显示，在这些配合物中，配合物 85 催化性能最佳，在 60℃，pH=9 条件下，以甲酸钠为氢供给体，催化剂的 TOF 值达 43 h^{-1}，ee 值超过 93%。

Xiao 等[230]将未经修饰的 Ru-TsDPEN 直接用于水相不对称转移氢化反应，研究发现水相反应活性比在 HCOOH/Et$_3$N 共沸体系中的高，并具有与有机相相当或略低的对映选择性。他们认为这可能与底物的疏水性有关，由于底物不溶于水，催化剂分配于底物相和水相中，但其在底物相中浓度更高，反应发生在底物/催化剂相中，从而提高了反应速率。2006 年他们又报道了[RuCl$_2$(p-cymene)]$_2$/β-氨基醇(88)/HCOONa 催化体系应用于水中芳香酮的还原[式(2-160)]。在 40℃反应 2~8 h，反应基本转化完全，但产物的 ee 值并不高，如邻氯苯乙酮，ee 值不足 40%[231]。随后 Collin 等[232]只改变反应体系中的配体，即用配体 89 取代 β-氨基醇同样催化水中芳香酮的加氢反应，不仅反应温度降低，而且产物的 ee 值提高到 99%。

$$\text{R}\underset{\text{}}{\overset{\text{O}}{\bigcirc}}\text{CH}_3 \xrightarrow[\text{HCO}_2\text{Na, H}_2\text{O, 40 °C}]{[\text{RuCl}_2(p\text{-cymene})]_2/88 \text{ 或 } 89} \text{R}\underset{\text{}}{\overset{\text{OH}}{\bigcirc}}\text{CH}_3 \qquad (2\text{-}160)$$

Espino 等[233]报道了采用 4,4'-二甲氧基-2,2'-二联吡啶(dmobpy) 为配体的有机钌 [(η^6-arene)RuCl($k2$-N,N-dmobpy)]X 催化这种反应，催化剂在此反应中表现了较高的活性，TON 值高达 200。

Joó 等[234]在 HCOONa 水溶液中用 10mol%[RuH$_2$(mtppms)$_x$][x = 3,4; mtppms = 间磺酰基苯基二甲基膦]催化剂催化邻甲氧基苯乙酮的选择性加氢反应生成 1-(2-甲氧基苯基)乙醇。如用配体 91 代替 90[235]，在相同的反应条件，可以使 α-苯基-β-卤代乙酮选择性加氢，生成卤代醇，ee 值为 84%~94%，而卤原子不受其影响[式 (2-161)]。Mao 和 Guo[236]报道在相同的催化体系下用配体 92，并在此反应体系中加入相转移催化剂 TBAB，此反应体系的优点是可以催化体积庞大的酮，如二茂铁乙酮，产物的化学选择性和对映选择性均在 90%以上。

$$\text{Ar}\underset{\text{}}{\overset{\text{O}}{\bigcirc}}\text{X} \xrightarrow[\text{H}_2\text{O, HCOONa, 30°C}]{2.5\% [\text{RuCl}_2(p\text{-cymene})]_2, 5\% 91} \text{Ar}\underset{\text{}}{\overset{\text{OH}}{\bigcirc}}\text{X} \qquad (2\text{-}161)$$

Deng 等[237]设计和合成了另一类型的水溶性手性二胺配体(TsDPEN-di-SO$_3$Na)。这一类水溶性配体与已报道的水溶性配体相比，因引入两个磺酸盐基团而具有更好的亲水性。以此二胺为配体的钌络合物作催化剂，以甲酸钠为氢源，芳香酮、杂芳香酮的水相不对称转移氢化反应，研究发现，在表面活性剂存在下，此催化体系能够获得高的转化率和对映选择性。另外，在阳离子表面活性剂存在下，催化剂可在水相反应中循环使用 21 次。

Fan 等[238]在研究聚乙二醇介质中不对称催化加氢反应的基础上，发展了 PEG/H$_2$O 为反应介质的不对称催化转移氢化反应新溶剂体系。在此体系中，以 HCOONa 为氢源，[RuCl$_2$(p-cymene)]$_2$/(S,S)-DPEN 催化苯乙酮的不对称转移氢化，因催化剂能保留在 PEG/H$_2$O 相中而不用萃取，催化剂可循环使用 14 次，且对映选择性保持不变。

Lin 等[239]利用他们自己发展的 SmI$_2$ 参与的偶联反应合成了一系列手性二胺，将其用于水相不对称转移氢化反应，发展了一种合成苯酞的新方法[240]；并发现配体 93 的苯环上取代基的体积对产物的对映选择性具有决定作用，叔丁基修饰的配体给出最好的结果。表面活性剂 CTAB 的加入有助于提高反应速率。在水介质中，各种取代底物都以高的收率和对映选择性转化为苯酞[式(2-162)]。

$$\text{配体 93}$$

$$R_2 \text{(苯环)} \text{—COOC}_2\text{H}_5,\ \text{—C(O)CH}_2R_1 \xrightarrow[\text{HCO}_2\text{Na, 4\%CTAB, H}_2\text{O, 40 ℃}]{[\text{RuCl}_2(p\text{-cymene})]_2/93} R_2\text{(苯酞)}R_1 \quad (2\text{-}162)$$

Chen 等[241]将水溶性手性二胺(S,S)-1,2-二苯基乙二胺二磺酸钠(S,S)-DPENDS 与钌膦配合物[RuCl$_2$(TPPTS)$_2$]$_2$ 原位生成的催化剂用于催化水相中苄叉丙酮的不对称转移氢化反应。在优化条件下，羰基加氢产物 4-苯基-3-丁烯-2-醇的选择性可达 96%，对映选择性可达 71.2%。经正己烷简单萃取后即可实现催化剂与加氢产物的分离，循环使用 5 次后，目标产物 4-苯基-3-丁烯-2-醇选择性和对映选择性没有明显下降。Zhou 等[242]报道将[RuCl$_2$(p-cymene)]$_2$/94 和[(C$_5$Me$_5$)IrCl$_2$]$_2$/94a 配体用于苯甲醛的还原，其中，94b 可以重复使用两次，94c 可以重复使用三次。

配体 94

94 a NR$_1$R$_2$R$_3$ = N(C$_2$H$_5$)$_3$
94 b NR$_1$R$_2$R$_3$ = N(CH$_2$CH$_2$CH$_2$CH$_3$)$_3$
94 c NR$_1$R$_2$R$_3$ = CH$_3$(CH$_2$)$_6$CH$_2$N(CH$_3$)$_2$

在 PEG-400-水介质中，以(1S,2S)-1,2-二苯基乙二胺的磺酸钠盐为手性修饰剂(95)，考察了水溶性三苯基磺酸钠膦稳定的有机金属钌催化苯乙酮及其衍生物的不对称加氢反应。结果表明，该催化剂体系具有良好的催化活性和对映选择性。在最佳反应条件下，苯乙酮转化率和对映选择性分别为100%和84.9%。经正己烷萃取后，催化剂即可从产物中分离出来，并可重复使用多次[243]。Liu 等[244]报道了水溶性的钌配合物 96 的合成，并将其在 N_2 氛围中于 60℃下水介质中不同醛的转移氢化反应。

最近 Singh 等[245]报道了一种新型水溶性钌配合物 97 的合成，采用此配合物催化水介质中羰基化合物的转移氢化反应。研究显示，用甲酸、柠檬水、抗坏血酸等作氢源时，催化剂受溶液的 pH 影响非常大；如用甘油作氢源，可以克服上述缺点。

Wills 等[246]首次报道 Ru(Ⅱ)/TsDPEN 催化水中富电子的芳酮，如氨基苯乙酮、甲氧基苯乙酮不对称氢化反应。这类酮在传统的不对称氢化反应条件下一般很难发生，但作者研究发现，催化剂 98～100 均可有效地催化对位含有给电子基团的芳酮的不对称还原反应，在 FA/TEA 存在下给出较高得率(96%～99%)和 ee 值(94%～98%)的还原产物，而在邻三氟甲基苯乙酮和环己酮的还原反应中，产物的得率虽然可达 99%，但对映选择性较低，分别为 69%和 75%。

2) 亚胺和亚胺离子的氢化反应

在水介质中,有机金属钌配合物也可以催化亚胺和亚胺离子的氢化反应。如 Deng 等[247]采用他们自己设计的水溶性手性二胺配体(TsDPEN-di-SO_3Na)101 与对-异丙基苯甲烷钌(Ⅱ)氯化物的二聚体络合物催化水介质中亚胺的不对称转移氢化反应,还原 3,4-二氢异喹啉、二氢-β-咔啉和 N-磺酰胺类化合物都获得了很好的得率和对映选择性[式(2-163)~式(2-166)]。以 N-磺酰亚胺为底物,研究了催化剂的循环。当反应结束后,以乙酸乙酯/正己烷混合溶剂萃取出产物,然后加入 1 当量甲酸,进行下一轮催化循环。同时,通过 ICP-MS 分析证实,萃取所流失的金属量少于 0.09%,催化剂可以循环使用 4 次且对映选择性没有显著的降低。有趣的是,无活性的亚胺可以通过形成亚胺慃离子而顺利进行不对称转移氢化反应。表面活性剂的使用不仅可以提高反应速率,还能够提高反应的对映选择性。加入阳离子表面活性剂 CTAB 获得了最好的结果。反应体系的 pH 也非常重要,当以 HCOOH 为氢源时,没有产物生成。随后,他们又合成了一系列修饰的 TSDPEN 配体,并将其与[(Cp*)Ru(Ⅱ)Cl_2]$_2$的络合物催化水介质硝基烯烃的不对称转移氢化反应,对各种取代底物显示出比较高的反应活性和对映选择性。当用 HCOOH-HCOONa 缓冲溶液调节溶液 pH 为 5.1~5.6 时,反应的化学选择性达到最佳,同时,配体的电子性质对反应的对映选择性也有非常重要的影响。

$$\text{NaO}_3\text{S} \quad \quad \text{SO}_3\text{Na}$$
$$101$$

$$\text{底物} \xrightarrow[\text{HCOONa, CTAB, H}_2\text{O}]{\text{RuCl}_2(p\text{-cymene})/101} \text{产物} \quad (2\text{-}163)$$

$$\xrightarrow[\text{HCOONa, CTAB, H}_2\text{O}]{\text{RuCl}_2(p\text{-cymene})/101} \quad (2\text{-}164)$$

$$\xrightarrow[\text{HCOONa, CTAB, H}_2\text{O}]{\text{RuCl}_2(p\text{-cymene})/101} \quad (2\text{-}165)$$

$$\text{H}_3\text{CO}\underset{\text{H}_3\text{CO}}{\bigcirc}\underset{R}{\overset{+}{N}}\text{-Bn}\quad\text{Br}^- \xrightarrow[\text{HCOONa, CTAB, H}_2\text{O}]{\text{RuCl}_2(p\text{-cymene})/101} \text{H}_3\text{CO}\underset{\text{H}_3\text{CO}}{\bigcirc}\underset{R}{\overset{*}{N}}\text{-Bn} \quad (2\text{-}166)$$

2. C—H 键活化反应

和其他 C—H 键相比，末端炔的 C—H 键具有弱酸性，且比较活泼，在水中容易被许多过渡金属活化形成金属炔化物或 π 配合物，这些配合物再与其他官能团作用形成新的 C—C 键[式(2-167)]。

$$R\text{≡}H \xrightarrow[\text{水}]{M=\text{Cu, Ag, Ru 等}} \begin{array}{c} R\text{≡}M \\ \text{或} \\ R\text{≡}H \\ \quad\vdots \\ M \end{array} \longrightarrow R\text{≡}CH_2\text{-}G \quad (2\text{-}167)$$

1995 年 Li 等[248]报道了配合物 $RuCl_2(PPh_3)_3$ 催化水中高烯丙醇化合物的官能团重排反应，但在此催化体系中会有副反应发生。后来也有报道[249]烯丙醇类化合物与醛发生的类似 Aldol 缩合反应，在反应过程中，烯丙醇的 C—H 键先被活化形成烯醇中间体，紧接着发生双键重排[式(2-168)]。

$$\underset{R}{\overset{O}{\parallel}}\text{H} + \overset{OH}{\underset{Me}{\diagup\!\!\diagdown}} \xrightarrow[\text{H}_2\text{O, 110℃, 5 h}]{RuCl_2(PPh_3)_3, 3\text{mol}\%} \underset{R}{\overset{OH}{\diagup}}\underset{Me}{\overset{O}{\diagdown}}\text{Me} \quad (2\text{-}168)$$

末端炔烃的加成偶联反应(Strauss 反应)是一个非常经典的反应，近来 Chen 等[250]报道催化剂 102 可以催化水中脂肪炔或芳香炔的加成偶联反应生成区域选择性和立体选择性均很高的反式烯炔产物[式(2-169)]。

$$\left[\begin{array}{c} \overset{PPh_2}{\diagup} \\ N\text{—}Ru\text{—}NCMe \\ \overset{\diagdown}{PPh_2} \\ PPh_2 \end{array}\right]^{\oplus} \quad OTf^{\ominus}$$

102

$$R\text{≡} \xrightarrow[\text{H}_2\text{O, 100℃}]{102} \underset{R}{\overset{H}{\diagup}}\text{≡}\underset{R}{\overset{H}{\diagdown}} \quad (2\text{-}169)$$

64%~86%

Arockiam 等[251]报道催化剂[Ru(O$_2$CR)$_2$(arene)]催化水中芳烃邻位进行活化再芳基化反应,反应无需加入表面活性剂[式(2-170)],且其在水中的催化效率明显优于有机溶剂 NMP。同时,作者也发展了催化剂[RuCl$_2$(p-cymene)]$_2$催化水中杂原子芳基苯衍生物功能化反应[式(2-171)]以及间三氯苯与联苯或稠环化合物的反应[式(2-172)]。

$$\text{2-吡啶基苯} + 2\ \text{PhCl} \xrightarrow[\substack{\text{KO}i\text{-Pr}\\ 3\ \text{eq K}_2\text{CO}_3\\ \text{H}_2\text{O, 100℃, 2 h}}]{[\text{Ru(O}_2\text{CR)}_2\text{(arene)}]} \text{邻芳基化产物} + \text{双邻芳基化产物} \quad (2\text{-}170)$$

$$\text{PhCH=N-} + \text{Het-X} \xrightarrow[\substack{20\ \text{mol\%KO}i\text{-Pr}\\ 3\ \text{equiv K}_2\text{CO}_3\\ \text{H}_2\text{O, 100 ℃, 2 h}}]{5\ \text{mol\%[RuCl}_2(p\text{-cymene)}]_2} \text{单 Het 产物} + \text{双 Het 产物} \quad (2\text{-}171)$$

$$(2\text{-}172)$$

Ackermann 等[252]报道了催化剂 103 能直接活化不活泼的 C—H 键进行芳基化反应,并给出化学选择性较高的产物[式(2-173)]。Singh[253]设计了两种钌催化剂 104 和 105 用于特定条件下的 C—H 活化或 2-吡啶苯和 N-咪唑基的功能化反应,在催化量的三甲基乙酸钾盐存在下,这两种双齿配体催化剂均能使芳烃与卤代苯进行两次芳基化反应。值得注意的是它们在水中 C—H 键活化以及芳烃芳基化反应的活性明显高于有机溶剂(甲基吡咯烷酮)介质中[式(2-174)~式(2-176)]。

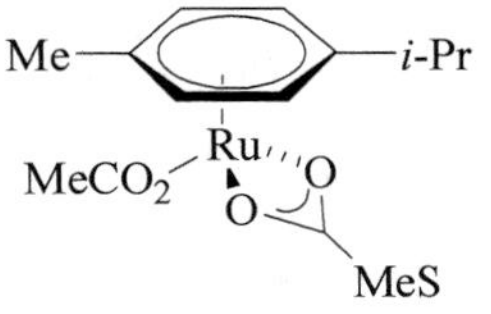

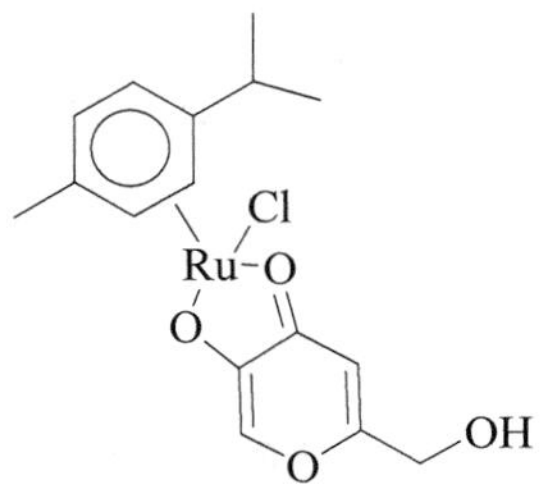

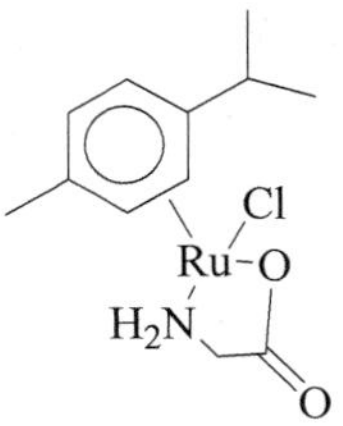

$$\text{103} \quad \text{104} \quad \text{105}$$

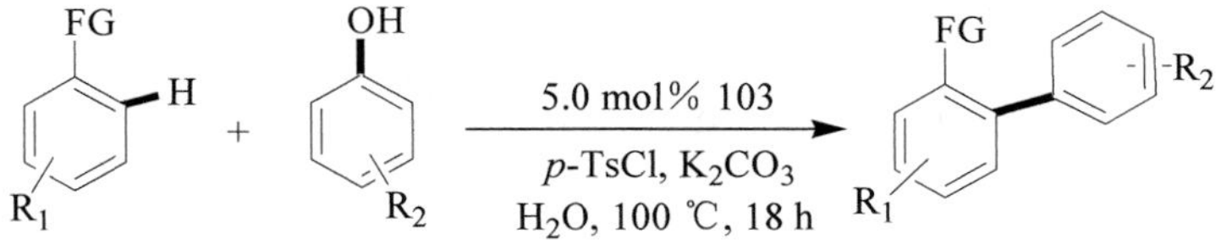

$$(2\text{-}173)$$

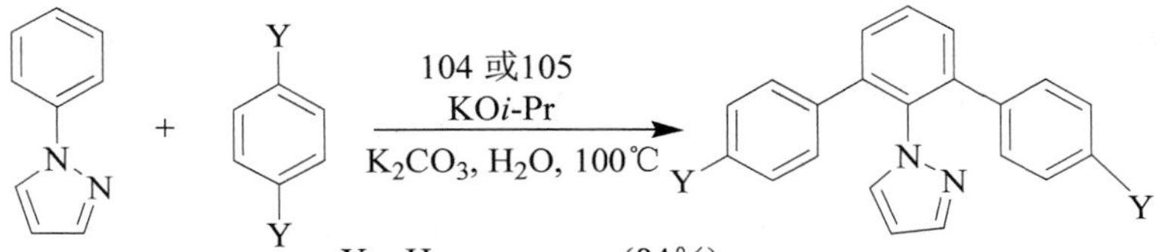

$$(2\text{-}174)$$

Y = H (84%)
Y = Me (85%)
Y = COMe (89%)
Y = CO$_2$Me (80%)

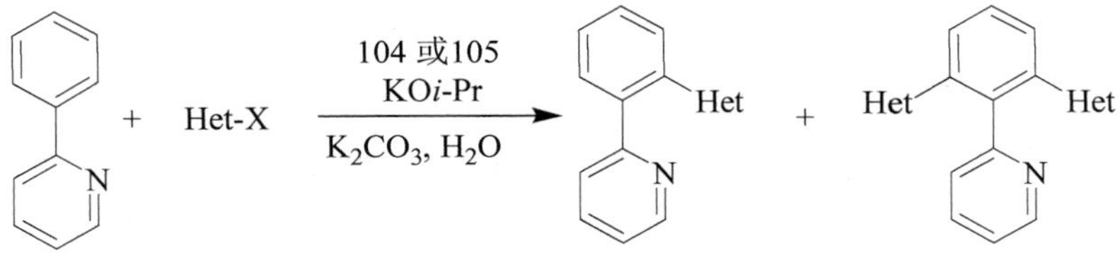

$$(2\text{-}175)$$

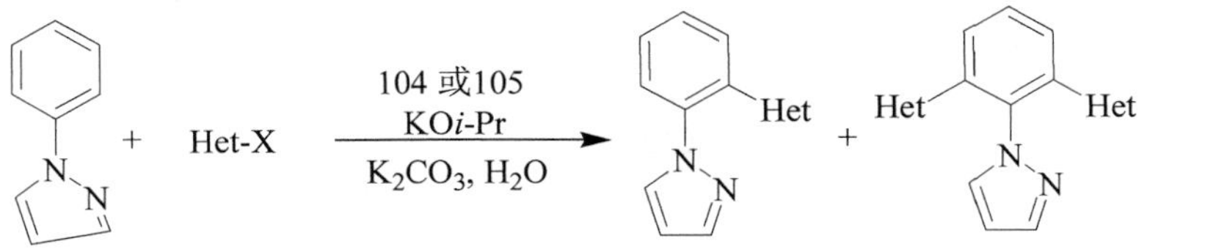

$$(2\text{-}176)$$

Het-X:

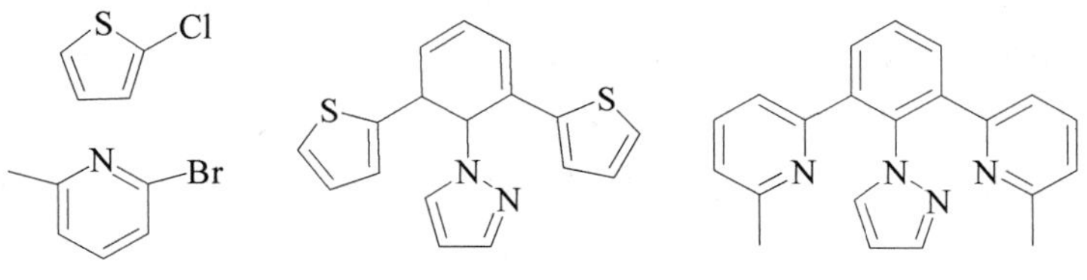

Beller 等[254]研究了催化剂[Ru(cod)Cl$_2$]催化水中 C—H 键的活化反应,他们将

该催化剂应用于卤代芳烃、CO 以及芳烃的羰基化偶联反应[式(2-177)]。随后 Tlili 等[255]报道将其应用于水中烯烃与 CO 和芳烃的反应[式(2-178)]，特别是在催化苯乙烯与(杂芳基)吡啶及其衍生物的反应中，选择性和原子经济性均达 100%。

$$\text{DG-C}_6\text{H}_4\text{-H} + \text{CO} + \text{X-C}_6\text{H}_4\text{-R} \xrightarrow[\text{H}_2\text{O}]{[\text{Ru(cod)Cl}_2]} \text{DG-C}_6\text{H}_3\text{-C(O)-C}_6\text{H}_4\text{-R} \quad (2\text{-}177)$$

$$\text{DG-C}_6\text{H}_4\text{-H} + \text{CO} + \text{CH}_2\text{=CH-R} \xrightarrow[\text{H}_2\text{O}]{[\text{Ru(cod)Cl}_2]} \text{DG-C}_6\text{H}_3\text{-C(O)-CH}_2\text{CH}_2\text{-R} \quad (2\text{-}178)$$

最近 Cai 等[256]报道了催化剂[RuCl$_2$(p-cymene)]$_2$ 应用于芳香羧酸与烯烃的烯基化反应，在 Cu(OAc)$_2$·H$_2$O 存在时，80℃下反应 16 h，可得到理想的邻羟甲基苯甲酸内酯目标产物，催化剂能循环使用 6 次[式(2-179)]。

$$\text{R}_1\text{-C}_6\text{H}_4\text{-COOH} + \text{CH}_2\text{=CH-R}_2 \xrightarrow[\substack{\text{Cu(OAc)}_2\cdot\text{H}_2\text{O (2.0 eq)} \\ \text{PEG-400/H}_2\text{O (3:2), 80℃, 16h}}]{[\text{RuCl}_2(p\text{-cycmene})]_2 \text{ (2 mol\%)}} \text{phthalide-CH}_2\text{R}_2 \quad (2\text{-}179)$$

3. 烯丙醇异构化反应

烯丙醇异构化反应[式(2-180)]是指通过烯醇中间体不可逆地自动转变成羰基化合物，它是制备许多具有生物功能天然产品的关键步骤[248]。早在 1994 年，Grubbs 和 McGrath[257]用[Ru(H$_2$O)$_6$](tos)$_2$ (tos 为对甲基苯磺酸盐)催化水介质烯丙醇和烯丙醚的异构化反应，效果优良。

$$\text{CH}_2\text{=CH-CH(OH)-R} \xrightarrow[\text{溶剂}]{\text{Ru 催化剂}} \text{CH}_3\text{-C(O)-CH}_2\text{-R} \quad (2\text{-}180)$$

2006 年 Cadierno 等[258]将合成的催化剂 106 用在水介质中 1-辛烯-3-醇的异构化生成 3-辛酮的反应。75℃下反应 45 min 即可完成，产物得率 100%、TOF 值为 750 h^{-1}，向体系中加入 Cs$_2$CO$_3$，反应时间可缩短到 15 min，TOF 值也提高到 2000 h^{-1}。Joó 等[259]发现，催化剂 [RuCp(mPTA)$_2$(OH$_2$O)]-(OSO$_2$CF$_3$)$_3$·(H$_2$O)(C$_4$H$_{10}$O)$_{0.5}$ 和 [RuClCp(mPTA)$_2$](OSO$_2$CF$_3$)$_2$ 均可催化水介质烯丙醇化合物的异构化反应，但 80℃下转化率不高，低于 50%。Gimeno 等[260]用催化剂 107 快速地催化水中异构化反应，催化剂在 1-辛烯-3-醇异构化反应中可重复使用 5 次。

Cadierno 等[261]将钌催化剂 108 应用在水中对烯丙基茴香醚的异构化反应,催化剂用量为 1 mol%时,80℃下微波辐射 5~30 min 可定量合成反式茴香醚,选择性高达 99%,这个反应可放大到 20.2 mmol,在 35℃下反应 6 h 可得到 2.89 g (96%)反式茴香醚。

Kathó 等[262]报道了催化剂 cis-[RuCl$_2$(DMSO)$_4$]催化水相中烯丙醇异构化反应,此反应对酮的选择性为 100%。Romerosa 等[263]研究了不同水溶性催化剂[RuClCp(PTA)$_2$]、[RuClCp(HPTA)$_2$]Cl$_2$·2H$_2$O、[RuCp(DMSO-kS)(PTA)$_2$]Cl、[RuCp(DMSO-kS)(PTA)$_2$](OSO$_2$CF$_3$)和[RuCp(DMSO-kS)(HPTA)$_2$]Cl$_3$·2H$_2$O 的催化性能,后三个催化剂几乎没有催化活性,而[RuClCp(PTA)$_2$]活性最高,产物得率为 69%~91%。

Romerosad 等[264]报道,虽然催化剂 109 水溶性不强,但仍然可以高效催化水中 1-辛烯-3-醇的异构化反应,受此启发[265],他们研究了催化剂 109 和双金属配合物 110 在水相 1-辛烯-3-醇、1-庚烯-3-醇和 1-己烯-3-醇异构化反应中的催化性能,获得良好的效果。

4. 其他反应

重氮乙酸酯与烯烃反应生成环丙化产物。在水中采用手性双羟甲基二氢噁唑基吡啶-钌催化剂,苯乙烯的不对称环丙化反应对映选择性高达 96%~97%,反-顺式立体选择性也可达 97%。Simonneaux 等[266]合成了一种新的水溶性钌卟啉配合物 111,并将其用于水介质重氮乙酸酯与烯烃的反应,生成具有光学活性的反式环丙基酯[式(2-181)],对映选择性达 86%,催化剂稳定性高,可回收重复使用。

$$\text{EtO}_2\text{C}\diagdown_{\text{H}}\diagup^{\text{N}_2} + \text{PhCH=CH}_2 \xrightarrow[\text{H}_2\text{O}]{111} \text{Ph}\diagup\triangle\diagdown\text{CO}_2\text{Et} + \text{Ph}\diagup\triangle\diagdown\text{CO}_2\text{Et} \qquad (2\text{-}181)$$

烯烃换位反应是形成不饱和碳碳键的重要方法，常见催化剂是亚苄基钌配合物，由于其对水具有相容性，适合于水介质反应。近来 Lynn 等报道[267]新型水溶性烷亚基钌配合物 112 和 113 催化水介质烯烃换位反应[式(2-182)]。

(2-182)

环合换位反应是构建中等和大环化合物的一个重要反应，催化剂 114 能够使 σ,ω-二烯的环合换位反应在水介质中顺利进行[268][式(2-183)]。

$$\text{底物(带 NMe}_3^+\text{Cl}^-\text{和 Ph 的二烯)} \xrightarrow{10 \text{ mol\% 114, H}_2\text{O}} \text{环戊烯产物(带 NMe}_3^+\text{Cl}^-\text{)} \quad (2\text{-}183)$$

早在 1990 年，Taqui 等[269]报道了水溶性钌配合物 $K[Ru^{III}(EDTA\text{-}H)Cl]_2 \cdot H_2O$ 催化水介质乙炔水合生成乙醛。通常，炔烃水合反应应遵循 Markovnikov 规则，生成相应的酮。但有报道在水/醇介质中，采用钌配合物为催化剂，能选择性地合成醛。近年来 Sears 等[270]发展了催化剂 115，可催化水介质中腈(芳香族，脂肪族或乙烯基型)反应，100℃下反应 7 h 可获得 43%~99%的酰胺[式(2-184)]。

$$R-C\equiv N \xrightarrow[\text{H}_2\text{O, 100℃, 空气气氛}]{5 \text{mol\% 115}} R-C(=O)NH_2 \quad (2\text{-}184)$$

Cadierno 等[271]发展了钌配合物 116 的合成并将其用于催化水介质中腈的水合反应中，在催化剂用量为 3mmol%时，无需添加其他任何试剂，反应在 100℃可顺利进行，短时间内可获得 78%以上的酰胺产物[式(2-184)]，同时对于不同底物，如芳香腈、脂肪族腈、α,β-不饱和腈等都具有优良的催化活性。

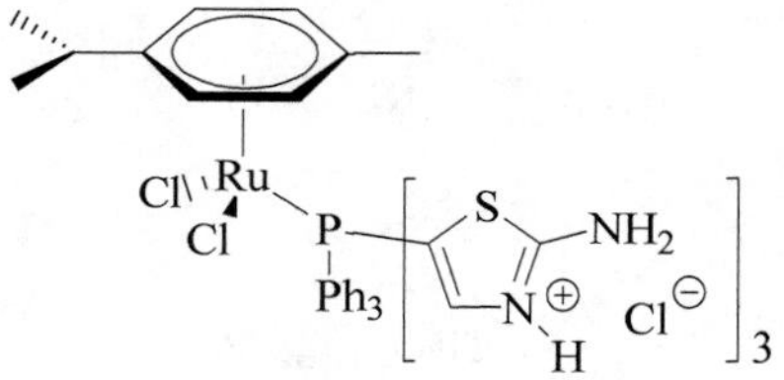

116

最近 Tomas-Mendivil 等[272]报道了一种新型钌配合物 117，不仅可用在环境温和友好介质中合成较高得率的抗癫痫药卢非酰胺[式(2-185)]，还能催化 α-羟基腈的水合反应[式(2-186)]，催化剂可重复使用 10 次，每次在 30 min 内就可达最高转化率。

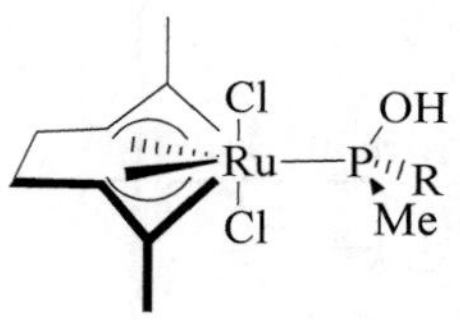

117

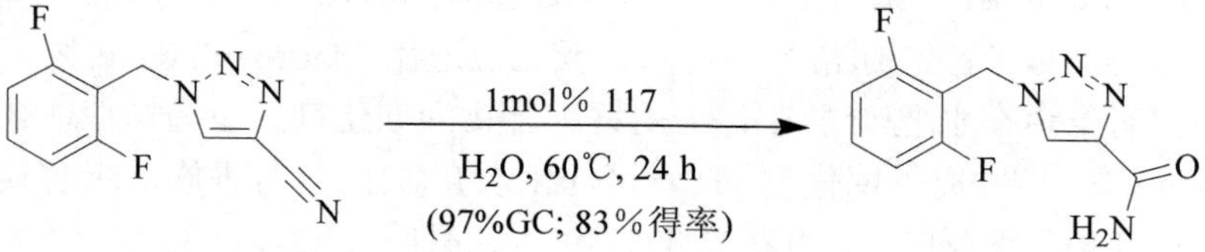

(2-185)

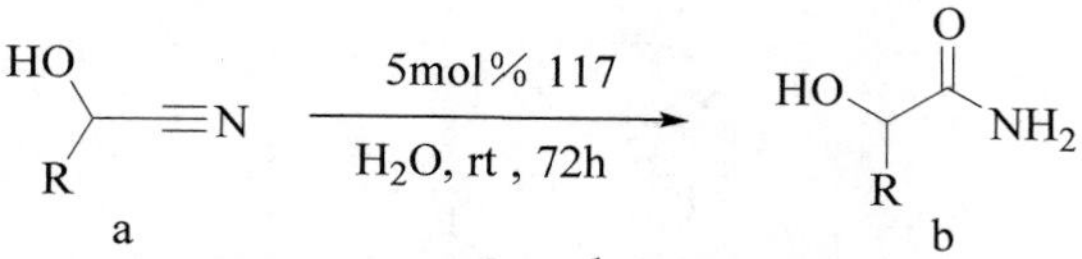

$1 mol\% \ 117$
$H_2O, 60℃, 24 h$
(97%GC; 83%得率)

(2-186)

$5 mol\% \ 117$
$H_2O, rt, 72h$

a
b
R = Me(a/b): > 99% yield by ^{1}H NMR (85%得率)
R = Et(a/b): > 99% yield by ^{1}H NMR (85%得率)

三、有机金属铑催化的反应

1. 还原反应

1) 羰基化合物的还原

Knowles 等率先报道含有手性膦配体的 Wilkinson 类铑膦络合物催化烯烃的不对称加氢反应，由于催化剂在水介质中易水解，且水也会破坏反应的过渡态和有机底物与催化剂的相互作用,通常在反应体系中需要加入相转移催化剂或加入两亲的表面活性剂以提高反应活性和选择性[273, 274]。近年来,采用水溶性手性催化剂应用于水介质不对称加氢的反应逐步开展,1990 年 Tóth 和 Hanson[275]将水溶性铑催化剂用于 α-乙酰胺基丙烯酯的还原[式(2-187)],获得产物的 ee 值高达 94%。

$$\underset{\text{NHCOR}_2}{\overset{\text{COOR}}{\text{Ph-CH=C}}} \xrightarrow{H_2, H_2O} \underset{\text{NHCOR}_2}{\overset{\text{COOR}}{\text{Ph-CH}_2-\text{CH}}} \quad (2\text{-}187)$$

Xiao 等[276]用[Rh Cp*Cl$_2$]$_2$/β-氨基醇催化剂催化苯乙酮还原成醇,如添加配体 118,虽然反应转化率略有下降,但产物的 ee 值可从 31%提高到 68%。随后他们又报道,Rh-TsDPEN 119 是水中酮不对称加氢反应的高效催化剂,且适用的底物范围较广,包括非功能化的芳香酮、杂环芳酮、酮酸酯和不饱和酮等,ee 值达到 99%,最佳反应条件下,TOF 值为 690 h^{-1}。在较高 S/C 下,催化反应在空气氛围下也可进行,如 2-乙酰基呋喃在 S/C = 100 时,空气氛围下反应 1 h,产物 ee 值也能达 99%[277]。

Deng 等[278]用 Rh-Cp*-TsDPEN 催化水介质 α, β-不饱和酮的加成反应[式(2-188)],反应底物中的 C═C 和 C═O 均被加成,产率和 ee 值均在 98%以上。

$$\text{R}\underset{\text{O}}{\overset{\text{O}}{\bigodot}} \xrightarrow[\substack{\text{HCOOH/HCOONa = 4/20} \\ \text{H}_2\text{O, 60°C, 3 h}}]{(S,S\text{-Cp*RhClTsDPEN, S/C = 100})} \text{R}\underset{\text{O}}{\overset{\text{OH}}{\bigodot}} \quad (2\text{-}188)$$

Kang 等[279]用水溶性配体 120 与[Cp*RhCl$_2$]$_2$ 构成高效催化体系,在苯乙酮加氢反应中,苯环对位无论是含有吸电子还是给电子基团,所得产物的 ee 值均在 90%以上,但如果在邻位上有吸电子基团如—F 时,产物的 ee 值只有 67%。以甲酸钠/水作为氢源,用[RhCl$_2$(Cp*)]$_2$ 催化水中苯乙酮的不对称加氢反应,当加入一定量的配体 120 时,室温下反应 0.45 h 即可完成,产物 ee 值达 96%。

Deng 等[280]发现手性二氨基二胺配体 121——铑络合物(Rh-TsDPEN-di-NH$_2$)能够高效催化水中多种不同类型底物(杂芳香酮、ω-溴代芳香酮、亚胺)的不对称转移氢化反应，反应都能在 10 h 以内完成，获得高得率产物。

Adolfsson 等[281]合成了一类长链烷烃修饰的表面活性剂型手性二胺配体，其与[(Cp*)RhCl$_2$]$_2$的络合物 122 在 SDS 存在下催化水相脂肪酮和芳香酮的不对称转移氢化反应。由于疏水作用，催化剂和底物插入 SDS 形成的胶束中，从而增加底物与催化剂的接触概率，提高反应活性。这种作用对于水溶性很差的长链脂肪酮表现得尤为明显，但这种相互作用对于对映选择性的提高不明显，原因在于胶束没有强刚性，不能对底物与催化剂的作用方式产生显著影响，对于甲基环己基酮的还原，ee 值可达到 37%~84%。

Salmain 等[282]以 2,2′-二联吡啶胺为配体合成了铑配合物 123，以甲酸盐为氢源，可高效催化水中芳香酮还原产生仲醇，甲酸盐的用量、pH 以及缓冲溶液的种类对反应速率和产物醇的得率有着重要的影响。

最近发现[283]，[Rh(COD)Cl]$_2$ 和配体 124 形成的水溶性有机金属铑催化剂 Rh(DHTANa)可高效催化水介质中环己烯酮的加氢反应[式(2-189)]，当底物/Rh 摩尔比为 500/1 时，转化率达 99%，环己酮选择性接近 96%，催化剂可循环使用多次，活性基本保持不变。

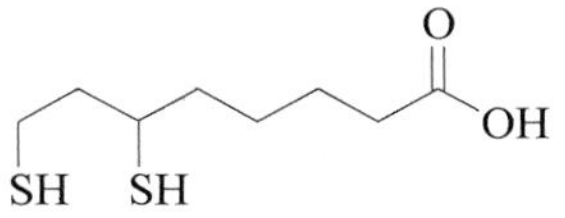

$$\text{DHTA}$$
124

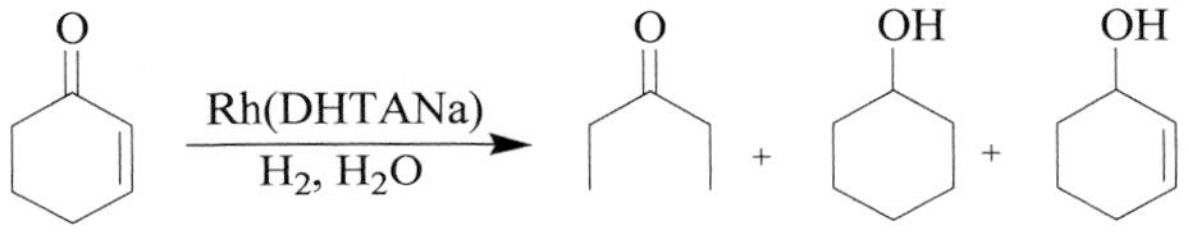

$$(2\text{-}189)$$

Deng 等[284]合成了一系列不同链长的两亲性配体，在水中与铑金属离子络合后可催化水中脂肪酮酯的不对称转移氢化还原反应。研究结果表明，只有当单长链催化剂边链碳数达到 8 个以上时，才能对脂肪酮酯的催化还原取得较好的反应活性和对映选择性。催化剂 125 在水介质催化体系中展现出了具有较广的酮酯底物普适性，不仅对各类 α-、β-、γ-、δ-和 ε-脂肪酮酯有很好的催化活性，而且对 α-乙酰氧基脂肪酮底物也具有很好的催化活性，同时能够获得高对映选择性(比值最高达到 99%)，反应主要是通过底物长链和催化剂长链之间的疏水作用来提高对映选择性的。此外，单长链表面活性剂型催化体系也能催化 α,β-不饱和脂肪酮以及卤代脂肪酮的不对称还原，并取得很好的催化活性。

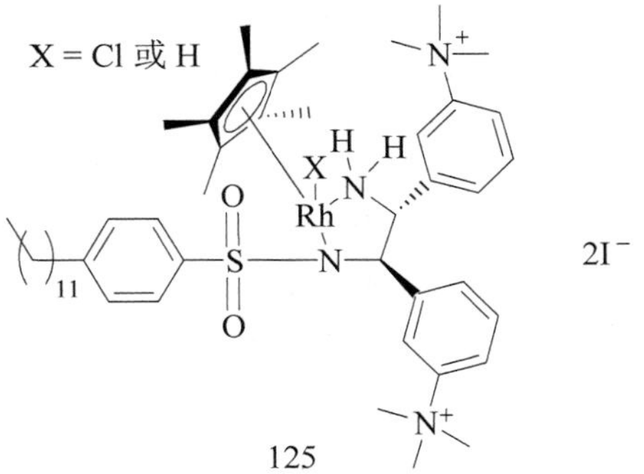

125

2) 亚胺的还原

有机金属铑催化剂不仅可以对羰基化合物进行催化加氢，还可催化水中亚胺和亚胺离子的催化加氢反应。例如，Wilkinson 等[285]采用 RhCl(PPh$_3$)$_3$ 和[Rh(PPh$_3$)$_2$(COD)]PF$_6$ 催化水相亚胺加氢反应。另外，在水相不对称转移氢化反应的底物扩展方面,最近也取得了令人振奋的成果。

Xiao 等[286]以 Noyori 类催化剂 126 成功实现了喹啉类化合物的水相不对称转移氢化反应[式(2-190)]。基于亚胺的不对称氢化需先形成亚胺离子,然后再被还原[287],他们将反应置于 pH = 5 的 HCOOH/HCOONa 缓冲体系中,在此条件下亚胺发生质子化而被活化,但结果并不理想。对于 2-甲基喹啉的还原,反应初速率很快,但 1 h 后反应几乎停止,转化率不到 50%,此时反应体系 pH 上升到 8。使用 pH = 5 的 HOAc-NaOAc 缓冲液,喹啉的水相不对称转移氢化反应 3 h,得到 95% 的转化率。研究还表明,金属前驱体 [(Cp*)RhCl$_2$]$_2$ 比 [(Cp*)IrCl$_2$]$_2$ 和 [(cymene)RuCl$_2$]$_2$ 表现出更好的活性和对映选择性。配体苯磺酰胺苯环基团上取代基的电子效应对反应的影响不大,而立体效应对收率和对映选择性有较大影响。对位叔丁基取代获得最好的效果,对各种取代喹啉的不对称转移氢化反应,均能获得很好的收率和对映选择性。而对于 2-芳基取代的喹啉,由于其碱性较弱,反应需要在 pH = 4 的缓冲液中才能顺利进行。

$$\underset{\text{N}}{\underset{R_1}{\bigodot}}\underset{R_2}{\bigodot} \xrightarrow[\text{HCO}_2\text{Na, 40°C, 6~24 h}]{\underset{\text{HOAc-NaOAc 缓冲液, pH = 5}}{126}} \underset{\text{N}}{\underset{R_1}{\bigodot}}\underset{H}{\underset{R_2}{\bigodot}} \quad (2\text{-}190)$$

Xu 等[288]发展了一种新型高效的 N-杂化化合物加氢方法[式(2-190)和式(2-191)],以甲酸盐为氢源,[Cp*RhCl$_2$]$_2$/bpy 可催化很多 N 杂环化合物,如喹喔啉、喹喔啉酮、喹啉和吲哚及其衍生物等的还原反应,双齿 N 配体联吡啶(bpy)起关键作用,溶液的 pH 也具有非常重要的影响。在最佳反应条件下,除 3-苯基喹啉给出的产物得率仅 69% 外,其他底物给出产物的得率均大于 90%,吲哚衍生物给出产物的得率也高于 80%,有的甚至高达 98.5%,如 5-甲基吲哚。

$$\underset{\text{H}}{\underset{R_1}{\bigodot}}\underset{R_3}{\overset{R_2}{\bigodot}} \xrightarrow[\text{H}_2\text{O, S/C = 200, 80°C}]{\underset{\text{HCOONa/HCOOH}}{[\text{Cp*RhCl}_2]_2/\text{bpy}}} \underset{\text{H}}{\underset{R_1}{\bigodot}}\underset{R_3}{\overset{R_2}{\bigodot}} \quad (2\text{-}191)$$

3) 硝基烯还原

Deng 等[289]合成了一种修饰的 TsDPEN 配体 127，与[(Cp*)RhCl$_2$]$_2$ 的络合物可催化硝基烯烃水相不对称转移氢化反应[式(2-192)]，不同于 Carreira 等[290]所用的水溶性催化剂，这类催化剂是非水溶性的，对各种取代底物都表现出高反应活性和对映选择性。研究表明，pH 对反应选择性有重要影响，当 HCOOH-HCOONa 缓冲液的 pH 在 5.1~5.6 时，选择性最高。同时，配体的电子性质对反应的对映选择性也有很大影响，机理研究表明，反应经历一个分步的 Michael 加成。

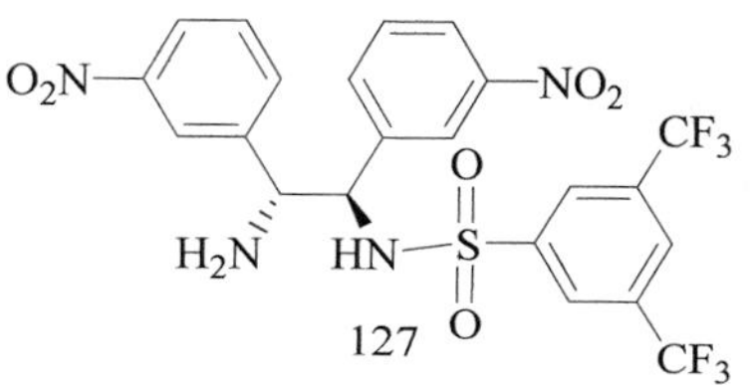

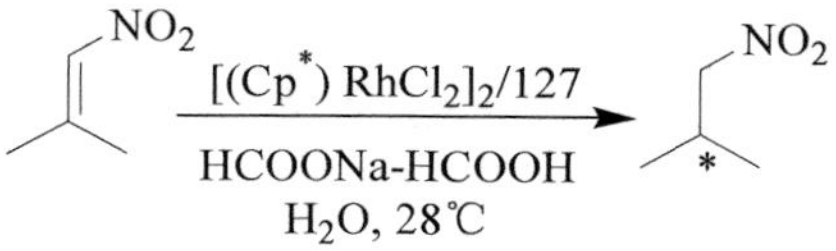

$$\text{(2-192)}$$

2. 加成反应

1997 年，Miyaura 等[291]首次报道了用[Rh(acac)(CO)$_2$](acac 为乙酰丙酮)与 1,4-双(二苯基膦基)丁烷的络合物能高效催化水介质芳基硼酸与 α, β-不饱和醛及链状和环状 α, β-不饱和酮的共轭加成反应[式(2-193)]，其中，环状 α, β-不饱和酮的产率较低。随后，他们用手性催化剂 128 催化水中芳基硼酸与 α, β-不饱和酮的共轭加成，在 KOH 和 Et$_3$N 等碱性条件下，对于环酮加成产物的 ee 值可达 77%以上，尤其是与环己酮的加成，其产物的 ee 值达到 99%以上[292]。Chan 等[293]同样用[Rh(acac)(C$_2$H$_4$)]催化剂，在反应体系中加入手型配体(3S)-4,4′-双(二苯基膦基)-2,2′-6,6′-四甲氧基-3,3′-联吡啶(S-P-Phos)129，催化 α, β-不饱和酮，实现与过量芳基硼酸的加成反应，产物的 ee 值达到 97%~99%。

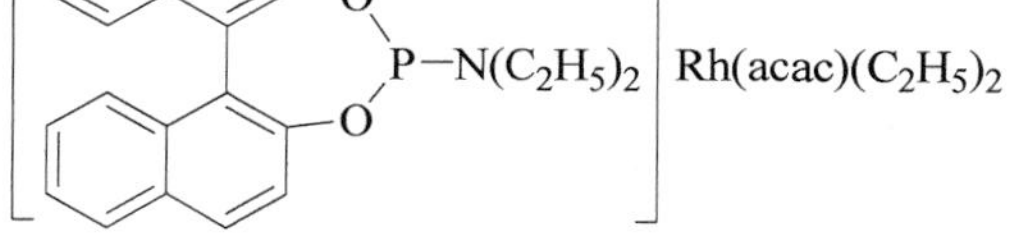

128

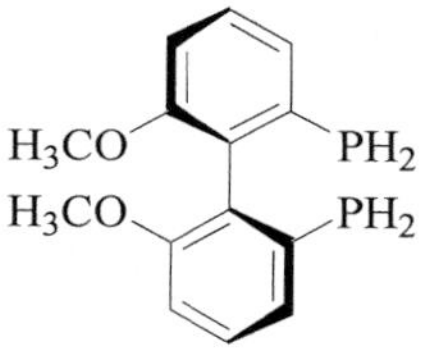

129

$$\underset{R_1}{\overset{O}{\diagdown}}\!\!\!\diagdown\!\!\!\underset{R_2}{\overset{}{\diagdown}} + R_3\text{-B(OH)}_2 \xrightarrow[\text{H}_2\text{O, 50 °C, 16 h}]{[\text{Rh(acac)(CO)}_2]/\text{dppb}} \underset{R_1}{\overset{R_3\ O}{\diagdown\!\!\!\diagdown\!\!\!\diagdown}}\underset{R_2}{} \quad (2\text{-}193)$$

Li 等[294]用铑催化剂, 如 $Rh_2(COD)_2Cl_2$ 和 $Rh(COD)_2BF_4$ 等催化了水中有机金属试剂与 α,β-不饱和羰基化合物的共轭加成[式(2-194)], 其中有机金属试剂可以是有机铋[295]、有机铅[296]、有机锡[297]等。

$$Rn\text{-}MXm + \underset{O}{\overset{R_1}{\diagdown\!\!\!\diagdown}}\!\!\!R_2 \xrightarrow[\text{空气/H}_2\text{O}]{\text{Rh(I) cat.}^①} \underset{R}{\overset{R_1}{\diagdown\!\!\!\diagdown}}\!\!\!\underset{O}{\overset{}{\diagdown}}\!\!\!R_2 \quad (2\text{-}194)$$

M=B, In, Si, Sn, Ge, Pb, As, Sb, Bi
X= R, Ar, Cl, Br, I, OH, OR

炔烃与有机硅烷的加成反应是一类重要的有机反应, 大多数反应在无水条件下进行。Oshima 等[298]报道了[RhCl(nbd)]$_2$/二(二苯基膦)丙烷催化水中炔与三乙基硅烷的选择性加成反应[式(2-195)], 加入十二烷基磺酸钠有助于 Rh—Cl 键的断裂形成阳离子铑物种, 而加入阳离子表面活性剂, 如甲基三辛基铵, 炔烃的加成反应则不能发生。

$$R\!\!\equiv\!\!H + (C_2H_5)_3SiH \xrightarrow[\text{SDS, H}_2\text{O, rt}]{[\text{RhCl(nbd)}]_2/\text{dppp}} \underset{SiC_2H_5}{\overset{H}{\diagdown\!\!\!=\!\!\!\diagdown}} \quad (2\text{-}195)$$

炔烃不仅可以与硅烷加成, 也能与有机金属氢化物加成, 如 Wu 和 Li 等报道了末端炔烃与氢锡化物及不饱和羰基化合物的串联反应[式(2-196)], 反应产物具有较高立体选择性[299]。

$$R\!\!\equiv\!\!H + Bu_3SnH \xrightarrow[\text{H}_2\text{O, rt, 20 min}]{\text{Rh(I)}} \underset{R_1}{\overset{EWG}{\diagdown\!\!\!=\!\!\!\diagdown}}\!\!R_2 \xrightarrow{100\text{°C, 24 h}} \underset{R_1}{\overset{R}{\diagdown}}\!\!\underset{R_2}{\overset{}{\diagdown}}\!\!EWG \quad (2\text{-}196)$$

Wender 等[300]报道了采用水溶性的二齿配体铑催化剂 $Rh(nbd)^+/130SbF_6^-$ 催化[5+3]环加成反应[式(2-197)], 由于催化剂为水溶性, 而反应底物在水中的溶解度很小, 避免了聚合反应, 环化产物的得率明显提高(90%), 催化剂可重复使用 8 次。

$$\xrightarrow[\text{H}_2\text{O, 12 h}]{\text{Rh(nbd)}^+/130SbF_6^-} \quad (2\text{-}197)$$

---

① cat.表示催化剂, 下同。

130

3. 甲酰化反应

烯烃与 CO 和 H$_2$ 反应生成多一个碳原子的醛或醇,称为氢甲酰化反应[式(2-198)],已成为当今世界最重要的有机化工工艺之一,应用在增塑剂、表面活性剂、药物中间体等精细化工合成领域。Rührchemie 公司用水溶性强的[HRh(CO)(Ph$_2$PPhSO$_3$Na)$_3$]催化丙烯氢甲酰化反应,成为生产醛的主要合成路线,年产量超过 10 万吨[301, 302]。Ajjou 和 Alper[303]首次用聚 4-戊烯酸和二[2-(二苯膦基)乙基]胺形成的水溶性配合物 Rh/PPA(Na$^+$)/DPPEA,催化水中氢甲酰化反应,尤其在催化苯基乙烯基醚为底物时,对产物醛的选择性接近 100%。Mieczynska 等[304]报道了水介质中 Rh(acac)(CO)$_2$ 催化 2-丁烯醇的氢甲酰化反应,加入水溶性配体 PNS(PNS = Ph$_2$PCH$_2$CH$_2$CONHC(CH$_3$)$_2$CH$_2$SO$_3$Li)可显著促进反应,也可催化水中 N-烯丙酰胺的氢甲酰化反应[式(2-199)],加入 TPPTS 配体后,TOF 值可达 3891 h^{-1},明显高于甲醇溶剂(1239 h^{-1}),与 Rh(acac)(CO)$_2$/PPh$_3$ 催化剂相比,其反应速率和选择性都有明显提高[305]。

$$R\text{—CH=CH}_2 + CO + H_2 \xrightarrow{\text{Rh-TPPTS}} R\text{—CH}_2\text{CH}_2\text{CHO} + R\text{—CH(CH}_3\text{)CHO} \quad (2\text{-}198)$$

$$\text{CH}_2\text{=CHCH}_2\text{NHAc} \xrightarrow[\text{H}_2\text{O}]{\text{Rh(acac)(CO)}_2, \text{TPPTS}} \text{OHC(CH}_2\text{)}_3\text{NHAc} + \text{OHC-CH(CH}_3\text{)CH}_2\text{NHAc} \quad (2\text{-}199)$$

对于长链烯烃,由于它在水中的溶解度很差,氢甲酰基化反应在两相体系中一般进行得很慢,选择性也很低。Monflier 等[306]报道在无有机溶剂的水介质中,加入部分甲基化的 β-环糊精(β-CD),铑催化的壬-1-烯的氢甲酰基化反应又可达到 100%的转化率和 95%的选择性[式(2-200)]。

$$\text{CH}_2\text{=CH(CH}_2\text{)}_6\text{CH}_3 \xrightarrow[\beta\text{-CD, H}_2\text{O}]{\text{Rh(acac)(CO)}_2} \text{OHC(CH}_2\text{)}_7\text{CH}_3 \quad (2\text{-}200)$$

随后报道了一系列有关磺化的 P 配体[P(m-C$_6$H$_4$SO$_3$Na)$_3$]络合的有机金属铑催化剂应用在氢甲酰化反应中,如 RhCl(CO)(TPPTS)$_2$-BISBIS[TPPTS=P(m-C$_6$H$_4$SO$_3$Na)$_3$];

BISBIS = sulfonated 1,1-*bis*(diphenylphosphinomethyl)-2,2'-biphenyl[307], RhCi(CO)(TPPTS)$_2$[308], HRh(CO)(CH$_3$CN)(TPPMS)$_3$[BF$_4$], [Rh(μ-Pz)(CO)(TPPTS)]$_2$[309], [Rh(*p*-Pz)(CO)(TPPMS)]$_2$, [Rh(CO)(μ-Pz)(TPPMS)]$_2$[310], 其中, RhCl(CO)(TPPTS)$_2$-BISBIS 可催化水介质中 1-十二烯的氢甲酰化反应, 不仅 TON 值高(2897)、立体选择性高(98%), 且可以重复使用 5 次。而[Rh(CO)(μ-Pz)(TPPMS)]$_2$ 催化正己烯[式(2-201)]和苯乙烯的氢甲酰化反应[式(2-202)], 催化剂可循环使用 7 次, 金属活性物种没有流失。

$$\text{烯烃} \xrightarrow[\text{H}_2, \text{CO}, \text{H}_2\text{O}]{\text{Rh cat.}} \text{醛} \quad (2\text{-}201)$$

$$\text{丁香酚} \xrightarrow[\text{H}_2, \text{CO}, \text{H}_2\text{O}]{\text{Rh cat.}} \text{醛} \quad (2\text{-}202)$$

最近, Liu 等[311]合成一种咪唑盐配体, 并由此制备了一种 Rh(Ⅲ)有机金属催化剂 131, 用于正辛烯的氢甲酰化反应。在配体 132 的存在下, 正辛烯的转化率和正壬醛的选择性均达到 98%, TON 值也达到 1920, 可能是由于 *P,N*-嵌入的阳离子与亲水性的 BF$_4^-$ 之间存在着协同作用。

131　　　132

四、其他金属催化剂

1. 有机金属金催化剂

炔烃水合反应[式(2-203)]是制备羰基化合物的经典绿色方法, 早在 1937 年, Frieman 等[312]报道了用 HgSO$_4$/H$_2$SO$_4$ 催化水介质乙炔水合反应。近期 Gautier 等[313]合成了水溶性氮杂环 Au(Ⅲ/Ⅰ)配合物 133 催化水介质中末端脂肪炔醇水合反应, 反应过程中不需加入 Brønsted 酸作共催化剂, 在 50~80℃下反应 18~40 h, 除了 2,2-二甲基丙炔醇给出 64%的产物得率外, 其他水合产物的得率均高于 94%,

有的接近100%。

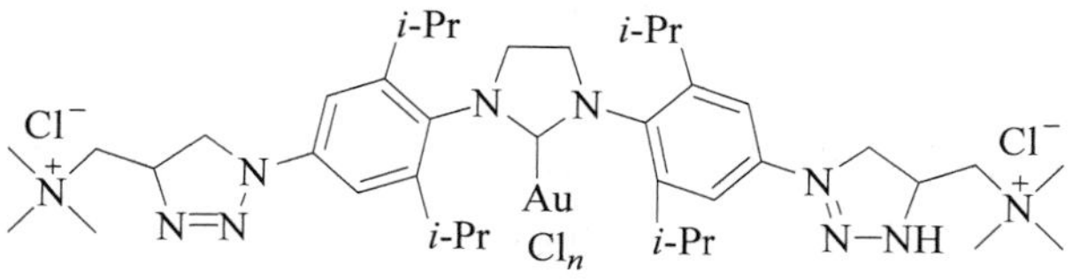

133 a: n = 1 [Au(Ⅰ)],
133 b: n = 3 [Au(Ⅲ)]
133

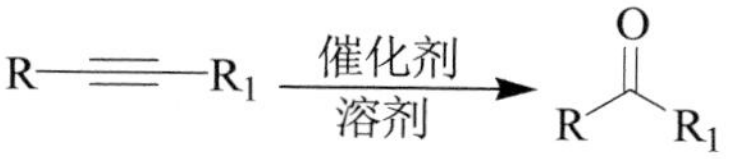

$$R\text{—}\!\!\!\equiv\!\!\!\text{—}R_1 \xrightarrow[\text{溶剂}]{\text{催化剂}} \underset{R}{\overset{O}{\|}}\!\!\text{—}R_1 \tag{2-203}$$

有机金属金不仅可催化炔与水的加成反应，还可以催化炔与醛、胺的 A^3-偶联反应[式(2-204)]。2003 年，Li 和 Wei[56]报道了用 $AuBr_3$ 高效催化水介质中炔烃、醛与胺的偶联反应，随后，Wong 等[314]报道了离子型催化剂 134 催化水中 A^3-偶联反应，在 40℃下即可得到高得率，若使用手性脯啉作为反应物，产物的 ee 值高达 99%，该偶联反应还可用于炔丙基胺功能化的青蒿素的合成[式(2-205)]。

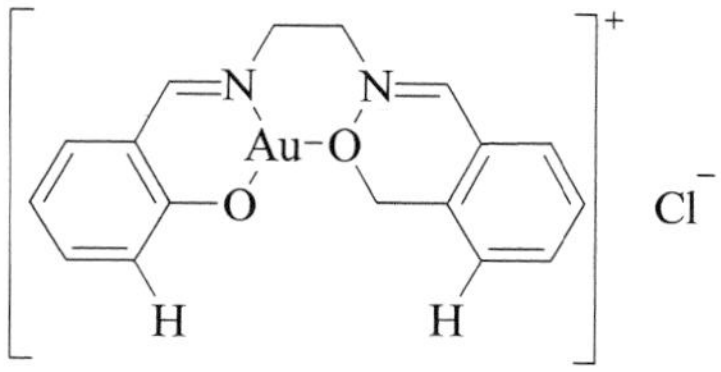

134

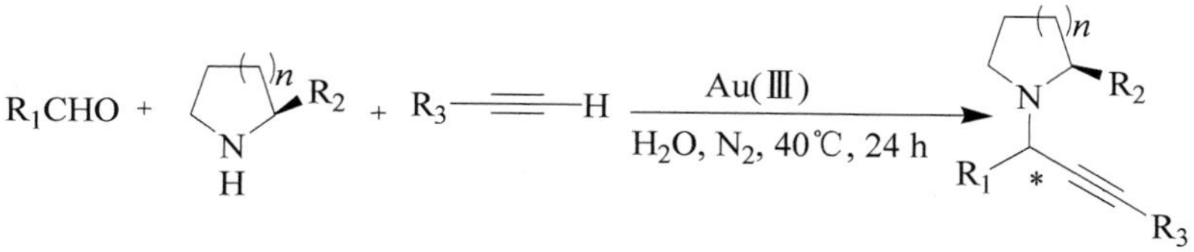

$$R_1CHO + \underset{H}{\overset{n}{\underset{N}{\bigcirc}}}\!R_2 + R_3\text{—}\!\!\!\equiv\!\!\!\text{—}H \xrightarrow[H_2O, N_2, 40℃, 24\text{ h}]{\text{Au(Ⅲ)}} \text{产物} \tag{2-204}$$

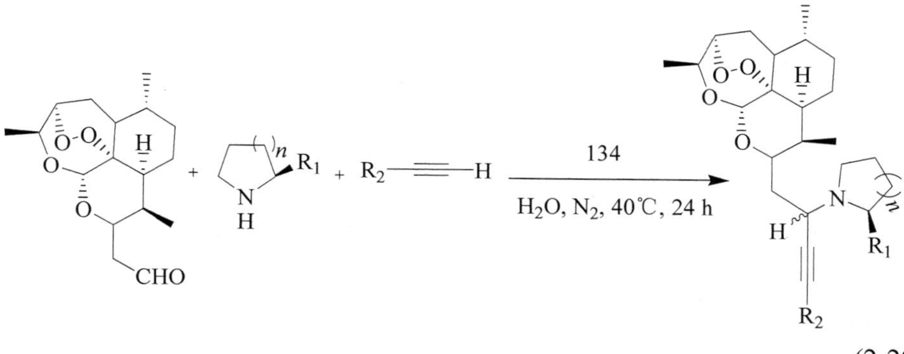

$$\text{(2-205)}$$

2009年，Wong 等[315]报道了[Au(C^N^C)Cl]135 催化水介质中的三组分偶联反应，在 40℃下反应 1 h 即可完成，可同时应用于芳香醛和脂肪醛偶联反应，产物的得率较高，ee 值也在 99%以上，催化剂可循环 10 次。Contel 等[316]合成了水溶性催化剂[AuCl(PR$_3$)](P = mC$_6$H$_4$SO$_3$Na$_2$)，该催化剂对空气比较稳定，在水介质 A^3-偶联反应中显示出较好的催化性能，可应用于抗菌剂和抗癌药品的合成。

[Au(C^N^C)Cl]
135

Belger 和 Krause 以商业化 3,5-二甲基-4-硝基-1-苯甲酸为原料，合成出一种水溶性的金配合物 136，在水介质中进行累积二烯烃[式(2-206)]和炔醇的环异构化反应[式(2-207)]中，展现出优良的催化性能。如在累积二烯烃的环异构化反应中，配合物 136 在 40℃下反应 24 h，目标产物的得率大于 82%，尤其是 136a，产物的得率为 98%。两种催化剂均可重复使用 5 次[317]。

136   a: R=Et
      b: R=n-Bu

$$\text{(2-206)}$$

$$\text{(2-207)}$$

Krause 和 Belger[318]报道了一系列催化剂 137 的合成，在催化水中炔基羧酸和炔基酰胺的环化反应时表现出高活性，产物分别是内酯[式(2-208)]和内酰胺[式(2-209)]，得率较高，在 pH = 7 时，催化剂 137ab 给出最佳活性，且可回收再使用。

		R_1 = Mes	Dipp
R_2=	CH_3	137ab	137ba
	C_2H_5	137ab	137bb
	nBu	137ac	137bc

$$\text{(2-208)}$$

$$\text{(2-209)}$$

Liu 等[319]报道了[Au(PPh$_3$)]Cl/Ag$_2$CO$_3$催化微波辐射下的水介质分子内末端炔的氢胺化反应[式(2-210)],对于C≡C 在末端的反应物,苯环无论含有吸电子基团还是给电子基团,均能在短时间内完成,为快速合成具有生物活性的吲哚-3-甲酰胺提供了一种新型绿色方法。

$$\text{(2-210)}$$

最近,Liu 等[320]用[(1,1′-二苯基-2-基)二叔丁基苯基膦氯化金 138 催化微波辐射下取代 2-(1H-吲哚基)烷基胺炔丙基羧酸间的串联反应[式(2-211)和式(2-212)],同时形成一个 C—C 键和两个 C—N 键,是一步合成含有多环吲哚的原子经济性反应,反应操作简单,产物得率较高。

138

$$\text{(2-211)}$$

$$\text{(2-212)} \quad n=1,2$$

Liu 等[321]报道了催化剂 139 和三氟乙酸共同催化水介质中 Michael 加成–分子内环化串联反应式[式(2-213)]，该反应在微波作用下能有效催化含有七元环的吲哚的合成。此方法操作简单，环境友好，可用于具有生物活性的吲哚类化合物的合成。

139

$$\text{(2-213)}$$

Li 等[322]首次报道将 i-PrAuNTf$_2$ 应用在水介质吲哚[式(2-214)]和炔丙基胺[式(2-215)]的分子内氧化加成，随后，他们又报道了该催化剂可以同时催化水介质中邻炔丙基苯胺环异构化与炔酰胺分子内的氧化反应，在 80℃下反应 1 h，产物得率为 43%~70%。

$$\text{(2-214)}$$

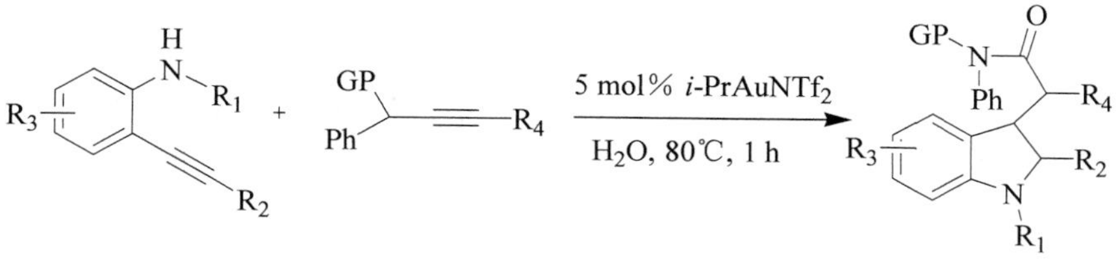

$$\text{(2-215)}$$

2. 有机金属镍催化剂

Arzoumanian 等[323]报道了在相转移催化剂存在下,镍催化 α-酮炔的羰基化反应,依据底物取代基,分别得到呋喃酮或不饱和羧酸,如 α-酮炔与 KCN 在水中的氰基化反应,给出不饱和羟基内酰胺[式(2-216)],而镍催化水中炔丙基卤化物和炔丙基醇获得吡喃酮,此为串联反应,涉及 CO 的羰基化反应及与 KCN 进行的氰基化反应[324][式(2-217)]。

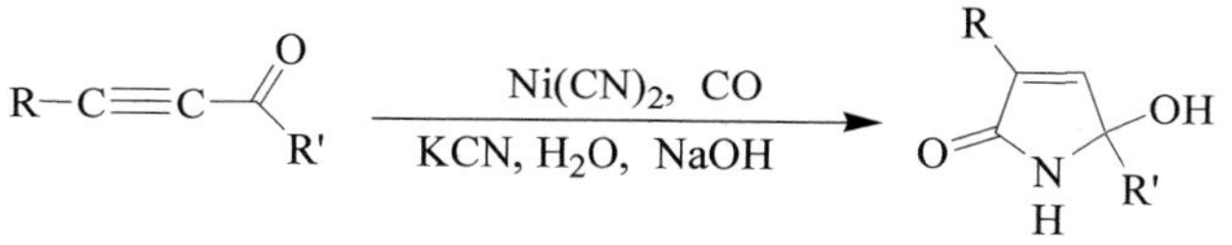

$$\text{(2-216)}$$

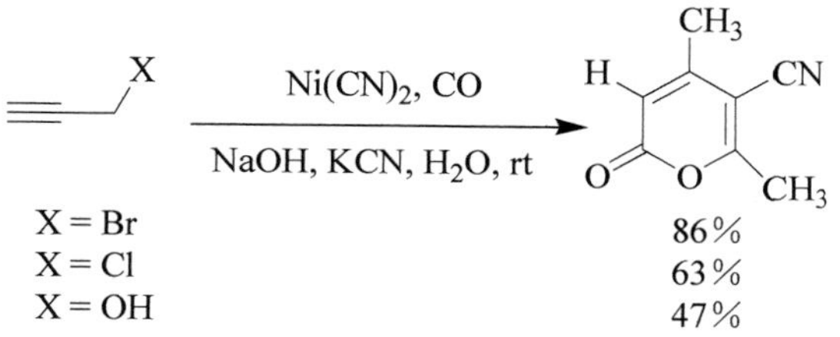

$$\text{(2-217)}$$

X = Br 86%
X = Cl 63%
X = OH 47%

也有报道室温下镍催化水中 6-氨基-1,3-二甲基尿嘧啶与取代炔基酮的反应,经离子化途径,以定量的产率获得具有潜在药理和生物活性的 2,4-二氧吡啶并[2,3-d]嘧啶衍生物[325][式(2-218)]。

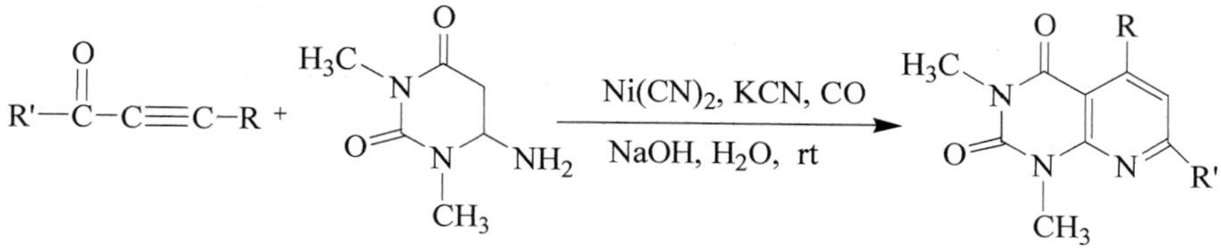

$$\text{(2-218)}$$

Sharma 等[326]报道用 Ni(CN)$_2$ 在温和条件下催化水介质 α-酮炔与 2-氨基-4(1H)-喹啉酮的缩合反应,获得苯并[b][1,8-]萘吡啶-5-酮衍生物[式(2-219)],得率为 69%~79%,在没有催化剂存在下,反应 24 h,得率低于 10%。

$$\text{(2-219)}$$

3. 有机金属钴催化剂

炔烃和烯烃的羰基化反应是合成各种羰基化合物的常见简易方法。早在 1977 年，Alper 和 Abbayes[327]采用相转移试剂和钴催化剂，在 NaOH 水溶液中实现苄基溴的羰基化反应[式(2-220)]，随后，Krafft 等[328]采用 CTAB 为表面活性剂，实现了钴催化水中分子间和分子内的 Pauson-Khand 反应[式(2-221)和式(2-222)]。

$$\text{PhCH}_2\text{Cl} + \text{CO} \xrightarrow[\text{PTC}]{\text{Co(CO)}_4^-,\ \text{aq NaOH}} \text{PhCH}_2\text{COONa} \quad (2\text{-}220)$$

$$\text{(2-221)}$$

$$\text{(2-222)}$$

4. 有机金属铱催化的反应

Li 等[329]用水溶性有机铱催化剂[Cp*Ir(H$_2$O)$_3$][OTf]$_2$(140)，在空气氛围下催化水中醛肟重排反应生成酰胺，可通过醛(脂肪醛、共轭不饱和醛、非共轭不饱和醛和芳香醛)、盐酸羟胺与碳酸钠的缩合重排，经"一锅"反应得到不同的酰胺，其优点是不需要添加膦配体、催化剂用量少、操作简单且得率高。Ma 等[330]报道了催化剂 140 应用于水中邻氨基苯甲酰胺与醛的脱氢缩合反应生产喹唑酮[式(2-223)]，在回流条件下，目标产物的得率为 75%~85%。

140

$$\text{R}_1\underset{\text{NH}_2}{\overset{\underset{\|}{\text{C(=O)NH}_2}}{\bigcirc}} + \text{R}_2\text{CHO} \xrightarrow[\text{2) 1 mol\% 140, 回流, 1 h}]{\text{1) H}_2\text{O, 回流, 2 h}} \text{R}_1\bigcirc\underset{\text{N}}{\overset{\text{NH}}{\bigcirc}}\text{R}_2 \qquad (2\text{-}223)$$

最近 Ni 等[331]首次报道在配体 141 存在下，有机金属铱催化水中 α-酮膦酸酯的不对称转移氢化反应，反应不需任何表面活性剂，获得产物 α-羟基膦酸酯[式(2-224)]，得率为 44%~78%，对映选择性大于 99%。

$$\text{配体 141}$$

$$\text{R}_1\underset{\text{R}_2\text{O}}{\overset{\text{O}}{\bigcirc}}\text{P(=O)OR}_2 \xrightarrow[\text{HCO}_2\text{Na, H}_2\text{O, rt}]{\text{mol\% [IrCl}_2\text{Cp}^*]\ 2\ \text{mol\% 141}} \text{R}_1\underset{\text{R}_2\text{O}}{\overset{\text{OH}}{\bigcirc}}\text{P(=O)OR}_2 \qquad (2\text{-}224)$$

第四节 有机小分子不对称催化剂

有机小分子催化剂因结构简单且易于制备而备受关注，近几年已有大量报道。总的来说，有机小分子催化剂可分为三类：氨基酸类、非氨基酸含氮有机碱类化合物、硫脲类，但有机小分子催化水中不对称有机合成报道较少，主要涉及 Aldol 缩合反应和 Michael 加成反应等。

一、Aldol 缩合反应

Aldol 反应即羟醛缩合反应[式(2-225)]，是有机化学中构建 C—C 键的最有效的反应之一。天然氨基酸可催化 Aldol 反应，除了 L-脯氨酸，很多有机小分子都可以催化不对称 Aldol 反应[332]。

$$\text{R}_1\underset{\text{R}_2}{\overset{\text{O}}{\bigcirc}} + \text{RCHO} \xrightarrow[\text{H}_2\text{O}]{\text{催化剂}} \text{R}_1\underset{\text{R}_2}{\overset{\text{O}}{\bigcirc}}\underset{}{\overset{\text{OH}}{\bigcirc}}\text{R} \qquad (2\text{-}225)$$

早在 2002 年，Janda 等[151, 333]报道了 3-吡咯基吡啶(142)催化水中丙酮与对氯苯甲醛的 Aldol 缩合反应，但对映选择性仅为 20%。他们认为水在催化反应中具有双重作用，一是水中的氢键活化了苯甲醛，有利于亲核试剂的进攻；二是水可以加速亚胺的水解给出产物，同时，促进催化剂再生，因此，在水中进行的缩合反应速率加快，而在有机溶剂中没有产物生成。Barbas 等[152]报道了脯氨酸及其类似物 143 催化的醛与酮的直接 Aldol 反应，反应介质由有机溶剂扩展到磷酸缓冲溶液，但获得外消旋产物。用质子化的二胺(144)作为催化剂[113]与 TFA 共同催化水中 Aldol 缩合反应，产物得率和 ee 值最高达 99%，而对于非环状酮，产物 ee 值不高。

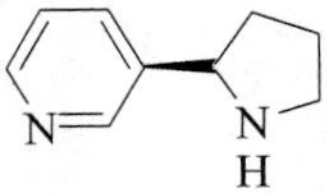

142 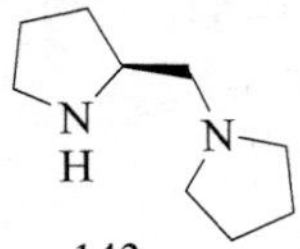　　143 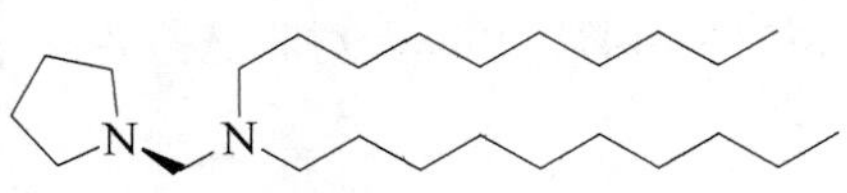　　144

Hayashi 等[334, 335]报道硅羟基脯氨酸(145)催化水中高对映选择性和非对映选择性的 Aldol 反应，给出 ee 值高于 99%的产物。

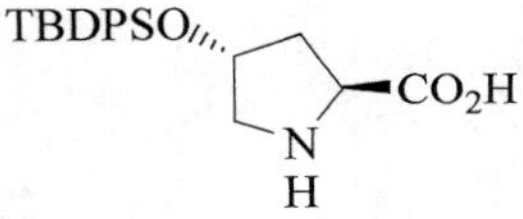

145

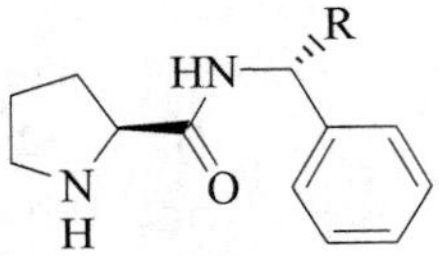

146
146a R= CH_3; 146b R= C_2H_5

Chimni 和 Mahajan 报道质子化的手性脯氨酰胺(146)催化水中 Aldol 反应[336]。他们发现在无水条件下，反应不能发生，在丙酮与对硝基苯甲醛反应中给出 ee 值最高为 55%。对于四取代的芳炔，也能给出较好的催化活性及立体选择性，尤其是催化剂 146b 催化的反应速率明显优于 145。Singh 等[337]发展了一种含有非极性的 γ-二苯基的脯氨酰胺催化剂 147、148、149，它们在 NaCl 水溶液中催化丙酮与醛的 Aldol 反应，具有较好的活性，并可推广到各类醛和酮[338,339]。与催化剂 147 和 148 相比，催化剂 149 在高活性的醛如对硝基苯甲醛和萘甲醛的 Aldol 反应中给出最高的立体选择性(anti：syn>99：1)和对映选择性(>99%)。研究也发现，盐析作用增加了疏水效应，使得率和 ee 值提高。Singh 和 Gandhi 通过叠氮离子与氮杂环丙烷的立体选择性和区域选择性的开环反应，合成了一系列脯氨酰胺催化剂 150，它不同于催化剂 147、148、149 结构中的 β-氨基醇被 β-氨基磺酰胺取代。催化剂 150 是水中 Aldol 反应的高效催化剂，无论是环酮还是非环酮，在盐水中仅用 2 mol%催化剂 150，得率和 ee 值均高于 99%[340]。

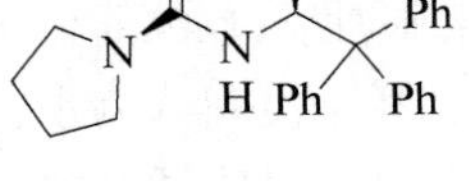

147 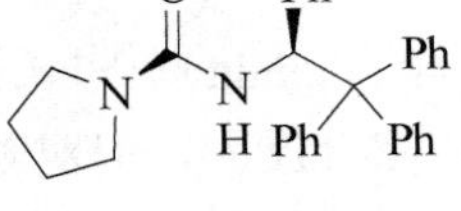　　148

Vilaivan 和 Sathapornvajana 用 2-氨基苯酚合成了脯氨酰胺催化剂 151，在脯氨酰胺结构中引入一个额外的氢键，在催化环己酮和芳醛的不对称 Aldol 反应中给出高得率、非对映选择性和对映选择性[341]。Chen 等报道了含有氨基和硫脲结构的不溶于水的双功能催化剂 152[342]。催化剂中樟脑的刚性结构主要有两个作用，一是有效的立体控制作用，二是在水中与疏水性的反应物作用，增加了催化剂的疏水性。该催化剂在水中的 Aldol 反应给出高得率、高非对映选择性和对映选择性的 anti-Aldol 产物。基于上述思路，Gryko 和 Saletra 合成了 L-脯氨硫酰胺(153)，用其催化水中环酮和芳醛的 Aldol 反应，给出高得率和高立体选择性[343]。

Chimni 等[344]合成了质子化的含有芳环的手性(S)-脯氨酰胺衍生物 154，催化水中不对称 Aldol 反应。催化剂中的芳基提供了疏水作用，而酰胺的亚氨基(NH)与底物形成氢键。在不同环酮与醛的反应中均给出高得率、高非对映选择性(dr 值 70%~99%)和对映选择性(ee 值 90%~98%)。Wang 等[345]合成了氟代-(S)-四氢吡咯磺酰胺催化剂 155，用于水相中不同酮与芳醛的缩合反应。氟的强吸电子作用增强了磺酰胺中亚氨基的酸性，使其与底物形成一个强的氢键。在 0℃反应时，产物的得率达到 92%，对映选择性也高达 98%。催化剂可重复使用 7 次。

Shao 等[346]合成了伯胺联萘催化剂 156，其在水相芳醛和环酮不对称 Aldol 反应中显示出高催化效率，给出高非对映选择性和对应选择性的产物，而在非环酮缩合反应中，产物的 ee 值和 de 值中等。

156　　　　　　　157

Benaglia 和 Nájera[347, 348]也报道了以硬脂酸为促进剂，1,1′-联萘-2,2′-二氨基-(S)-脯氨酰胺(157)为催化剂，催化水中不对称 Aldol 反应。对于环己酮或非环酮如 2-辛酮、2-丁酮与芳醛反应，产物具有高的 ee 值和 de 值，这可能是由于硬脂酸在反应体系中形成了一种疏水环境。Jiang 等[349]研究表明，用 TFA 作添加剂，用 NOBIN-脯氨酰胺(158)催化水中 Aldol 反应，可给出 99%得率的产物，且非对映立体选择性和对映选择性也分别达到 99%和 97%。如用甲基保护催化剂 159 上的羟基，也能获得相似的结果，这表明羟基对过渡态的形成没有明显的影响。

158　　　　　　　159

Lu 等[350]用羟基脯氨酸为原料合成了疏水性伯氨基脯氨酸衍生物 160 催化剂，L-苏氨酸的羟基被甲硅氧基取代，其催化水相环酮与芳醛反应给出高 ee 值和 de 值的反式产物，而非环酮如被保护，催化羟基丙酮与芳醛反应则给出高 de 值和 ee 值的顺式产物。催化剂 160 催化脂肪醛的 Aldol 反应没有明显活性，可能是由于底物中的芳环增强了疏水性，而脂肪醛却不能增强疏水性。Teo[351]采用甲硅氧基-L-丝氨酸 161 催化水中环酮和芳醛的不对称 Aldol 缩合反应，催化剂适用于底物的范围广，并给出高 ee 值和 de 值。如用 L-丝氨酸直接催化该反应，反应 2 天也检测不到产物，这归因于催化剂 161 中含有一个疏水性的空腔，有利于对映选择性反应。

160　　　　　　　161

Barbas 等[352]报道了含有 γ-二苯基的伯-*O-t*Bu-L-thr 基酰胺催化剂 162,将其应用于水中羟基丙酮的不对称顺式 Aldol 反应,在催化被保护羟基丙酮的反应中给出较高得率和立体选择性,而在未保护的羟基丙酮的反应中没有活性,这可能是由于羟基丙酮具有亲水性。同时,Gong 等[353]报道了含有 γ-二苯基的伯氨基酸基酰胺催化剂 163 应用于水中脂肪酮与醛的 Aldol 反应,在对硝基苯甲酸的盐水中,很多醛与酮反应均能得到高 ee 值和 de 值的目标产物,这可能是由于水分子中的羟基与小分子催化剂中的氧形成了一个分子间的氢键,提高了小分子催化剂中酰胺的 NH 酸性,从而增强了氢与醛的作用,但是随着脂肪酮中碳链增长,ee 值也随之降低,而 1,3-二羟基丙酮与醛在水中则不能发生反应。

Singh 等[354]合成了一系列含有芳环的伯叔二胺催化剂 164,并将其用于环酮和非环酮与醛的不对称 Aldol 反应,主要基于催化剂中的芳环可以与反应底物形成一个疏水性的空腔,使 Aldol 反应在空腔中进行。研究表明,在三氟乙酸(TFA)存在下,反应可以给出更高的对映选择性和非对映选择性。作者也将催化剂 164/TFA 应用于不同环酮和醛的缩合反应中,获得高对映选择性和非对映选择性的反式 Aldol 产物,以环己酮作为电子给体,可获得 ee 值为 88%的顺式 Aldol 产物,该催化体系还可催化水中两个分别作为给体和受体的酮的 Aldol 反应[354, 355],如环己酮和苯乙醛酸酯的反应,在催化量(10mol%)的催化剂 164/TFA 存在下,室温下生成四取代的碳化合物[354],具有高得率以及优良的非对映选择性和对映选择性。

Chimni 等[356]发现质子化的手性脯氨酰胺可催化水中不对称 Aldol 反应,给出高得率和高对映选择性的产物。有趣的是,两个非对映立体选择性的 165 和 166 在水相 Aldol 反应中的反应活性和对映选择性有明显的不同。165 具有较高的对映选择性,但反应活性较低,而 166 虽然反应活性较高,但对映选择性较低。Cai 等[357]报道了以 D-樟脑磺酸(D-CSA)为共催化剂,用 L-脯氨酸催化水中芳香醛与不同的酮的缩合反应,给出产物得率为 61%~83%,顺式异构体的 ee 值为 43.6%~99%,反式异构体的 ee 值为 63.4%~99%。

Hayashi 等[358]首次报道将合成的一系列化合物 167 用于水中不同醛的交叉 Aldol 反应。研究发现，含有羟基的脯氨酸衍生物的催化效率最高，如反-硅羟基-L-脯氨酸和顺-硅羟基-D-脯氨酸。此外，水对反应的立体选择性也具有非常重要的作用。在含有少量有机溶剂或干燥的有机溶剂中，反应活性非常低，而在最佳反应条件(1mol%催化剂 167，0.2 eq 酮和 3 eq 醛)下，大多数水介质反应均能给出高活性和立体选择性的缩合产物。他们还设计了一种脯氨酸-表面活性剂类型的小分子催化剂 168，用于水中两种不同醛的交叉 Aldol 反应，反应过程中不需要添加有机溶剂和酸，即可获得高选择性和得率的交叉缩合产物[359]。

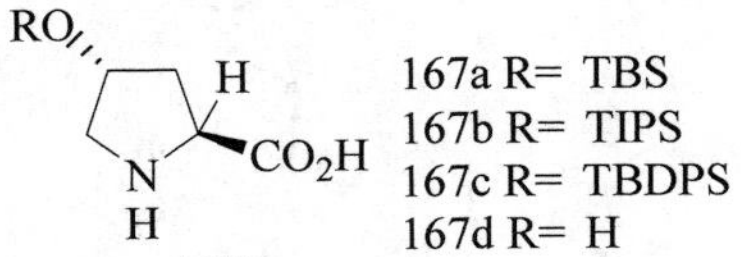
RO
H
N
H
CO_2H
167a R= TBS
167b R= TIPS
167c R= TBDPS
167d R= H

167

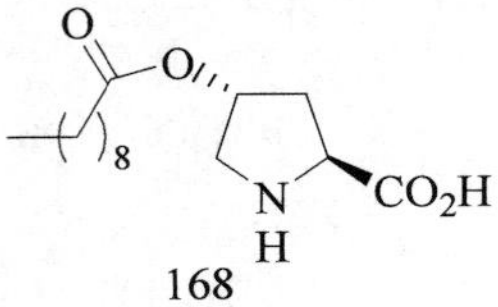
O
O
8
N
H
CO_2H

168

2006 年，Lu 等[360]提出了用简单有机小分子催化 Aldol 反应。随后，Amedjkouh 用未修饰的天然氨基酸催化水中不对称 Aldol 反应[361]，考察了各种氨基酸的催化性能，结果显示，L-色氨酸和 L-苯丙氨酸为催化剂时，获得最高得率、非对映选择性和对映选择性的目标产物。和脯氨酸相比，芳香氨基酸具有较高的活性，主要归因于其具有疏水性，使其在水中的溶解性降低，产生一种非均相混合物。他们又以 L-色氨酸催化水中环酮和芳香醛的 Aldol 缩合反应，获得高对映选择性的目标产物。作者认为，这个反应首先经过亚胺进攻醛形成一个过渡态，然后再形成立体异构体。色氨酸的芳香环有利于水中疏水核与另一个醛的形成，从而促进反应的进行。Tripathi 等[362]用天然氨基酸 D-葡萄糖胺催化水中不同醛与酮的 Aldol 反应，获得高得率的目标产物，但在大多数情况，产物 ee 值不到 10%。3,4-二甲氧基苯甲醛和丙酮反应获得最高 ee 值(54%)。

2005 年有报道[363]单独用脯氨酸催化水中醛-醛、醛-酮的 Aldol 反应，反应时间较长，且催化剂用量达到 30 mol%。后来，Benaglia 等[347]开发了催化剂 169，用于水中不对称 Aldol 反应，体系中需加入长链脂肪酸，如硬脂酸提供一个疏水环境，从而提高催化活性和立体选择性。

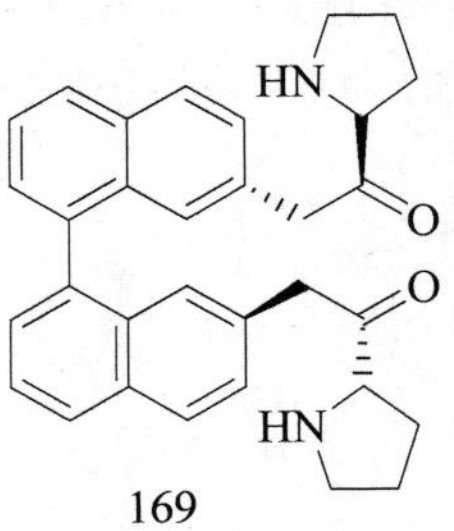
HN
O
O
HN

169

Li 等[364]报道非环状氨基酸催化室温不对称 Aldol 反应,考察了多种非环状氨基酸的催化性能,结果显示,在 5 mol%的二硝基苯酚存在下,L-异亮氨酸的催化性能最佳,β-羟基酮得率最高,且 ee 值也达到 99%以上。

2007 年,Hayashi 等[365]首次报道了均相催化剂脯氨酰胺 170 应用于水相不对称 Aldol 反应,获得高 de 值和 ee 值。Darbre 等[366]报道了锌和未经修饰的 L-脯氨酸形成的配合物[(L)-Proline]₂Zn 催化水中丙酮与对硝基苯甲醛的 Aldol 反应,赖氨酸和精氨酸也可替代 L-脯氨酸。反应在室温下进行,产物的对映选择性为 56%。另外,将手性胺 171 和 DBSA 混合,可得到一种高活性和高选择性的小分子催化体系[367]。

Noto 等[368]研究了一系列脯氨酸衍生物 172 催化的水介质 Aldol 反应,发现水对产物 ee 值有促进作用,当用 2 mol%催化剂时,无需添加剂,在室温下就能获得高 ee 值和 de 值。Xiao 等[369]用伯胺叔胺-Brønsted 酸 173 催化水中环酮和芳香醛的 Aldol 反应,无需添加剂就能获得高得率和高对映选择性的缩合产物。另有报道用吗啉和反-4-二甲基叔丁基硅氧基-L-脯氨酸 174 催化水中 3-吡啶甲醛与环己酮的缩合反应,研究获得了反应温度影响催化性能的规律[370]。

有报道超低温下,用 5 mol%或 10 mol%的脯氨酰胺 175 和 176 催化盐水溶液中丙酮和芳醛的 Aldol 反应,立体选择性达到 99%,对映选择性为 99%[339]。

Wu 等[371]报道用 N-聚磺酰胺合成了小分子催化剂 177,其在催化环酮与不同芳醛的 Aldol 反应中表现出优异的催化性能,不需加入添加剂,只需 3 mol%催化剂,产物得率高于 99%,且 ee 值和 de 值均大于 99%。磺酰胺类型的催化剂 178

同样也能有效地催化 Aldol 反应[372]。

Zhang 等[373]合成了 4-酚氧基取代的脯氨酰胺苯酚小分子催化剂 179,其可有效地催化水中不同醛和酮的不对称 Aldol 反应。在最佳反应条件下,得率和 dr 值均在 99% 以上,ee 值也能达到 97%。也有报道称,全氟苯基取代的脯氨酰胺小分子催化剂 179 在极性和非极性如盐水中均能催化 Aldol 反应,获得较好的立体选择性[374]。Singh 和 Gandh 合成了一系列脯氨酰胺小分子催化剂,其中用 β-氨基磺酰胺取代 β-氨基醇的催化剂 180 在盐水中催化环酮或非环酮的 Aldol 反应均显示活性,当催化剂用量为 2 mol% 时,产物得率和对映选择性均高达 99%[340]。Chimmi 等[375]证实质子化的脯氨酰胺 181 可有效催化水中不对称 Aldol 反应,溶液 pH 影响较大,当 pH = 4~5 时,dr 值和 ee 值均在 96% 以上。

Nugent 等[376]合成了仅有一个立体中心的小分子催化剂 PicAm-2(182),可快速催化水中 Aldol 反应 [式(2-226)和式(2-227)],用 2 mol%~5 mol% 催化剂,酮很容易与邻位、间位和对位取代的芳醛反应,即使是弱亲电性的对甲基苯甲醛,也给出产物的 ee 值为 95%~99%,功能化的环酮底物在 5 mol% 催化剂用量下,能给出最高得率和对映选择性。

(2-226)

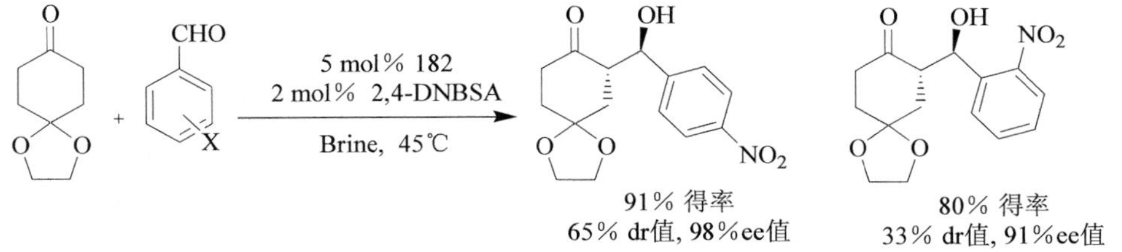

$$\text{(2-227)}$$

Fu 等[377]以三氟乙酸为反应介质,室温下通过丝氨酸与酰氯反应合成 O-芳基丝氨酸小分子催化剂 183,在催化水中不同醛与酮的反应中给出产物的得率和对映选择性均高于 99%,催化剂可循环使用 5 次,同时催化剂 183 可用于较大规模的 Aldol 反应,且对映选择性不受影响[378]。

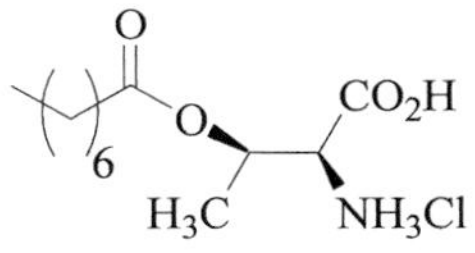

183

Juaristi 和 Hernández[379]以 5 mol%苯甲酸为添加剂,用(S)-脯氨酸-(S)-色氨酸甲基酯 184 催化水中酮与不同芳香醛的 Aldol 缩合反应,产物的得率为 80%~94%,立体选择性大于 98%,对映选择性大于 98%。Altenbach 等[380]研究了一系列 L-脯氨酸衍生物催化 Aldol 反应,发现含有三氟甲基的磺酰胺基团的小分子催化剂 185,在用量为 10 mol%时,获得非对映立体选择性为 97%,对映选择性为 97%,催化剂可循环使用。Plusquellec 等[381]用 2 mol%(R)-3-羟基吡咯烷催化水中缩合反应,发现添加 1 mol/L 水溶性的两亲糖苷 186,反应活性和立体选择性均明显提高。

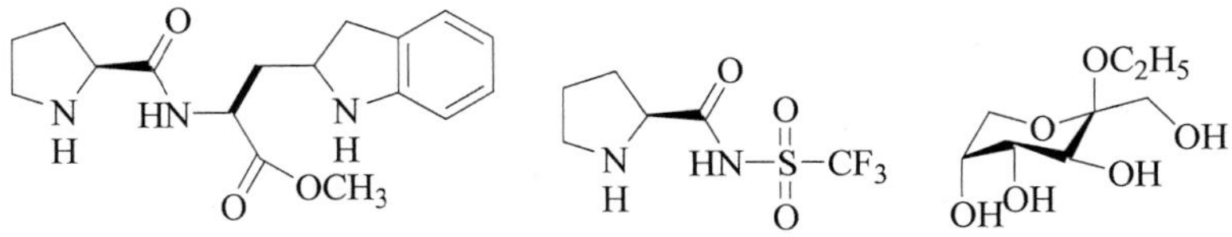

184 **185** **186**

Itoh[382]合成了一系列氟代磺酰胺用于 NaCl 水溶液中 Aldol 缩合反应,其中催化剂 187 性能最佳,给出反式缩合产物的对映选择性为 99%。催化剂用硅胶通过固相萃取,无需纯化就能直接进行重复循环使用。Schafmeister 等[383]合成了一系列含有疏水性基团的 4-羟基脯氨酸衍生物催化水中 Aldol 反应。研究结果显示,将芳环引入脯氨酸结构中所得的小分子催化剂 188 催化活性更高。

Fu等[384]用不同的二胺盐直接催化水中大规模化学计量的Aldol反应。结果表明，催化剂189性能优良，产物的对映选择性达99%。Chimni等[385]考察了含有伯胺和叔胺的不同二胺化合物催化水中靛蓝与丙酮或环己酮的反应，发现190催化剂性能最佳，当催化剂用量为1mol%时就能有效催化反应，获得高得率和对映选择性的3-取代-3-羟基-2-羟基吲哚衍生物。

Li等[386]发展了脯氨酸-胆固醇和薯蓣皂苷配基的双亲小分子催化剂191，应用于水中芳醛和环酮的不对称Aldol反应，在不加添加剂时，水溶剂可提高产物得率和立体选择性，明显优于传统有机溶剂，如DMSO、DMF、THF等。Zlotin等[387]报道了用酰胺硝基上含有(R)-α-苯乙基的(S)-缬氨酰胺(192)催化水中环酮和芳醛的不对称Aldol缩合反应，发现含有伯胺α-氨基酸酰胺单元的小分子催化剂具有较高的催化效率和立体选择性。

Zhao等[388]用N-Cbz-(S)-脯氨酸作为手性源合成了2,2′-二苯酚基的脯氨酰胺的C2-和C1-对称小分子催化剂193，C1-对称小分子催化剂在水介质Aldol反应中表现出高催化效率，产物得率为100%，de值和ee值分别为97%和98%。

Li 等[389]采用缩二氨酸(194)催化不同醛和酮的缩合反应,产物得率在 94%以上,de 值和 ee 值分别为 99%和 97%。Nisco 等[390]发现,二肽 195 是一种具有较好应用前景的环境友好的小分子催化剂,催化水中 Aldol 反应,产物得率为 96%,de 和 ee 值分别达到 97%和 90%。Da 等[391]报道了以二硝基苯酚为酸添加剂,脯氨酰胺类型的催化剂 196 可催化水中酮与醛的不对称 Aldol 反应,当催化剂用量为 1 mol%时,de 和 ee 值分别为 98%和 97%。

Kokotos 等[392]以脯氨酸、苯基丙氨酸及氨基酸叔丁基酯为原料合成了三肽(Pro-Phe-AA-Ot-Bu),并将其用于催化水相不同醛和酮的不对称 Aldol 反应,产物得率高于 99%,非对映立体选择性大于 97%,对映选择性为 99%以上。研究发现,氨基酸的 C 端决定了该小分子催化剂的催化活性,当用二叔丁基天冬氨酸与脯氨酸及苯基丙氨酸制得的三肽 197 催化水中不对称 Aldol 反应时,产物的得率和选择性都比较高。而用叔丁基-(R)-苯基甘氨酸盐代替二叔丁基天冬氨酸制得的三肽 198,活性则较差。

Kokotos 等[393]报道了各种多肽的叔丁基酯用于催化水中酮和醛的对映选择性 Aldol 反应,其中 Pro-Glu(Ot-Bu)-Ot-Bu(199)活性最佳,给出高得率、对映选择性和非对映选择性。

Wang 等[394]采用不对称方法合成了氮杂环丁二烯-2-羧酸酰胺对映体小分子催化剂 200,它能顺利催化盐水中含有不同取代基的苯甲醛与丙酮的 Aldol 反应,产物 β-羟基酮的 ee 值超过 96%。在苯甲醛与环酮的反应中,产物的 ee 值和 de 值均超过 99%。

Bhowmick 等[395]用 4-羟基-L-脯氨酸和枞酸合成了一种新型小分子催化剂 201，可催化水相不同芳香醛与酮的 Aldol 缩合反应，在添加剂 TFA 存在时，用 1mol%催化剂，反应在短时间内就能完成，产物的得率在 99%以上，且立体和对映选择性分别高于 94%和 99.9%。

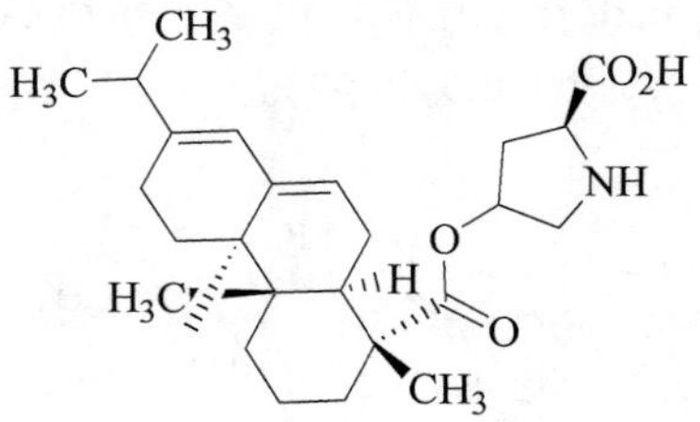

201

Zhao 等[396]设计了一系列含有二芳基的双功能小分子催化剂，用于水中 Aldol 反应合成各种 3-烷基-3-羟基-2-氧代吲哚衍生物，其中催化剂 202 在靛红与对映异构体的酮的反应中活性最佳，产物得率、立体选择性和对映选择性均高于 99%。

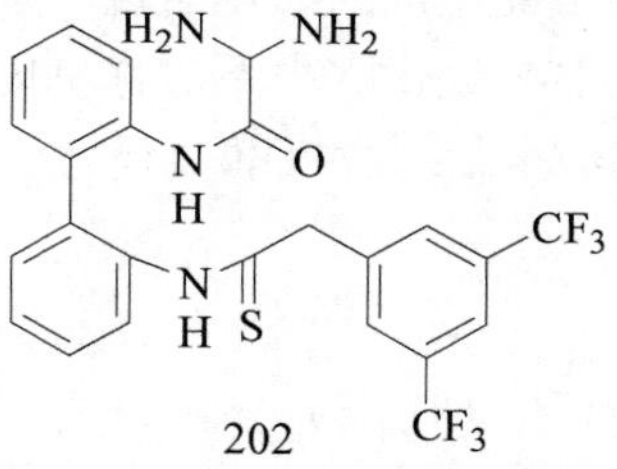

202

Yilmaz 等[397]从 L-脯氨酸出发，合成了两种新的对叔丁基杯[4]芳烃基手性小分子催化剂 203，在最佳反应条件下，催化水中环己酮和芳醛的 Aldol 反应，产物得率为 95%，对映选择性为 90%，非对映选择性为 65%。

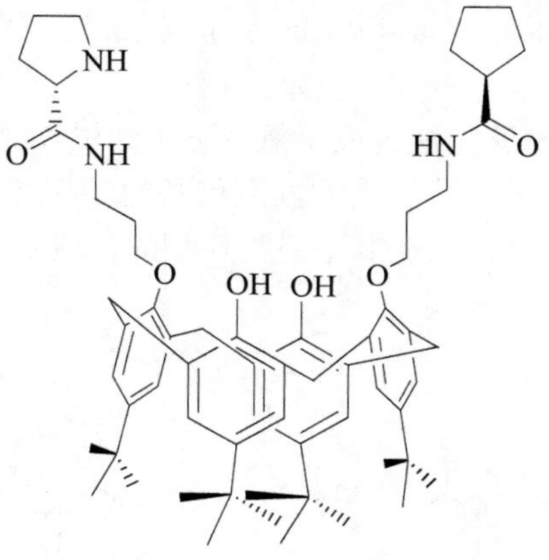

203

Miura 和 Machinami 报道了用含有配糖的水溶性小分子催化剂 204 和 205 催化水中不对称 Aldol，催化剂中的脯氨酰基和 α-手性醛的构象对产物的立体选择性具有明显的影响，催化剂很容易从反应化合物中分离、回收和重复使用[398]。

Itoh 等[399]报道了光催化水中不对称 Aldol 反应，首先，苄醇在一水蒽醌-2-磺酸钠(AQN-2-SO$_3$Na·H$_2$O)于空气中和有机光催化剂的催化下，被氧化成苯甲醛，然后在小分子催化剂 206 催化下与酮发生 Aldol 反应，给出中等得率的产物，对映选择性大于 94%。

Jimeno 等[400]合成了一种新型、高效的阳离子型小分子胍催化剂 207 应用于不对称 Aldol 反应，展现出优良的催化性能和立体选择性，这归因于小分子催化剂 207 中含有一个酰基的胍结构，有利于形成分子内氢键。而在 2-辛酮与对硝基苯甲醛的 Aldol 反应中，产物得率仅为 55%，对映选择性为 75%。

Yilmaz 等[401]以疏水性的杯芳烃合成了一种脯氨酸基酯、酰胺和酸组成的小分子催化剂 208。在水介质环己酮与醛的 Aldol 反应中，催化剂 208 具有较好的活性和立体选择性，给出产物得率、对映选择性和立体选择性分别为 82%、81% 和 91%。

二、Michael 加成反应

在各种 C—C 键的形成反应中,1,4-共轭加成反应在有机合成中起到非常重要的作用[402]。在所有 Michael 受体中,硝基烯非常活泼,产物中的硝基可以进一步转变成腈的氧化物、酮、酰胺或羧酸等,因而其在有机合成中具有非常重要的意义。特别是酮和醛与硝基烯的 Michael 加成反应,生成 γ-硝基羰基化合物[402-404]。有关水中小分子催化的 Michael 反应也有很多报道,如 Barbas 等[405]采用二胺 209/TFA 催化水中醛和酮与硝基烯的 Michael 加成反应,给出高对映选择性(97%)的加成产物。实验证实,在盐水中能避免胺聚合,提高得率和 ee 值。

Cheng 等[406]合成了表面活性剂类型的不对称小分子(STAO)催化剂 210,并将其用于水相醛和酮与硝基烯的加成反应。在无添加剂和有机溶剂的条件下,获得较高的对映选择性(>98%)和非对映选择性(99%)。STAO 催化剂在反应体系中具有双功能作用,既是加速反应进行的不对称催化剂,也作为表面活性剂帮助有机底物完成溶剂化。Wang 等[407]报道用 F-(S)-四氢吡咯磺酰胺小分子催化剂 211 催化水中酮和醛与硝基烯在水中的 Michael 加成反应,产物得率为 98%,对映选择性为 95%,非对映选择性为 50%,催化剂可重复使用 6 次。

所有上述水相小分子催化的 Michael 反应都有一定的局限性,如反应底物范围窄,催化剂用量大,反应时间长,且需加入添加剂等。2007 年 Xiao 等[408]报道了一种结构可调的吡啶-硫脲双功能小分子催化剂 212,并将其用于水中酮与硝基烯的不对称 Michael 加成反应[式(2-228)]。在最佳条件下,给出高得率(88%~90%)和对映及立体选择性(均为 99%)的产物,硝基烯的芳环上取代基的性质对活性没有明显影响。

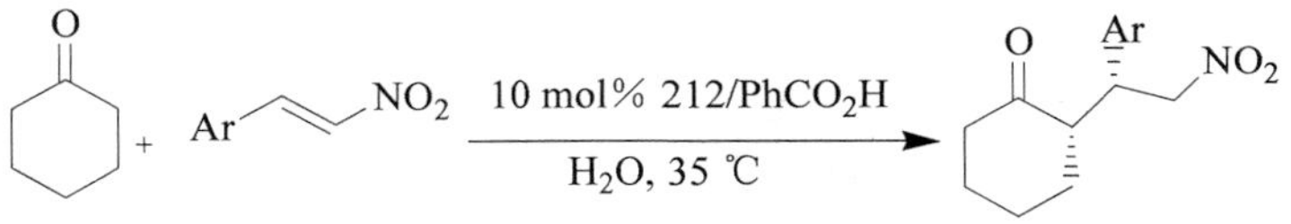

$$\text{环己酮} + Ar\text{—CH=CH—}NO_2 \xrightarrow[H_2O,\ 35\ ^\circ C]{10\ \text{mol}\%\ 212/PhCO_2H} \text{产物} \qquad (2\text{-}228)$$

Singh 和 Maya 合成了一种新型水溶性的联萘基手性二胺小分子催化剂 213，用于盐水中酮与硝基烯的 Michael 加成反应。在较大范围的 β-硝基烯(芳香硝基和脂肪硝基烯)和环酮的 Michael 加成反应中，均给出较高立体选择性(99%)和对映选择性(99%)，这可归因于小分子催化剂结构中的联萘基团提供了一种疏水性的环境，使有机受体和给体分子在水中的溶解性降低，从而提高了反应活性和立体选择性[409]。

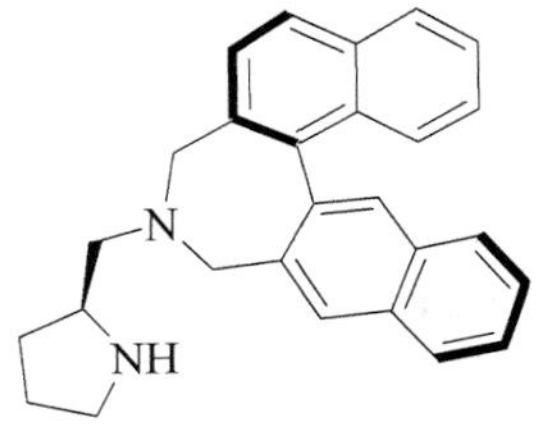

213

Ma 等[410-412]合成了 OTMS 取代的联苯脯氨醇小分子催化剂 214，并将其用于催化水中醛与 β-硝基丙烯酸酯的不对称 Michael 加成反应，不仅具有非常高的催化活性和立体选择性(>98%)及对映选择性(>99%)，也适用于更多的醛和 β-硝基丙烯酸酯参与的 Michael 反应。后来，该催化剂也被用在简单的芳香硝基烯和脂肪硝基烯与醛的反应。对于芳香硝基烯，仅用 0.5mol%催化剂就能给出最佳的结果，产物得率为 90%，对映选择性大于 99%，且在丙二腈与 α,β-不饱和醛的反应中，也能给出非常好的结果。

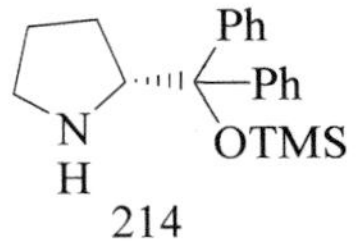

214

Jørgenson 等[413]同样也报道了用三甲基硅保护的二芳基脯氨醇 215 催化水中酯和 α,β-不饱和醛的 Michael 反应，较大范围的酯和醛均能给出高得率和对映选择性的产物。这种 Michael 产物可用于合成光学活性的环己烯酮和具有生物活性的化合物如哌啶衍生物。

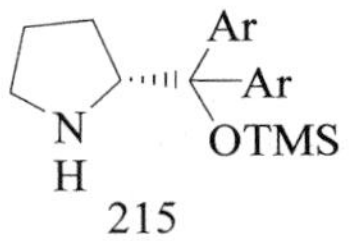

215

Palomo 等[414]合成了一系列新型脯氨醇基小分子催化剂 216，其在水中硝基甲烷与烯醛的 Michael 加成反应[式(2-229)]中表现高活性，很多反应底物在此催化体系中均能给出高得率和立体选择性的目标产物。这个催化体系也适用于丙二酸酯和醛的共轭加成[式(2-230)]以及醛与烯醛的 Michael 加成反应[式(2-231)]，同样能给出高得率和对映选择性的目标产物。

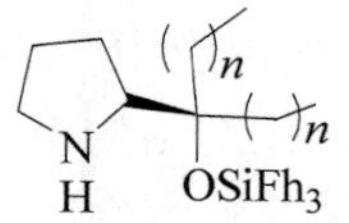
N
H
OSiFh3
n
n

216

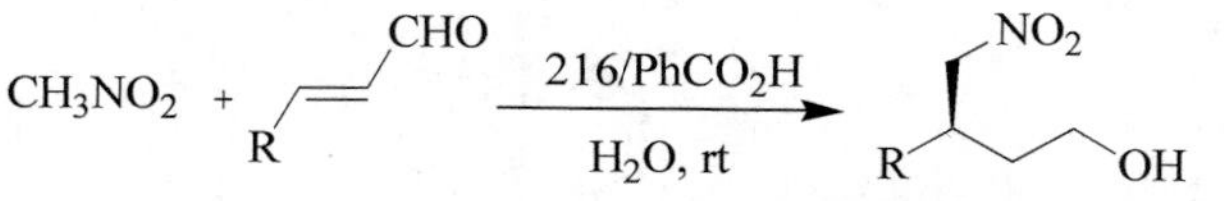
CH3NO2
CHO
R
216/PhCO2H
H2O, rt
NO2
R
OH

70% 得率, 97% ee值

(2-229)

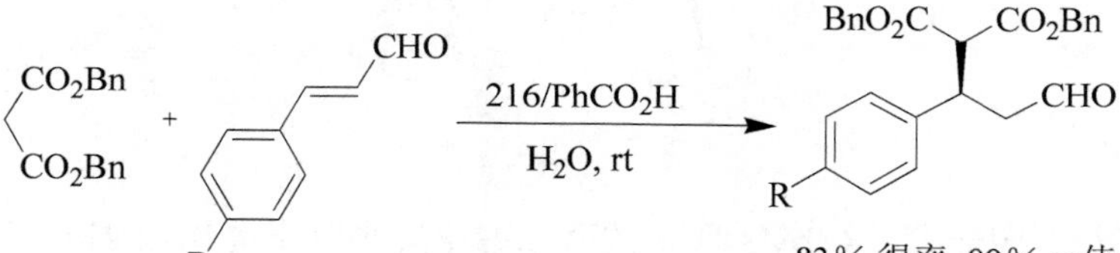
CO2Bn
CO2Bn
CHO
R
216/PhCO2H
H2O, rt
BnO2C
CO2Bn
CHO
R

83% 得率, 99% ee值

(2-230)

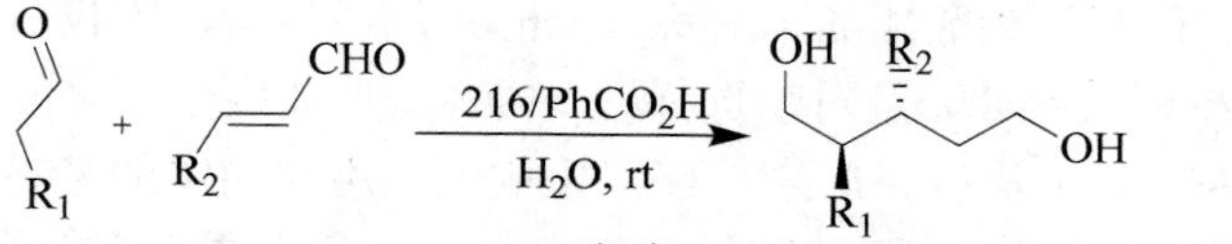
O
R1
CHO
R2
216/PhCO2H
H2O, rt
OH
R2
OH
R1

68%得率, 98∶2(*anti*∶*syn*), 98% ee值

(2-231)

2009 年 Wu 等[415]在吡咯烷支链引入一个疏水性基团，合成了吡咯烷基小分子催化剂 217，在水相不同硝基烯与醛的 Michael 加成反应中显示出高活性。Chandrasekhar 等[416]合成了一种新的羟基邻苯二甲酰亚胺-偶合三唑-脯氨酸衍生物 218，其可作为一种高效的、高立体选择性的催化剂，催化水中硝基烯与酮的不对称 Michael 加成反应，在苯甲酸存在下，给出高得率和高选择性。

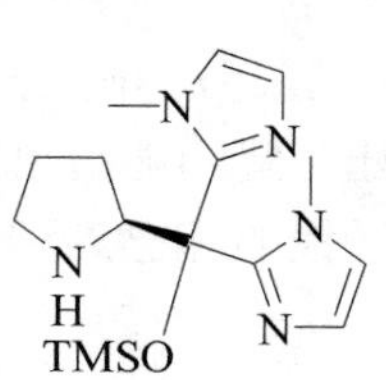
N
N
N
H
N
N
TMSO

217

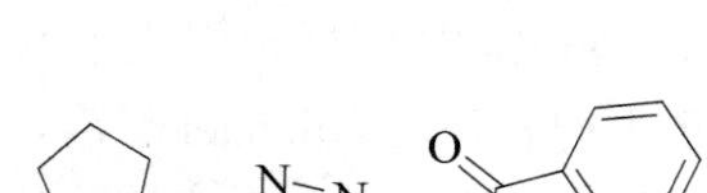
HN
N
N
N
O
O
N
O

218

Huang 等[417]成功合成了含长链的(*S*)-脯氨醇甲硅氧基醚小分子催化剂，其在室温下催化含有少量甲酸的水溶液中各种醛与反式硝基烯的不对称 Michael 加成

反应，给出高得率的目标产物，dr 值和 ee 值均大于 99%。Moon 和 Kim 发展了一种含有手性伯胺的金鸡纳生物碱衍生物的催化剂 219，通过水中 α, β-不饱和酮和 α-硝基羧酸酯的对映选择性共轭加成，合成手性 γ-硝基酮[式(2-232)]。此催化剂也适用于水中 δ-羰基硝基烷和 δ-羰基酯的 Michael 反应，同样也给出高得率(81%)和对映选择性(89%)的不对称加成产物[418]。

$$R_1\text{-CH=CH-C(O)-}R_2 + \text{CO}_2\text{C}_2\text{H}_5\text{CH}_2\text{NO}_2 \xrightarrow[\text{H}_2\text{O, rt}]{\substack{2\text{ mol\% 219}\\40\text{ mol\% PhCO}_2\text{H}}} \text{产物} \quad (2\text{-}232)$$

Loh 等[419]首次报道了小分子催化水中乙烯砜与醛的 Michael 加成反应[式(2-233)]，该反应在水中非常难进行，但他们报道的 220/DMAP 可成功催化该反应进行，给出中等得率和对映选择性的产物，而脯氨酸和其他氨基酸催化剂均不能催化这个反应。

$$\text{RCH}_2\text{CHO} + \text{CH}_2\text{=C(SO}_2\text{Ph)}_2 \xrightarrow[\text{2) NaBH}_4, \text{CH}_3\text{OH}]{\text{1) 10 mol\% 220, DMAP}} \text{HOCH}_2\text{CHR-CH}_2\text{CH(SO}_2\text{Ph)}_2 \quad (2\text{-}233)$$

Ni 等[420]合成了类似的小分子催化剂 221，同样也考察了其在醛与硝基烷中的催化性能。研究表明，催化剂 221 不仅能给出较高得率(99%)和立体选择性(syn : anti = 98 : 2，99% ee)的产物，且催化剂中 Brønsted 酸和二甲基胺形成的铵盐不仅增强了催化剂在水中的溶解能力，而且也提供了一种容易回收循环使用的催化剂。

221

Lin 等[421]合成了一种可用于水中不对称 Michael 加成反应的催化剂 222，该催化剂结构中含有四氢吡啶和砜的单元，在硝基烯与环己酮的不对称加成反应中显示了高催化活性，无需添加任何试剂，产物得率高达 90%~99%，且也有非常高的立体选择性(>99%)和对映选择性(>98%)。

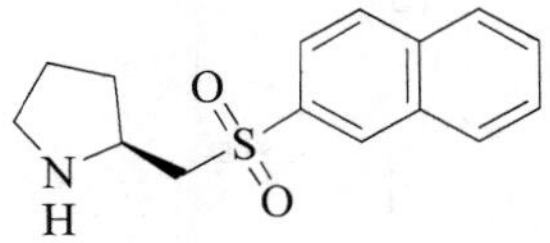

222

Moorthy 等[422]合成了脯氨酸衍生物小分子催化剂 223 并将其应用于盐水中硝基烯与酮的不对称 Michael 加成反应，给出高得率的产物(>97%)，且 ee 值也达 95%，这可归因于催化剂的甲苯磺酰基与硝基烯的芳环的 π-π 电子堆积作用。

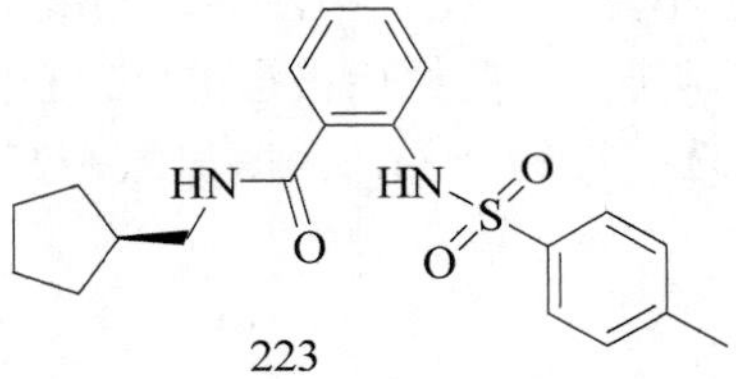

223

Luo 和 Du 报道了用二萘磺酰胺催化剂 224 催化水中酮和硝基烯的 Michael 加成，产物得率中等，对映选择性为 93%~97%，非对映选择性大于 99%。催化剂 224 也可催化异丁醛和硝基烯的不对称 Michael 加成，产物得率中等，ee 值为 83%[423]。

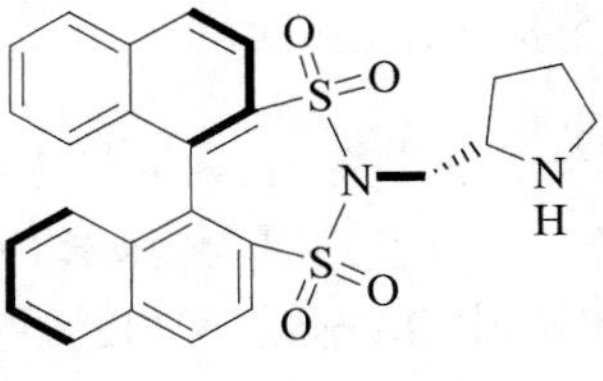

224

Liu 等[424]报道将 10 mol%含有长链的表面活性剂类型双功能硫脲小分子催化剂 225 与 10 mol%苯甲酸催化体系用于硝基烯和环酮的不对称 Michael 加成反应，在室温下给出目标加成产物，得率为 80%~93%，立体选择性为 95%~99%，对映选择性为 92%。Sebesta 等[425]报道了 O-十二碳酰基-4-羟基脯氨酸小分子催化剂 226 催化水中硝基烯与醛的对映选择性 Michael 加成反应，产物得率为 78%~94%，ee 值为 83%~98%。

Ni 等[426]报道了采用二芳基脯氨醇甲硅氧基醚 227/苯甲酸共催化盐水中硝基烯和乙醛的不对称 Michael 加成反应，不需加入任何有机溶剂，各种硝基烯均能和醛反应且均能给出较好得率(30%~61%)和高的对映选择性(80%~97%)。

Li 等合成了一种由脯氨酸合成双亲含硫咪唑小分子催化剂 228，其与对硝基苯甲酸共同催化水中硝基烯与酮的不对称 Michael 加成反应，给出对映和立体选择性均为 99%[427]。

Kumar 等[428]报道了一种由脯氨酸和羟基酰胺合成的新型吡咯烷-O-代酰胺小分子催化剂 229，它在无添加剂的水中，于室温下有效地催化酮和硝基烯的不对称 Michael 加成反应，给出较好得率和高选择性的目标产物。

Guo 等[429]设计了一种小分子催化剂 230 用于水中环己酮和不同硝基烯的不对称 Michael 加成反应，产物得率为 99%，对映选择性为 94%，立体选择性大于 99%。

对于丙二酸酯与 α,β-不饱和酮的 Michael 加成反应也有报道[430]，采用含有长链烷基的伯胺仲胺二胺催化剂 231，可高效催化酯与不同 α,β-不饱和酮的 Michael 对映选择性加成[式(2-234)]。Du 等[431]采用二萘磺酰胺 232 于室温下催化水中酮和烯基丙二酸酯的不对称 Michael 加成反应，产物得率为 98%，立体选择性为 99%，对映选择性为 99%。

$$R_1OOC\text{-}CH_2\text{-}COOR_1 + Ph\text{-}CH=CH\text{-}CO\text{-}R_2 \xrightarrow[\text{H}_2\text{O, 50℃}]{\text{20 mol\% 231}, \text{20 mol\% TFA}} R_1OOC\text{-}C(COOR_1)\text{-}CH(Ph)\text{-}CH_2\text{-}CO\text{-}R_2 \quad (2\text{-}234)$$

Miura 等[432]用苯基丙氨酸合成了 β-氨基磺酰胺小分子催化剂 233 和 234，并将其用于催化水中苯基丙二酸酯与 α,β-不饱和酮的共轭加成反应[式(2-235)]。结果表明，不需添加剂就可获得高得率和优良对映选择性(80%~99%)的目标产物，这可归因于全氟取代基可为反应提供一种疏水性环境以及强的吸电子能力，且可通过简单的固相萃取就能完成回收并重复使用。

$$R_1\text{-}CH=CH\text{-}CO\text{-}R_2 + CH_2(CO_2Bn)_2 \xrightarrow[\text{H}_2\text{O, rt}]{0.1\text{ eq 233 或 234}} BnO_2C\text{-}CH(CO_2Bn)\text{-}CH(R_1)\text{-}CH_2\text{-}CO\text{-}R_2 \quad (2\text{-}235)$$

Bae 和 Song 采用金鸡纳基方酰胺小分子催化剂 HQN-SQA(235)催化丙二酸酯和不同硝基烯的 Michael 加成反应。实验证实，该反应在水中的反应速率明显大于传统的有机溶剂，这可能是由于疏水的水合效应有利于金鸡纳基催化剂的 C-3 取代基的疏水性，另外，水在反应中有利于形成非共价的氢键，从而提高反应的对映选择性，因此水是这个反应的最佳溶剂。使用 0.01 mol%的催化剂足以催化 β-酮酸酯和硝基烯的加成反应，产物得率和立体选择性均大于 99%，对映选择性高达 99%[433]。

三、Mannich 反应

有机小分子催化酮和芳香胺、芳醛的不对称三组分的不对称 Mannich 反应是合成光学活性的 α- 或 β-氨基酸衍生物和 γ-氨基醇的重要方法[434]。另外，Mannich 加合物可用来合成生物活性分子，水中小分子催化的 Mannich 反应为合成生物活性分子提供了一种环境友好的合成路线[435]。

Hayashi 等[436]合成了甲硅氧基四唑杂化的催化剂 236，它与甲硅氧基脯氨酸联合催化水中不对称 Mannich 反应，二甲氧基乙醛与对甲氧基苯胺和酮反应生成高得率和高对映选择性的顺式产物[式(2-236)]。甲硅氧基脯氨酸的钠盐 237 可成功催化 α-亚胺乙基乙醛酸[式(2-237)]，这归因于催化剂中的钠离子与羧基具有相同的作用。

Teo 等[437]合成了一种高效的甲硅氧基色氨酸小分子催化剂 238，用于催化水中三组分的不对称 Mannich 反应，很多环酮和非环酮均能作为给体与受体醛和乙基乙醛酸酯反应，给出高立体化学选择性的顺式 Mannich 产物。

以上小分子催化的 Mannich 反应均给出了顺式产物，而反式产物的合成仍是一个挑战。为了在水中获得反式产物，Lu 等[438]以苏氨酸合成了催化剂 239，其在邻苄基羟基丙酮与对甲氧基苯胺、脂肪醛或芳香醛的反应[式(2-238)]中，获得反式产物，具有高得率和对映选择性，同样，含有线型和支链的脂肪醛反应也能获得反式产物，具有高得率和对映选择性。

$$(2\text{-}238)$$

Amedjkouh 和 Brandberg 报道了一种 Mannich 加合物自催化水中不对称 Mannich 反应。该加合物具有双功能催化作用，显示出较高的活性和立体选择性。它在 pH=7 的缓冲溶液中可以得到比较好的催化效率[439]。Bhandari 和 Srinivas 报道了 L-脯氨酸催化水中 Mannich 反应形成 C—C—N 键的方法，适用于很多种酮的 Mannich 反应，获得高得率的 Mannich-Aldol 类型产物，这种产物在碱性条件下很容易发生脱甲酰化反应，从而得到目标 Mannich 产物[440]。

Tao 等[441]报道了异甜菊醇-氨基酸的双亲小分子催化剂 240，其能够高效催化水中环酮和芳香胺、芳醛的不对称三组分 Mannich 反应，获得顺式 Mannich 产物，立体选择性为 98%，对映选择性大于 99%。

Šebesta 等[442]从 N-丙基磺酰胺合成 4-叔丁基二甲基硅氧基取代的催化剂 241，催化水中环酮与亚胺乙醛酸酯的 Mannich 反应，获得顺式非对映异构体，ee 值大于 99%。该催化剂也能有效地催化水中四氢-2H-吡喃-2,6-二醇与亚胺乙醛酸酯的反应，获得相应的 2,3-二取代四氢吡啶，dr 和 ee 值分别为 95%和 98%。

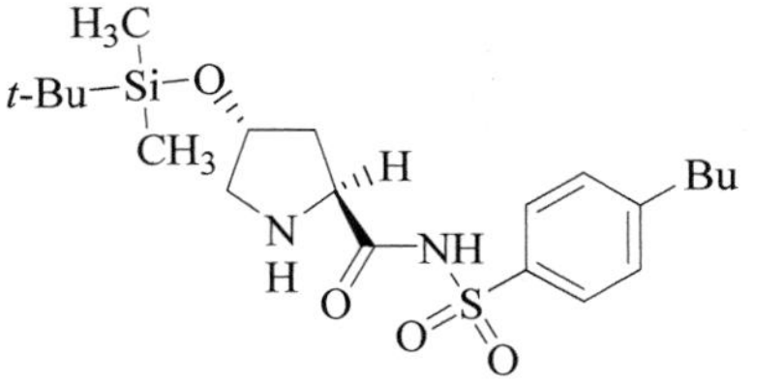

241

四、环加成反应

MacMillan 和 Northrup 报道了催化剂 242 与 HClO₄ 联合催化水中二烯和 α, β-不饱和酮的 Diels-Alder 反应[式(2-239)和式(2-240)][443]。后来，Ogilvie 等[444]报道环状酰肼催化剂 243 与 HClO₄ 或 TfOH 以及 244 与 TfOH 联合催化环戊酮和 α, β-不饱和醛的 Diels-Alder 反应，获得高得率和非对映选择性及对映选择性的环加成产物。Bonini 等[445]用 245 与 HClO₄ 联合催化相同的反应，但产物的对映选择性低到中等。

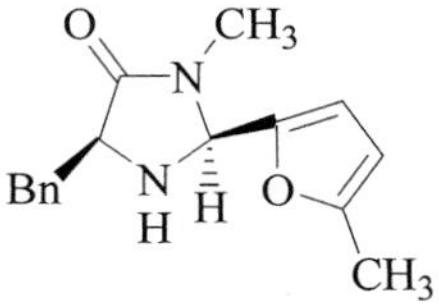

242

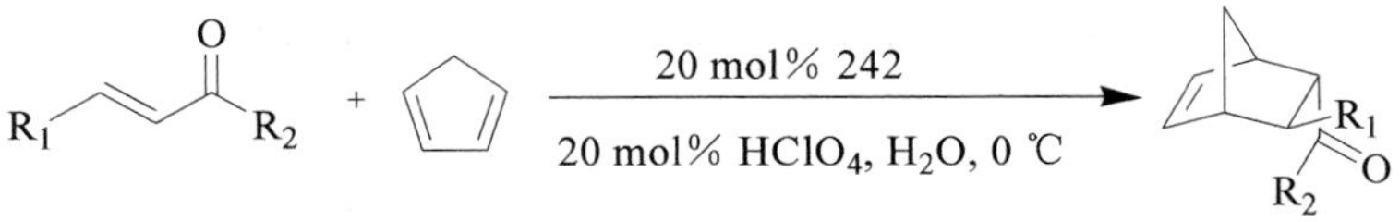

$$R_1\text{-CH=CH-CO-}R_2 + \text{cyclopentadiene} \xrightarrow[20\text{ mol\% HClO}_4, H_2O, 0\ ℃]{20\text{ mol\% 242}} \quad (2\text{-}239)$$

$$\text{cyclopentadiene} + \text{cyclic ketone} \xrightarrow[20\text{ mol\% HClO}_4, H_2O, 0\ ℃]{20\text{ mol\% 242}} \quad (2\text{-}240)$$

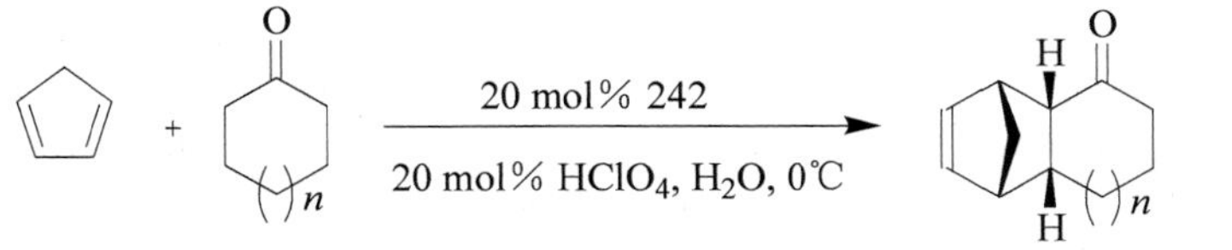

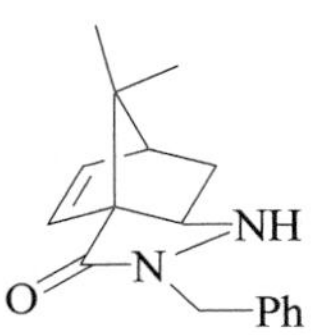

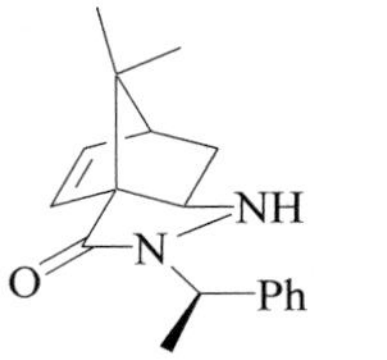

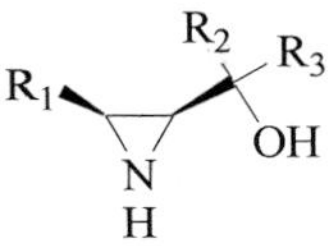

	243	244	245
得率	71%~96%	78%~94%	16%~88%
exo∶endo	1.2∶1~2.6∶1	1.1∶1~4∶1	1∶1.8~1.6∶1
ee值	85%~94%ee(exo)	85%~96%ee(exo)	10%~66%ee(exo)
		80%~93%ee(endo)	11%~48%ee(endo)

Hayashi 等[446]报道了二芳基脯氨酸甲硅氧基醚盐 246 催化水中肉桂醛和戊二烯的 Diels-Alder 反应[式(2-241)]，获得的产物具有高 exo-选择性和优良的对映选择性，水溶剂可加速反应速率并提高对映选择性。

246

$$Ph-CH=CH-CHO + \text{环戊二烯} \xrightarrow{5 \text{ mol\% } 246}{H_2O, \text{rt}} \text{产物} \quad (2\text{-}241)$$

95% 得率; exo ∶ endo = 85 ∶ 15
99% ee(exo); 99% ee(endo)

参 考 文 献

[1] Li C J. Organic reactions in aqueous media-with a focus on carbon-carbon bond formation. Chemical Reviews, 1993, 93: 2023-2035.

[2] Li C J. Organic reactions in aqueous media with a focus on carbon-carbon bond formations: A decade update. Chemical Reviews, 2005, 105: 3095-3166.

[3] Hollis T K, Robinson N P, Bosnich B. Homogeneous catalysis.[Ti(Cp*)$_2$(H$_2$O)$_2$]$_2$$^+$: An air-stable, water-tolerant Diels-Alder catalyst. Journal of the American Chemical Society, 1992, 114: 5464-5466.

[4] Lubineau A, Meyer E. Water-promoted organic reactions. aldol reaction of silyl enol ethers with carbonyl compounds under atmospheric pressure and neutral conditions. Tetrahedron, 1988, 44: 6065-6070.

[5] Kobayashi S, Nagayama S, Busujima T. Lewis acid catalysts stable in water. correlation between catalytic activity in water and hydrolysis constants and exchange rate constants for substitution of inner-sphere water ligands. Journal of the American Chemical Society, 1998, 120: 8287-8288.

[6] Kobayashi S, Hachiya I, Araki M, Ishitani H. Scandium trifluoromethanesulfonate (Sc(OTf)$_3$). A novel reusable catalyst in the Diels-Alder reaction. Tetrahedron Letters, 1993, 34: 3755-3758.

[7] Chen W H, Yuan D Q, Fujita K. Regiospecifically multifunctional β-cyclodextrins with two or three glucose residues bearing imidazolyl groups at the C$_3$ positions. Tetrahedron Letters, 1997, 38: 4599-4602.

[8] Kobayashi S, Wakabayashi T. Scandium trisdodecylsulfate (STDS). A new type of lewis acid that forms stable dispersion systems with organic substrates in water and accelerates aldol reactions much faster in water than in organic solvents. Tetrahedron Letters, 1998, 39: 5389-5392.

[9] Manabe K, Mori Y, Wakabayashi T, Nagayama S, Kobayashi S. Organic synthesis inside particles in water: Lewis acid-surfactant-combined catalysts for organic reactions in water using colloidal dispersions as reaction media. Journal of the American Chemical Society, 2000, 122: 7202-7207.

[10] Kobayashi S, Busujima T, Nagayama S. On indium(Ⅲ) chloride-catalyzed aldol reactions of silyl enol ethers with aldehydes in water. Tetrahedron Letters, 1998, 39: 1579-1582.

[11] Tian H Y, Chen Y J, Wang D, Bu Y P, Li C J. The effects of aromatic and aliphatic anionic surfactants on Sc(OTf)$_3$-catalyzed Mukaiyama aldol reaction in water. Tetrahedron Letters, 2001, 42: 1803-1805.

[12] Mori Y, Manabe K, Kobayashi S. Catalytic use of a boron source for boron enolate mediated stereoselective aldol reactions in water. Angewandte Chemie International Edition, 2001, 40: 2815-2818.

[13] Nagayama S, Kobayashi S. A novel chiral lead(Ⅱ) catalyst for enantioselective aldol reactions in aqueous media. Journal of the American Chemical Society, 2000, 122: 11531-11532.

[14] Kobayashi S, Hamada T, Nagayama S, Manabe K. Lanthanide trifluoromethanesulfonate-catalyzed asymmetric aldol reactions in aqueous media. Organic Letters, 2001, 3: 165-167.

[15] Yang B Y, Chen X M, Deng G J, Zhang Y L, Fan Q H. Chiral dendritic bis(oxazoline) copper(Ⅱ) complexes as Lewis acid catalysts for enantioselective aldol reactions in aqueous media. Tetrahedron Letters, 2003, 44: 3535-3538.

[16] Li H J, Tian H Y, Chen Y J, Wang D, Li C J. Chiral anionic surfactants for asymmetric Mukaiyama Aldol-type reaction in water. Journal of Chemical Research, 2003, 2003: 153-156.

[17] Zhang T, Zhang Y, Li Z, Song Z, Liu H, Tao J. Direct asymmetric aldol reaction co-catalyzed by amphiphilic prolinamide phenol and Lewis acidic metal on water. Chinese Journal of Chemistry, 2013, 31: 247-255.

[18] Loh T P, Wei L L. Novel one-pot Mannich-type reaction in water: indium trichloride-catalyzed condensation of aldehydes, amines and silyl enol ethers for the synthesis of β-amino ketones and esters. Tetrahedron Letters, 1998, 39: 323-326.

[19] Loncaric C, Manabe K, Kobayashi S. Mannich-type reactions catalyzed by neutral salts in water. Advanced Synthesis & Catalysis, 2003, 345: 1187-1189.

[20] Ishimaru K, Kojima T. Addition of water to zinc triflate promotes a novel reaction: stereoselective Mannich-type reaction of chiral aldimines with 2-silyloxybutadienes. The Journal of Organic Chemistry, 2003, 68: 4959-4962.

[21] Eder U, Sauer G, Wiechert R. New type of asymmetric cyclization to optically active steroid CD partial structures. Angewandte Chemie International Edition in English, 1971, 10: 496-497.

[22] Hajos Z G, Parrish D R. Synthesis and conversion of 2-methyl-2-(3-oxobutyl)-1,3-cyclopentanedione to the isomeric racemic ketols of the[3.2.1] bicyclooctane and of the perhydroindane series. The Journal of Organic Chemistry, 1974, 39: 1612-1615.

[23] Harada N, Sugioka T, Uda H, Kuriki T. Efficient preparation of optically pure Wieland-Miescher ketone and confirmation of its absolute stereochemistry by the CD exciton chirality method. Synthesis, 1990, 1: 53-56.

[24] Sussangkarn K, Fodor G, Karle I, George C. Ascorbic acid as a Michael donor : Part Ⅱ. reaction with alicyclic enones. Tetrahedron, 1988, 44: 7047-7054.

[25] Keller E, Feringa B L. Ytterbium triflate catalyzed Michael additions of β-ketoesters in water. Tetrahedron Letters, 1996, 37: 1879-1882.

[26] Kobayashi S, Kakumoto K, Mori Y, Manabe K. Chiral Lewis acid-catalyzed enantioselective Michael reactions in water. Israel Journal of Chemistry, 2001, 41: 247-250.

[27] Firouzabadi H, Iranpoor N, Nowrouzi F. The facile and efficient Michael addition of indoles and pyrrole to α,β-unsaturated electron-deficient compounds catalyzed by aluminium dodecyl sulfate trihydrate[Al(DS)$_3$]·3H$_2$O in water. Chemical Communications, 2005: 789-791.

[28] Ueno M, Kitanosono T, Sakai M, Kobayashi S. Chiral Sc-catalyzed asymmetric Michael reactions of thiols with enones in water. Organic & Biomolecular Chemistry, 2011, 9: 3619-3621.

[29] Bonollo S, Lanari D, Pizzo F, Vaccaro L. Sc(Ⅲ)-catalyzed enantioselective addition of thiols to α,β-unsaturated ketones in neutral water. Organic Letters, 2011, 13: 2150-2152.

[30] Aplander K, Ding R, Krasavin M, Lindström U M, Wennerberg J. Asymmetric Lewis acid catalysis in water: α-Amino acids as effective ligands in aqueous biphasic catalytic Michael additions. European Journal of Organic Chemistry, 2009, 2009: 810-821.

[31] Veisi H, Maleki B, Eshbala F H, Veisi H, Masti R, Ashrafi S S, Baghayeri, M. In situ generation of iron(Ⅲ) dodecyl sulfate as Lewis acid-surfactant catalyst for synthesis of *bis*-indolyl, *tris*-indolyl, di(*bis*-indolyl), tri(*bis*-indolyl), tetra(*bis*-indolyl)methanes and 3-alkylated indole compounds in water. RSC Advances, 2014, 4: 30683-30688.

[32] Zhuang W, Jorgensen K A. Friedel-Crafts reactions in water of carbonyl compounds with heteroaromatic compounds. Chemical Communications, 2002: 1336-1337.

[33] Manabe K, Aoyama N, Kobayashi S. Friedel-Crafts-type conjugate addition of indoles using a Lewis acid-surfactant-combined catalyst in water. Advanced Synthesis & Catalysis, 2001, 343: 174-176.

[34] Ding R, Zhang H B, Chen Y J, Liu L, Wang D, Li C J. In(OTf)$_3$-catalyzed Friedel-Crafts reaction of aromatic compounds with methyl trifluoropyruvate in water. Synlett, 2004, 2004: 555-557.

[35] Rideout D C, Breslow R. Hydrophobic acceleration of Diels-Alder reactions. Journal of the American Chemical Society, 1980, 102: 7816-7817.

[36] Otto S, Boccaletti G, Engberts J B F N. A chiral Lewis-acid-catalyzed Diels-Alder reaction. water-enhanced enantioselectivity. Journal of the American Chemical Society, 1998, 120: 4238-4239.

[37] Otto S, Bertoncin F, Engberts J B F N. Lewis acid catalysis of a Diels-Alder reaction in water. Journal of the American Chemical Society, 1996, 118: 7702-7707.

[38] Otto S, Engberts J B F N. A systematic study of ligand effects on a Lewis-acid-catalyzed Diels-Alder reaction in water. water-enhanced enantioselectivity. Journal of the American Chemical Society, 1999, 121: 6798-6806.

[39] Larsen S D, Grieco P A. Aza Diels-Alder reactions in aqueous solution: cyclocondensation of dienes with simple iminium salts generated under Mannich conditions. Journal of the American Chemical Society, 1985, 107: 1768-1769.

[40] Grieco P A, Parker D T, Cornwell M, Ruckle R. Retro aza Diels-Alder reactions: Acid catalyzed heterocycloreversion of 2-azanorbornenes in water at ambient temperature. Journal of the American Chemical Society, 1987, 109: 5859-5861.

[41] Yu L, Li J, Ramirez J, Chen D, Wang P G. Synthesis of azasugars via lanthanide-promoted aza Diels-Alder reactions in aqueous solution. The Journal of Organic Chemistry, 1997, 62: 903-907.

[42] Lubineau A, Augé J, Lubin N. Hetero Diels-Alder reaction in water. synthesis of α-hydroxy-γ-lactones. Tetrahedron Letters, 1991, 32: 7529-7530.

[43] Grieco P A, Henry K J, Nunes J J, Matt J E. Total synthesis of (+/-)-sesbanimide A and B. Journal of the Chemical Society, Chemical Communications, 1992, 4: 368-370.

[44] Zhang J H, Li C J. InCl$_3$-catalyzed domino reaction of aromatic amines with cyclic enol ethers in water: A highly efficient synthesis of new 1,2,3,4-tetrahydroquinoline derivatives. The Journal of Organic Chemistry, 2002, 67: 3969-3971.

[45] Kobayashi S. Sc(OTf)$_3$-Catalyzed three-component reactions of aldehydes, amines and allyltributylstannane in micellar systems. facile synthesis of homoallylic amines in water. Chemical Communications, 1998: 19-20.

[46] Krishnaveni N S, Surendra K, Kumar V P, Srinivas B, Suresh Reddy C, Rao K R. β-Cyclodextrin promoted allylation of aldehydes with allyltributyltin under supramolecular catalysis in water. Tetrahedron Letters, 2005, 46: 4299-4301.

[47] Li G L, Zhao G. Three-component synthesis of homoallylic amines promoted by carboxylic acids. Synthesis, 2006, 2006: 3189-3194.

[48] Xie Z, Li G, Zhao G, Wang J. Three-component synthesis of homoallylic amines catalyzed by phosphomolybdic Acid in water. Chinese Journal of Chemistry, 2009, 27: 925-929.

[49] Manabe K, Kobayashi S. Facile synthesis of α-amino phosphonates in water using a Lewis acid-surfactant-combined catalyst. Chemical Communications, 2000: 669-670.

[50] Ye Y, Ding Q, Wu J. Three-component reaction of 2-alkynylbenzaldehyde, amine, and nucleophile using Lewis acid-surfactant combined catalyst in water. Tetrahedron, 2008, 64: 1378-1382.

[51] Wang S, William R, Georgina Estelle Seah K K, Liu X W. Lewis acid-surfactant-combined catalyzed synthesis of 4-aminocyclopentenones from glycals in water. Green Chemistry, 2013, 15: 3180-3183.

[52] Amantini D, Fringuelli F, Pizzo F, Vaccaro L. Bromolysis and iodolysis of α,β-epoxycarboxylic acids in water catalyzed by indium halides. The Journal of Organic Chemistry, 2001, 66: 4463-4467.

[53] Fringuelli F, Pizzo F, Tortoioli S, Vaccaro L. Thiolysis of alkyl-and aryl-1,2-epoxides in water catalyzed by $InCl_3$. Advanced Synthesis & Catalysis, 2002, 344: 379-384.

[54] Procopio A, Gaspari M, Nardi M, Oliverio M, Rosati O. Highly efficient and versatile chemoselective addition of amines to epoxides in water catalyzed by erbium(Ⅲ) triflate. Tetrahedron Letters, 2008, 49: 2289-2293.

[55] Li C J, Wei C M. Highly efficient Grignard-type imine additions via C-H activation in water and under solvent-free conditions. Chemical Communications, 2002, 6: 268-269.

[56] Wei C M, Li C J. A highly efficient three-component coupling of aldehyde, alkyne, and amines via C-H activation catalyzed by gold in water. Journal of the American Chemical Society, 2003, 125: 9584-9585.

[57] Jiang N, Li C J. Novel 1,3-dipolar cycloaddition of diazocarbonyl compounds to alkynes catalyzed by $InCl_3$ in water. Chemical Communications, 2004, 35: 394-395.

[58] Khurana J M, Chaudhary A, Nand B, Lumb A. Aqua mediated indium(Ⅲ) chloride catalyzed synthesis of fused pyrimidines and pyrazoles. Tetrahedron Letters, 2012, 53: 3018-3022.

[59] Manabe K, Sun X M, Kobayashi S. Dehydration reactions in water. surfactant-type Brønsted acid-catalyzed direct esterification of carboxylic acids with alcohols in an emulsion system. Journal of the American Chemical Society, 2001, 123: 10101-10102.

[60] Manabe K, Iimura S, Sun X M, Kobayashi S. Dehydration reactions in water. Brønsted acid-surfactant-combined catalyst for ester, ether, thioether, and dithioacetal formation in water. Journal of the American Chemical Society, 2002, 124: 11971-11978.

[61] Aubele D L, Lee C A, Floreancig P E. The "aqueous" Prins reaction. Organic Letters, 2003, 5: 4521-4523.

[62] Peng L C, Lin L, Zhang J H, Zhuang J P, Zhang B X, Gong Y. Catalytic conversion of cellulose to levulinic acid by metal chlorides. Molecules, 2010, 15: 5258-5261.

[63] Rasrendra C B, Fachri B A, Makertihartha I G B N, Adisasmito S, Heeres H J. Catalytic conversion of dihydroxyacetone to lactic acid using metal salts in water. ChemSusChem, 2011, 4: 768-777.

[64] Shen G S, Zhou H F, Sui Y, Liu Q X, Zou K. $FeCl_3$-catalyzed tandem condensation/intramolecular nucleophilic addition/C-C bond cleavage: A concise synthesis of 2-substitued quinazolinones from 2-aminobenzamides and 1,3-diketones in aqueous media. Tetrahedron Letters, 2016, 57: 587-590.

[65] Jalil P A. Investigations on polyethylene degradation into fuel oil over tungstophosphoric acid supported on MCM-41 mesoporous silica. Journal of Analytical and Applied Pyrolysis, 2002, 65: 185-195.

[66] Zong Y, Zhao Y, Luo W, Yu X H, Wang J K, Pan Y. Highly efficient synthesis of 2,3-dihydroquinazolin-4(1H)-ones catalyzed by heteropoly acids in water. Chinese Chemical Letters, 2010, 21: 778-781.

[67] Chen X, She J, Shang Z, Wu J, Wu H, Zhang P. Synthesis of pyrazoles, diazepines, enaminones, and enamino esters using 12-tungstophosphoric acid as a reusable catalyst in water. Synthesis, 2009, 40: 3478-3486.

[68] Singh P, Kumar P, Katyal A, Kalra R, Dass S. K, Prakash S, Chandra R. Phosphotungstic acid: an efficient catalyst for the aqueous phase synthesis of bis-(4-hydroxycoumarin-3-yl)methanes. Catalysis Letters, 2010, 134: 303-308.

[69] Huang T K, Shi L, Wang R, Guo X Z, Lu X X. Keggin type heteropolyacids-catalyzed synthesis of quinoxaline derivatives in water. Chinese Chemical Letters, 2009, 20: 161-164.

[70] Heravi M M, Sadjadi S. Recent developments in use of heteropolyacids, their salts and polyoxometalates in organic synthesis. Journal of the Iranian Chemical Society, 2009, 6: 1-54.

[71] Heravi M M, Ghods A, Derikvand F, Bakhtiari K, Bamoharram F F. $H_{14}[NaP_5W_{30}O_{110}]$ catalyzed one-pot three-component synthesis of dihydropyrano[2,3-c] pyrazole and pyrano[2,3-d] pyrimidine derivatives. Journal of the Iranian Chemical Society, 2010, 7: 615-620.

[72] Li L, Xu L W, Ju Y D, Lai G Q. Asymmetric direct aldol reactions catalyzed by a simple chiral primary diamine-Brønsted acid catalyst in/on water. Synthetic Communications, 2009, 39: 764-774.

[73] Hasaninejad A, Zare A, Zolfigol M A, Shekouhy M. Zirconium Tetrakis(dodecyl sulfate)[Zr(DS)$_4$] as an efficient Lewis acid-aurfactant combined catalyst for the synthesis of quinoxaline derivatives in aqueous media. Synthetic Communications, 2009, 39: 569-579.

[74] Shirakawa S, Kobayashi S. Carboxylic acid catalyzed three-component aza-Friedel-Crafts reactions in water for the synthesis of 3-substituted indoles. Organic Letters, 2006, 8: 4939-4942.

[75] Halimehjani A Z, Farvardin M V, Zanussi H P, Ranjbari M A, Fattahi M. Friedel-Crafts alkylation of *N,N*-dialkylanilines with nitroalkenes catalyzed by heteropolyphosphotungstic acid in water. Journal of Molecular Catalysis A: Chemical, 2014, 381: 21-25.

[76] Manabe K, Iimura S, Sun X M, Kobayashi S. Dehydration reactions in water. Brønsted acid-surfactant-combined catalyst for ester, ether, thioether, and dithioacetal formation in water. Journal of the American Chemical Society, 2002, 124: 11971-11978.

[77] Azizi N, Arynasab F, Saidi M R. Efficient Friedel-Crafts alkylation of indoles and pyrrole with enones and nitroalkene in water. Organic & Biomolecular Chemistry, 2006, 4: 4275-4277.

[78] Ogawa C, Kobayashi S. Catalytic asymmetric C-C bond formation in aqueous medium. Comprehensive Inorganic Chemistry II, 2013, 6: 605-624.

[79] Xu X, De Almeida C P, Antal M J. Mechanism and kinetics of the acid-catalyzed formation of ethene and diethyl ether from ethanol in supercritical water. Industrial & Engineering Chemistry Research, 1991, 30: 1478-1485.

[80] Azizi N, Torkian L, Saidi M R. Highly efficient synthesis of *bis*(indolyl)methanes in water. Journal of Molecular Catalysis A: Chemical, 2007, 275: 109-112.

[81] Shirakawa S, Kobayashi S. Surfactant-type Brønsted acid catalyzed dehydrative nucleophilic substitutions of alcohols in water. Organic Letters, 2007, 9: 311-314.

[82] Manabe K, Kobayashi S. Mannich-type reactions of aldehydes, amines, and ketones in a colloidal dispersion system created by a Brønsted acid-surfactant-combined catalyst in water. Organic Letters, 1999, 1: 1965-1967.

[83] Manabe K, Mori Y, Kobayashi S. A Bronsted acid-surfactant-combined catalyst for Mannich-type reactions of aldehydes, amines, and silyl enolates in water. Synlett, 1999: 1401-1402.

[84] Akiyama T, Takaya J, Kagoshima H. Brønsted acid-catalyzed Mannich-type reactions in aqueous media. Advanced Synthesis & Catalysis, 2002, 344: 338-347.

[85] Akiyama T, Takaya J, Kagoshima H. HBF$_4$ catalyzed Mannich-type reaction in aqueous media. Synlett, 1999, (07): 1045-1048.

[86] Akiyama T, Takaya J, Kagoshima H. One-pot Mannich-type reaction in water: HBF$_4$ catalyzed condensation of aldehydes, amines, and silyl enolates for the ayntheisis of β-amino carbonyl compounds. Synlett, 1999, 1999: 1426-1428.

[87] Akiyama T, Takaya J, Kagoshima H. A highly stereo-divergent Mannich-type reaction catalyzed by Brønsted acid in aqueous media. Tetrahedron Letters, 2001, 42: 4025-4028.

[88] Akiyama T, Matsuda K, Fuchibe K. HCl-catalyzed stereoselective Mannich reaction in H$_2$O-SDS system. Synlett, 2005, 32: 322-324.

[89] Azizi N, Torkiyan L, Saidi M R. Highly efficient one-pot three-component Mannich reaction in water catalyzed by heteropoly acids. Organic Letters, 2006, 8: 2079-2082.

[90] Heravi M M, Bakhtiari K, Zadsirjan V, Bamoharram F F, Heravi O M. Aqua mediated synthesis of substituted 2-amino-4*H*-chromenes catalyzed by green and reusable Preyssler heteropolyacid. Bioorganic & Medicinal Chemistry Letters, 2007, 17: 4262-4265.

[91] Chen C, Zhu X Y, Wu Y, Sun H M, Zhang G F, Zhang Q, Gao Z W. 5-sulfosalicylic acid catalyzed direct Mannich reaction in pure water. Journal of Molecular Catalysis A: Chemical, 2014, 395: 124-127.

[92] Chang T, He L Q, Bian L, Han H Y, Yuan M X, Gao X R. Bronsted acid-surfactant-combined catalyst for the Mannich reaction in water. RSC Advances, 2014, 4: 727-731.

[93] Mukhopadhyay C, Datta A, Butcher R J. Highly efficient one-pot, three-component Mannich reaction catalysed by boric acid and glycerol in water with major syn-diastereoselectivity. Tetrahedron Letters, 2009, 50: 4246-4250.

[94] Javanshir S, Ohanian A, Heravi M M, Naimi-Jamal M R, Bamoharram F F. Ultrasound-promoted, rapid, green, one-pot synthesis of 2′-aminobenzothiazolomethylnaphthols via a multi-component reaction, catalyzed by heteropolyacid in aqueous media. Journal of Saudi Chemical Society, 2014, 18: 502-506.

[95] Akiyama T, Takaya J, Kagoshima H. Brønsted acid-catalyzed aza Diels-Alder reaction of Danishefsky's diene with aldimine generated in situ from aldehyde and amine in aqueous media. Tetrahedron Letters, 1999, 40: 7831-7834.

[96] Ai M. Oxidative dehydrogenation of isobutyric acid on V_2O_5-P_2O_5 catalysts. Journal of Catalysis, 1986, 98: 401-410.

[97] Azizi N, Saidi M R. Highly efficient ring opening reactions of epoxides with deactivated aromatic amines catalyzed by heteropoly acids in water. Tetrahedron, 2007, 63: 888-891.

[98] Shul'pina S L, Veghini D, Kudinov R A, Shul'pin B G. Oxidation of alcohols with hydrogen peroxide catalyzed by soluble iron and osmium derivatives. Reaction Kinetics and Catalysis Letters, 2006, 88: 157-163.

[99] Jiang H S, Mao X, Xie S J, Zhong B K. Partially reduced heteropoly compound catalysts for the selective oxidation of propane. Journal of Molecular Catalysis A: Chemical, 2002, 185: 143-149.

[100] Kolesnik I G, Zhizhina E G, Matveev K I. Catalytic oxidation of 2,6-dialkylphenols to the corresponding 2,6-dialkyl-1,4-benzoquinones by molecular oxygen in the presence of P-Mo-V heteropoly acids. Journal of Molecular Catalysis A: Chemical, 2000, 153: 147-154.

[101] Rosatella A A, Afonso C A M. Brønsted acid-catalyzed dihydroxylation of olefins in aqueous medium. Advanced Synthesis & Catalysis, 2011, 353: 2920-2926.

[102] Jin T S, Zhang J S, Guo T T, Wang A Q, Li T S. One-pot clean synthesis of 1,8-dioxo-decahydroacridines catalyzed by p-dodecylbenezenesulfonic acid in aqueous media. Synthesis, 2005, 36: 2001-2005.

[103] Alimohammadi K, Sarrafi Y, Tajbakhsh M. $H_6P_2W_{18}O_{62}$: A green and reusable catalyst for the synthesis of 3,3-diaryloxindole derivatives in water. Monatshefte für Chemie-Chemical Monthly, 2008, 139: 1037-1039.

[104] Khodabakhshi S, Karami B, Eskandari K, Rashidi A. A facile and practical p-toluenesulfonic acid catalyzed route to dicoumarols containing an aroyl group. South African Journal of Chemistry, 2015, 68: 53-56.

[105] Amrollahi M A, Kheilkordi Z. $H_3PW_{12}O_{40}$-catalyzed one-pot synthesis of *bis*(indole) derivatives under silent and ultrasonic irradiation conditions in aqueous media. Journal of the Iranian Chemical Society, 2016, 13: 925-929.

[106] Hekmatshoar R, Sajadi S, Sadjadi S, Heravi M M, Beheshtiha Y S, Bamoharram F F. Multifunctional catalysis of heteropoly acid: One-pot synthesis of quinolines from nitroarene and various aldehydes in the presence of hydrazine. Journal of the Chinese Chemical Society, 2008, 55: 1195-1198.

[107] Yadav J S, Subba Reddy B V, Narayana Kumar G G K S, Aravind S. The 'aqueous' prins cyclization: A diastereoselective synthesis of 4-hydroxytetrahydropyran derivatives. Synthesis, 2008, 39: 395-400.

[108] Oltmanns J U, Palkovits S, Palkovits R. Kinetic investigation of sorbitol and xylitol dehydration catalyzed by silicotungstic acid in water. Applied Catalysis A: General, 2013, 456: 168-173.

[109] Yang W J, Yang Z G, Xu L J, Zhang L L, Xu X, Miao M Z, Ren H J. Surfactant-type Brønsted acid catalyzed stereoselective synthesis of trans-3-alkenyl indazoles from triazenylaryl allylic alcohols in water. Angewandte Chemie International Edition, 2013, 52: 14135-14139.

[110] Ballini R, Bosica G. Nitroaldol reaction in aqueous media: An important improvement of the Henry reaction. The Journal of Organic Chemistry, 1997, 62: 425-427.

[111] Ayed T B, Amri H. A convenient synthesis of α-functional alkyl vinyl ketones. Synthetic Communications, 1995, 25: 3813-3819.

[112] Wang Z T, Xue H Q, Wang S C, Yuan C G. 'Green' synthesis of nitro alcohols catalyzed by solid-supported tributylammonium chloride in aqueous medium. Chemistry & Biodiversity, 2005, 2: 1195-1199.

[113] Mase N, Nakai Y, Ohara N, Yoda H, Takabe K, Tanaka F, Barbas C F. Organocatalytic direct asymmetric aldol reactions in water. Journal of the American Chemical Society, 2006, 128: 734-735.

[114] Shrikhande J J, Gawande M B, Jayaram R V. Cross-aldol and Knoevenagel condensation reactions in aqueous micellar media. Catalysis Communications, 2008, 9: 1010-1016.

[115] Wang G W, Zhang Z, Dong Y W. Environmentally friendly and efficient process for the preparation of β-hydroxyl ketones. Organic Process Research & Development, 2004, 8: 18-21.

[116] Sobhani S, Parizi Z P. An eco-friendly procedure for one-pot synthesis of β-phosphonomalonates: micellar solution of sodium stearate catalyzes tandem Knoevenagel-phospha-Michael reaction of aldehydes, malonitrile, and phosphites in aqueous media. Tetrahedron, 2011, 67: 3540-3545.

[117] Brahmachari G. Room temperature one-pot green synthesis of coumarin-3-carboxylic acids in water: A practical method for the large-scale synthesis. ACS Sustainable Chemistry & Engineering, 2015, 3: 2350-2358.

[118] Lu G P, Cai C. An odorless, one-pot synthesis of thioesters from organic halides, thiourea and benzoyl chlorides in water. Advanced Synthesis & Catalysis, 2013, 355: 1271-1276.

[119] Ahmed N, Brahmbhatt K G, Singh I P, Bhutani K K. Efficient chemoselective alkylation of quinoline 2,4-diol derivatives in water. Journal of Heterocyclic Chemistry, 2011, 48: 237-240.

[120] Fringuelli F, Pani G, Piermatti O, Pizzo F. Condensation reactions in water of active methylene compounds with arylaldehydes. One-pot synthesis of flavonols. Tetrahedron, 1994, 50: 11499-11508.

[121] Amarasekara A S, Singh T B, Larkin E, Hasan M A, Fan H J. NaOH catalyzed condensation reactions between levulinic acid and biomass derived furan-aldehydes in water. Industrial Crops and Products, 2015, 65: 546-549.

[122] Tamaddon F, Tayefi M, Hosseini E, Zare E. Dolomite ($CaMg(CO_3)_2$) as a recyclable natural catalyst in Henry, Knoevenagel, and Michael reactions. Journal of Molecular Catalysis A: Chemical, 2013, 366: 36-42.

[123] Iwamura M, Gotoh Y, Hashimoto T, Sakurai R. Michael addition reactions of acetoacetates and malonates with acrylates in water under strongly alkaline conditions. Tetrahedron Letters, 2005, 46: 6275-6277.

[124] Mudaliar C D, Nivalkar K R, Mashraqui S H. Michael reactions in aqueous micelles of cetyltrimethylammounium bromide. Organic Preparations and Procedures International, 1997, 29: 584-587.

[125] Miranda S, López-Alvarado P, Giorgi G, Rodriguez J, Avendaño C, Menéndez J C. Practical and user-friendly procedure for Michael reactions of α-nitroketones in water. Synlett, 2003, 2003: 2159-2162.

[126] Murata J, Tamura M, Sekiya A. Selective synthesis of fluorinated ethers by addition reaction of alcohols to fluorinated olefins in water. Green Chemistry, 2002, 4: 60-63.

[127] Shimizu S, Suzuki T, Shirakawa S, Sasaki Y, Hirai C. Water-soluble calixarenes as new inverse phase-transfer catalysts. Their scope in aqueous biphasic alkylations and mechanistic implications. Advanced Synthesis & Catalysis, 2002, 344: 370-378.

[128] Cherng Y J. Efficient nucleophilic substitution reaction of aryl halides with amino acids under focused microwave irradiation. Tetrahedron, 2000, 56: 8287-8289.

[129] Semple G, Skinner P J, Cherrier M C, Webb P J, Sage C R, Tamura S Y, Chen R, Richman J G, Connolly D T. 1-Alkyl-benzotriazole-5-carboxylic acids are highly selective agonists of the human orphan G-protein-coupled receptor GPR109b. Journal of Medicinal Chemistry, 2006, 49: 1227-1230.

[130] Degani I, Fochi R, Regondi V. S,S-Dialkyl dithiocarbonates as a convenient source of alkanethiolate anions in phase-transfer-catalysis systems: An improved synthesis of organic sulfides. Synthesis, 1983, 1983: 630-632.

[131] Peng Y Q, Song G H. Simultaneous microwave and ultrasound irradiation: A rapid synthesis of hydrazides. Green Chemistry, 2001, 3: 302-304.

[132] Peng Y Q, Song G H, Dou R L. Surface cleaning under combined microwave and ultrasound irradiation: Flash synthesis of 4H-pyrano[2,3-c] pyrazoles in aqueous media. Green Chemistry, 2006, 8: 573-575.

[133] Ju Y, Varma R S. Aqueous N-alkylation of amines using alkyl halides: direct generation of tertiary amines under microwave irradiation. Green Chemistry, 2004, 6: 219-221.

[134] Pironti V, Colonna S. Microwave-promoted synthesis of β-hydroxy sulfides and β-hydroxy sulfoxides in water. Green Chemistry, 2005, 7: 43-45.

[135] Wang S L, Wang X, Tu M S, Jiang B, Tu S J. Tandem one-pot synthesis of polysubstituted 1,3-thiazine derivatives. Chinese Journal of Chemistry, 2011, 29: 2411-2415.

[136] Kidwai M, Saxena S, Rahman Khan M K, Thukral S S. Aqua mediated synthesis of substituted 2-amino-4H-chromenes and in vitro study as antibacterial agents. Bioorganic & Medicinal Chemistry Letters, 2005, 15: 4295-4298.

[137] Quesada E, Taylor R J K. Tandem Horner-Wadsworth-Emmons olefination/Claisen rearrangement/hydrolysis sequence: Remarkable acceleration in water with mcrowave irradiation. Synthesis, 2005, 2005: 3193-3195.

[138] Ju Y, Varma R S. An efficient and simple aqueous N-heterocyclization of aniline derivatives: Microwave-assisted synthesis of N-aryl azacycloalkanes. Organic Letters, 2005, 7: 2409-2411.

[139] Ju Y, Varma R S. Aqueous N-heterocyclization of primary amines and hydrazines with dihalides: Microwave-Assisted syntheses of N-azacycloalkanes, isoindole, pyrazole, pyrazolidine, and phthalazine derivatives. The Journal of Organic Chemistry, 2006, 71: 135-141.

[140] Ju Y, Varma R S. Microwave-assisted cyclocondensation of hydrazine derivatives with alkyl dihalides or ditosylates in aqueous media: Syntheses of pyrazole, pyrazolidine and phthalazine derivatives. Tetrahedron Letters, 2005, 46: 6011-6014.

[141] Barnard T M, Vanier G S, Collins M J. Scale-up of the green synthesis of azacycloalkanes and isoindolines under microwave irradiation. Organic Process Research & Development, 2006, 10: 1233-1237.

[142] Schmitt M, Bourguignon J J, Wermuth C G. (E)-ethyl β-formylacrylate dimethylhydrazone methiodide: A reactive and convenient precursor of (E)-ethyl β-formylacrylate. Tetrahedron Letters, 1990, 31: 2145-2148.

[143] Russell M G, Warren S. Wittig reactions in water. synthesis of new water-soluble phosphonium salts and their reactions with substituted benzaldehydes. Tetrahedron Letters, 1998, 39: 7995-7998.

[144] Lu X, Li Z, Gao F. Base-catalyzed reactions in NH_3-enriched near-critical water. Industrial & Engineering Chemistry Research, 2006, 45: 4145-4149.

[145] Hao W J, Jiang B, Tu S J, Cao X D, Wu S S, Yan S, Zhang X H, Han Z G, Shi F. A new mild base-catalyzed Mannich reaction of hetero-arylamines in water: Highly efficient stereoselective synthesis of β-aminoketones under microwave heating. Organic & Biomolecular Chemistry, 2009, 7: 1410-1414.

[146] González-Cruz D, Tejedor D, deArmas P, García-Tellado F. Dual reactivity pattern of allenolates "on water": The chemical basis for efficient allenolate-driven organocatalytic systems. Chemistry-A European Journal, 2007, 13: 4823-4832.

[147] Gonzalez-Cruz D, Tejedor D, de Armas P, Morales E Q, Garcia-Tellado F. Organocatalysis "on water". Regioselective[3+2] -cycloaddition of nitrones and allenolates. Chemical Communications, 2006: 2798-2800.

[148] Zhang Y, Wei B W, Lin H, Zhang L, Liu J X, Luo H Q, Fan X L. "On water" direct catalytic vinylogous Henry (nitroaldol) reactions of isatins for the efficient synthesis of isoxazole substituted 3-hydroxyindolin-2-ones. Green Chemistry, 2015, 17: 3266-3270.

[149] Zhang Y, Wei B W, Zou L N, Kang M L, Luo H Q, Fan X L. "On water" direct vinylogous Henry (nitroaldol) reactions of 3,5-dimethyl-4-nitroisoxazole with aldehydes and trifluoromethyl ketones. Tetrahedron, 2016, 72: 2472-2475.

[150] Lu G P, Cai C. A facile, one-pot, green synthesis of polysubstituted 4H-pyrans via piperidine-catalyzed three-component condensation in aqueous medium. Journal of Heterocyclic Chemistry, 2011, 48: 124-128.

[151] Dickerson T J, Janda K D. Aqueous aldol catalysis by a nicotine metabolite. Journal of the American Chemical Society, 2002, 124: 3220-3221.

[152] Cordova A, Notz W, Barbas Ⅲ C F. Direct organocatalytic aldol reactions in buffered aqueous media. Chemical Communications, 2002, 24: 3024-3025.

[153] Peng Y Y, Ding Q P, Li Z, Wang P G, Cheng J P. Proline catalyzed aldol reactions in aqueous micelles: An environmentally friendly reaction system. Tetrahedron Letters, 2003, 44: 3871-3875.

[154] Ko K, Nakano K, Watanabe S, Ichikawa Y, Kotsuki H. Development of new DMAP-related organocatalysts for use in the Michael addition reaction of β-ketoesters in water. Tetrahedron Letters, 2009, 50: 4025-4029.

[155] Pande A, Ganesan K, Jain A K, Gupta P K, Malhotra R C. A novel eco-friendly process for the synthesis of 2-chlorobenzylidenemalononitrile and its analogues using water as a solvent. Organic Process Research & Development, 2005, 9: 133-136.

[156] Pang W, Zhu S F, Jiang H F, Zhu S Z. A novel synthesis of 5-perfluorophenyl 4,5-dihydro-1H-pyrazoles in THF or water. Journal of Fluorine Chemistry, 2007, 128: 1379-1384.

[157] Tullberg M, Grøtli M, Luthman K. Efficient synthesis of 2,5-diketopiperazines using microwave assisted heating. Tetrahedron, 2006, 62: 7484-7491.

[158] Jaiswal P K, Biswas S, Singh S, Samanta S. An organocatalytic highly efficient approach to the direct synthesis of substituted carbazoles in water. Organic & Biomolecular Chemistry, 2013, 11: 8410-8418.

[159] Heravi M M, Mousavizadeh F, Ghobadi N, Tajbakhsh M. A green and convenient protocol for the synthesis of novel pyrazolopyranopyrimidines via a one-pot, four-component reaction in water. Tetrahedron Letters, 2014, 55: 1226-1228.

[160] Mosslemin M H, Hzarenezhad E, Shams N, Rad M N S, Anaraki-Ardakani, H, Fayazipoor R. Green synthesis of 5-aryl-(1H,3H,5H,10H)-pyrimido[4,5-b] quinoline-2,4-diones catalysed by 1,4-diazabicyclo[2.2.2] octane in water. Journal of Chemical Research, 2014, 38: 169-171.

[161] Khurana J M, Nand B, Saluja P. 1,8-Diazabicyclo[5.4.0] undec-7-ene, a highly efficient catalyst for one-pot synthesis of substituted tetrahydro-4H-chromenes, tetrahydro[b] pyrans, pyrano[d] pyrimidines, and 4H-pyrans in aqueous medium. Journal of Heterocyclic Chemistry, 2014, 51: 618-624.

[162] Zakeri M, Nasef M M, Abouzari-Lotf E, Moharami A, Heravi M M. Sustainable alternative protocols for the multicomponent synthesis of spiro-4H-pyrans catalyzed by 4-dimethylaminopyridine. Journal of Industrial and Engineering Chemistry, 2015, 29: 273-281.

[163] Rimaz M, Aali F. An environmentally-friendly base organocatalyzed one-pot strategy for the regioselective synthesis of novel 3,6-diaryl-4-methylpyridazines. Chinese Journal of Catalysis, 2016, 37: 517-525.

[164] Biswas S, Majee D, Dagar A, Samanta S. Metal-free one-pot protocol for the preparation of unexpected carbazole derivative in water: Easy access to pyrimidocarbazole and pyridocarbazolone derivatives. Synlett, 2014, 25: 2115-2120.

[165] Biswas S, Dagar A, Srivastava A, Samanta S. Access to substituted carbazoles in water by a one-pot sequential reaction of α,β-substituted nitro olefins with 2-(3-formyl-1H-indol-2-yl)acetates. European Journal of Organic Chemistry, 2015, 46: 4493-4503.

[166] Cho S H, Kim J Y, Kwak J, Chang S. Recent advances in the transition metal-catalyzed two fold oxidative C-H bond activation strategy for C-C and C-N bond formation. Chemical Society Reviews, 2011, 40: 5068-5083.

[167] Lindström U M. Stereoselective organic reactions in water. Chemical Reviews, 2002, 102: 2751-2772.

[168] Brink G J T, Arends I W C E, Sheldon R A. Green, catalytic oxidation of alcohols in water. Science, 2000, 287: 1636-1639.

[169] Casalnuovo A L, Calabrese J C. Palladium-catalyzed alkylations in aqueous media. Journal of the American Chemical Society, 1990, 112: 4324-4330.

[170] Kostas I D, Coutsolelos A G, Charalambidis G, Skondra A. The first use of porphyrins as catalysts in cross-coupling reactions: A water-soluble palladium complex with a porphyrin ligand as an efficient catalyst precursor for the Suzuki-Miyaura reaction in aqueous media under aerobic conditions. Tetrahedron Letters, 2007, 48: 6688-6691.

[171] Zhang J, Yang F, Ren G, Mak T C W, Song M, Wu Y. Ultrasonic irradiation accelerated cyclopalladated ferrocenylimines catalyzed Suzuki reaction in neat water. Ultrasonics Sonochemistry, 2008, 15: 115-118.

[172] Li H, Wu Y. Water-soluble 2-arylnaphthoxazole-derived palladium (II) complexes as phosphine-free catalysts for the Suzuki reaction in aqueous solvent. Applied Organometallic Chemistry, 2008, 22: 233-236.

[173] Bai S Q, Hor T S A. Isolation of an[SNS] Pd(II) pincer with a water ladder and its Suzuki coupling activity in water. Chemical Communications, 2008: 3172-3174.

[174] Zhou J, Li X Y, Sun H J, Zhou J, Li X Y. An efficient and recyclable water. Journal of Organometallic Chemistry, 2010, 695: 297-303.

[175] Güelcemal S, Kani I, Yilmaz F, Cetinkaya B. N,N,N',N'-tetrakis(2-hydroxyethyl)ethylenediamine palladium(Ⅱ) complex as efficient catalyst for the Suzuki/Miyaura reaction in water. Tetrahedron, 2010, 66: 5602-5606.

[176] Conelly-Espinosa P, Morales-Morales D.[Pd(HQS)$_2$] (HQS=8-hydroxyquinoline-5-sulfonic acid) a highly efficient catalyst for Suzuki-Miyaura cross couplings in water. Inorganica Chimica Acta, 2010, 363: 1311-1315.

[177] Tu T, Feng X K, Wang Z X, Liu X Y. A robust hydrophilic pyridine-bridged *bis*-benzimidazolylidene palladium pincer complex: Synthesis and its catalytic application towards Suzuki-Miyaura couplings in aqueous solvents. Dalton Transactions, 2010, 39: 10598-10600.

[178] Tang Y Q, Lv H, Lu J M, Shao L X. Palladium(Ⅱ)-*N*-heterocyclic carbene complex derived from proline towards Suzuki-Miyaura coupling reaction in water at room temperature. Journal of Organometallic Chemistry, 2011, 696: 2576-2579.

[179] Banik B, Tairai A, Shahnaz N, Das P. Palladium(Ⅱ) complex with a potential N$_4$-type schiff-base ligand as highly efficient catalyst for Suzuki-Miyaura reactions in aqueous media. Tetrahedron Letters, 2012, 53: 5627-5630.

[180] Hanhan M E, Cetinkaya C, Shaver M P. Effective binuclear Pd(Ⅱ) complexes for Suzuki reactions in water. Applied Organometallic Chemistry, 2013, 27: 570-577.

[181] Zhang G F, Luan Y X, Han X W, Wang Y, Wen X, Ding C R. Pd(l-proline)$_2$ complex: An efficient catalyst for Suzuki-Miyaura coupling reaction in neat water. Applied Organometallic Chemistry, 2014, 28: 332-336.

[182] Charbonneau M, Addoumieh G, Oguadinma P, Schmitzer A R. Support-free palladium-NHC catalyst for highly recyclable heterogeneous Suzuki-Miyaura coupling in neat water. Organometallics, 2014, 33: 6544-6549.

[183] Dewan A, Bora U, Borah G. A simple and efficient tetradentate Schiff base derived palladium complex for Suzuki-Miyaura reaction in water. Tetrahedron Letters, 2014, 55: 1689-1692.

[184] Gao H, Xu C, Li H M, Lou X H, Wang Z Q, Fu W J. Dinuclear triphenylphosphine-cyclopalladated 2,4-dichloropyrimidine complex with a Pd-Pd bond: Synthesis, crystal dstructure, and application in Suzuki reaction of 3-(hydroxymethyl)phenylboronic acid in water. Bulletin of the Korean Chemical Society, 2015, 36: 2355-2358.

[185] Tsang M Y, Viñas, Teixidor F, Planas J G, Conde N, San Martin R, Herrero M T, Domínguez E, Lledós A, Vidossich P, Choquesillo-Lazarte D. Synthesis, structure, and catalytic applications for ortho-and meta-carboranyl based NBN pincer-Pd complexes. Inorganic Chemistry, 2014, 53: 9284-9295.

[186] Hanhan M E, Senemoglu Y. Microwave-assisted aqueous Suzuki coupling reactions catalyzed by ionic palladium(Ⅱ) complexes. Transition Metal Chemistry, 2012, 37: 109-116.

[187] Li Q, Zhang L M, Bao J J, Li H X, Xie J B, Lang J P. Suzuki-Miyaura reactions promoted by a PdCl$_2$/sulfonate-tagged phenanthroline precatalyst in water. Applied Organometallic Chemistry, 2014, 28: 861-867.

[188] Kapdi A, Gayakhe V, Sanghvi Y S, Garcia J, Lozano P, da Silva I, Perez J, Serrano J L. New water soluble Pd-imidate complexes as highly efficient catalysts for the synthesis of C$_5$-arylated pyrimidine nucleosides. RSC Advances, 2014, 4: 17567-17572.

[189] Zhang G F, Zhang W, Luan Y X, Han X W, Ding C R. PEG cick-triazole palladacycle: An efficient precatalyst for palladium-catalyzed Suzuki-Miyaura and copper-free Sonogashira reactions in neat water. Chinese Journal of Chemistry, 2015, 33: 705-710.

[190] Inés B, SanMartin R, Churruca F, Domínguez E, Urtiaga M K, Arriortua M I. A nonsymmetric pincer-type palladium catalyst in Suzuki, Sonogashira, and Hiyama couplings in neat water. Organometallics, 2008, 27: 2833-2839.

[191] Zhao X M, Hao X Q, Wang K L, Liu J R, Song M P, Wu Y J. Catalysis of the Suzuki-Miyaura coupling reaction in water by heteroannular cyclopalladated ferrocenylimine complexes. Transition Metal Chemistry, 2009, 34: 683-688.

[192] Marziale A N, Faul S H, Reiner T, Schneider S, Eppinger J. Facile palladium catalyzed Suzuki-Miyaura coupling in air and water at ambient temperature. Green Chemistry, 2010, 12: 35-38.

[193] Marziale A N, Jantke D, Faul S H, Reiner T, Herdtweck E, Eppinger J. An efficient protocol for the palladium-catalysed Suzuki-Miyaura cross-coupling. Green Chemistry, 2011, 13: 169-177.

[194] Qu G R, Xin P Y, Niu H Y, Jin X, Guo X T, Yang X N, Guo H M. Microwave promoted palladium-catalyzed Suzuki-Miyaura cross-coupling reactions of 6-chloropurines with sodium tetraarylborate in water. Tetrahedron, 2011, 67: 9099-9103.

[195] Firouzabadi H, Iranpoor N, Gholinejad M. 2-Aminophenyl diphenylphosphinite as a new ligand for heterogeneous palladium-catalyzed Heck-Mizoroki reactions in water in the absence of any organic co-solvent. Tetrahedron, 2009, 65: 7079-7084.

[196] Sabounchei S J, Hosseinzadeh M, Panahimehr M, Nematollahi D, Khavasi H R, Khazalpour S. A palladium-phosphine catalytic system as an active and recyclable precatalyst for Suzuki coupling in water. Transition Metal Chemistry, 2015, 40: 657-663.

[197] Godoy F, Segarra C, Poyatos M, Peris E. Palladium catalysts with sulfonate-functionalized-NHC ligands for Suzuki-Miyaura cross-coupling reactions in water. Organometallics, 2011, 30: 684-688.

[198] Zhong R, Pothig A, Feng Y K, Riener K, Herrmann W A, Kuhn F E. Facile-prepared sulfonated water-soluble PEPPSI-Pd-NHC catalysts for aerobic aqueous Suzuki-Miyaura cross-coupling reactions. Green Chemistry, 2014, 16: 4955-4962.

[199] Li L Y, Wang J Y, Zhou C S, Wang R H, Hong M C. pH-Responsive chelating N-heterocyclic dicarbene palladium(II) complexes: Recoverable precatalysts for Suzuki-Miyaura reaction in pure water. Green Chemistry, 2011, 13: 2071-2077.

[200] Tan J J, Cui J L, Ding G Q, Deng T S, Zhu Y L, Li Y W. Efficient aqueous hydrogenation of levulinic acid to γ-valerolactone over a highly active and stable ruthenium catalys. Catalysis Science & Technology, 2016, 6: 1469-1475.

[201] Kinzhalov M A, Luzyanin K V, Boyarskiy V P, Haukka M, Kukushkin V Y. ADC-based palladium catalysts for aqueous Suzuki-Miyaura cross-coupling exhibit greater activity than the most advantageous catalytic systems. Organometallics, 2013, 32: 5212-5223.

[202] Vlassa M, Ciocan-Tarta I, Mărgineanu F, Oprean I. Synthesis of diynes by phase transfer catalysis in the presence of a Pd(0) catalyst. Tetrahedron, 1996, 52: 1337-1342.

[203] Bhattacharya S, Sengupta S. Palladium catalyzed alkynylation of aryl halides (Sonogashira reaction) in water. Tetrahedron Letters, 2004, 45: 8733-8736.

[204] Kamali T A, Bakherad M, Nasrollahzadeh M, Farhangi S, Habibi D. Synthesis of 6-substituted imidazo[2,1-b] thiazoles via Pd/Cu-mediated Sonogashira coupling in water. Tetrahedron Letters, 2009, 50: 5459-5462.

[205] Luo F T, Lo H K. Short synthesis of *bis*-NHC-Pd catalyst derived from caffeine and its applications to Suzuki, Heck, and Sonogashira reactions in aqueous solution. Journal of Organometallic Chemistry, 2011, 696: 1262-1265.

[206] Ahammed S, Dey R, Ranu B C. Palladium and copper catalyzed one-pot Sonogashira reaction of 2-nitroiodobenzenes with aryl acetylenes and subsequent regioselective hydration in water: synthesis of 2-(2-nitrophenyl)-1-aryl ethanones. Tetrahedron Letters, 2013, 54: 3697-3701.

[207] Park K, Bae G, Park A, Kim Y, Choe J, Song K H, Lee S. Synthesis of symmetrical diarylalkyne from palladium-catalyzed decarboxylative couplings of propiolic acid and aryl bromides under water. Tetrahedron Letters, 2011, 52: 576-580.

[208] Lipshutz B H, Chung D W, Rich B. Sonogashira couplings of aryl bromides: Room temperature, water only, no copper. Organic Letters, 2008, 10: 3793-3796.

[209] Bakherad M, Keivanloo A, Bahramian B, Hashemi M. Copper-free Sonogashira coupling reactions catalyzed by a water-soluble Pd-salen complex under aerobic conditions. Tetrahedron Letters, 2009, 50: 1557-1559.

[210] Wang X Z, Zhang J, Wang Y Y, Liu Y. Amphiphilic ionic palladium complexes for aqueous-organic biphasic Sonogashira reactions under aerobic and CuI-free conditions. Catalysis Communications, 2013, 40: 23-26.

[211] Zhang C Y, Liu J H, Xia C G. Easily prepared water soluble Pd-NHC complex as an efficient, phosphine-free palladium catalyst for the Sonogashira reaction. Chinese Journal of Catalysis, 2015, 36: 1387-1390.

[212] Sarkar S, Pal R, Chatterjee N, Dutta S, Naskar S, Sen A K. A green approach for highly regioselective syntheses of furo[3,2-h] quinolines in aqueous medium. Tetrahedron Letters, 2013, 54: 3805-3809.

[213] Zhou R, Wang W, Jiang Z J, Fu H Y, Zheng X L, Zhang C C, Chen H, Li R X. Pd/tetraphosphine catalytic system for Cu-free Sonogashira reaction "on water". Catalysis Science & Technology, 2014, 4: 746-751.

[214] Voronova K, Homolya L, Udvardy A, Benyei A C, Joo F. Pd-tetrahydrosalan-type complexes as catalysts for Sonogashira couplings in water: Efficient greening of the procedure. ChemSusChem, 2014, 7: 2230-2239.

[215] Gazvoda M, Virant M, Pevec A, Urankar D, Bolje A, Kocevar M, Kosmrlj J. Mesoionic *bis*(Py-tzNHC) palladium(II) complex catalyses "green" Sonogashira reaction through an unprecedented mechanism. Chemical Communication, 2016, 52: 1571-1574.

[216] Manabe K, Kobayashi S. Palladium-catalyzed, carboxylic acid-assisted allylic substitution of carbon nucleophiles with allyl alcohols as allylating agents in water. Organic Letters, 2003, 5: 3241-3244.

[217] Chen L, Li C J. A remarkably efficient coupling of acid chlorides with alkynes in water. Organic Letters, 2004, 6: 3151-3153.

[218] Wolf C, Lerebours R. Palladium-phosphinous acid-catalyzed NaOH-promoted cross-coupling reactions of arylsiloxanes with aryl chlorides and bromides in water. Organic Letters, 2004, 6: 1147-1150.

[219] Lerebours R, Wolf C. Palladium(II)-catalyzed conjugate addition of arylsiloxanes in water. Organic Letters, 2007, 9: 2737-2740.

[220] Arkady K, Christophe D, Lipshutz B H. Zn-mediated, Pd-catalyzed cross-couplings in water at room temperature without prior formation of organozinc reagents. Journal of the American Chemical Society, 2009, 131: 15592-11593.

[221] Nishikata T, Lipshutz B H. Cationic Pd(II)-catalyzed Fujiwara-Moritani reactions at room temperature in water. Organic Letters, 2010, 12: 1972-1975.

[222] Tang Y Q, Chu C Y, Zhu L, Qian B, Shao L X. N-Heterocyclic carbene-Pd(II) complex derived from proline for the Mizoroki-Heck reaction in water. Tetrahedron, 2011, 67: 9479-9483.

[223] Yang S C, Hsu Y C, Gan K H. Direct palladium/carboxylic acid-catalyzed allylation of anilines with allylic alcohols in water. Tetrahedron, 2006, 62: 3949-3958.

[224] Shen X B, Gao T T, Lu J M, Shao L X. Imidazole-coordinated monodentate NHC-Pd(II) complex derived from proline and its application to the coupling reaction of arylboronic acids with carboxylic acid anhydrides in water at room temperature. Applied Organometallic Chemistry, 2011, 25: 497-501.

[225] Lin X F, Li Y, Li S Y, Xiao Z K, Lu J M. NHC-Pd(II)-Im (NHC = N-heterocyclic carbene, Im = 1-methylimidazole) complex catalyzed coupling reaction of arylboronic acids with carboxylic acid anhydrides in water. Tetrahedron, 2012, 68: 5806-5809.

[226] Zhang Y, Yin S C, Lu J M. *N*-Heterocyclic carbene-palladium(II)-1-methylimidazole complex catalyzed allyl-aryl coupling of allylic alcohols with arylboronic acids in neat water. Tetrahedron, 2015, 71: 544-549.

[227] Joó F, Bényei A. A biphasic reduction of unsaturated aldehydes to saturated alcohols by ruthenium complex-catalyzed hydrogen transfer. Journal of Organometallic Chemistry, 1989, 363: C19-C21.

[228] Darensbourg D J, Joo F, Kannisto M, Katho A, Reibenspies J H, Daigle D J. Water-soluble organometallic compounds: Catalytic-hydrogenation of aldehydes in an aqueous 2-phase solvent system using a 1,3,5-triaza-7-phosphaadamantane complex of ruthenium. Inorganic Chemistry, 1994, 33: 200-208.

[229] Jérôme C, Gael L, Helen S E, Georg S F. Water-soluble arene ruthenium complexes containing a trans-1,2-diaminocyclohexane ligand as enantioselective transfer hydrogenation catalysts in aqueous solution. European Journal of Inorganic Chemistry, 2005, 2005: 4493-4500.

[230] Wu X F, Li X G, Hems W, King F, Xiao J L. Accelerated asymmetric transfer hydrogenation of aromatic ketones in water. Organic & Biomolecular Chemistry, 2004, 2: 1818-1821.

[231] Wu X F, Li X H, McConville M, Saidi O, Xiao J L. *β*-Amino alcohols as ligands for asymmetric transfer hydrogenation of ketones in water. Journal of Molecular Catalysis A-Chemical, 2006, 247: 153-158.

[232] Zeror S, Collin J, Fiaud J C, Zouioueche L A. A recyclable multi-substrates catalytic system for enantioselective reduction of ketones in water. Journal of Molecular Catalysis A: Chemical, 2006, 256: 85-89.

[233] Aliende C, Perez-Manrique M, Jalon F A, Manzano B R, Rodriguez A M, Espino G A. Ruthenium complexes as versatile catalysts in water in both transfer hydrogenation of ketones and oxidation of alcohols. Selective deuterium labeling of rac-1-phenylethanol. Organometallics, 2012, 31: 6106-6123.

[234] Rossin A, Kovács G, Ujaque G, Lledós A, Joó F. The active role of the water solvent in the rregioselective CO hydrogenation of unsaturated aldehydes by[RuH$_2$(mtppms)$_x$] in basic media. Organometallics, 2006, 25: 5010-5023.

[235] Zeror S, Collin J, Fiaud J C, Zouioueche L A. Evaluation of ligands for ketone reduction by asymmetric hydride transfer in water by multi-substrate screening. Advanced Synthesis & Catalysis, 2008, 350: 197-204.

[236] Mao J, Guo J. Chiral amino amides for the ruthenium(II)-catalyzed asymmetric transfer hydrogenation reaction of ketones in water. Chirality, 2009, 22: 173-181.

[237] Ma Y P, Liu H, Chen L, Cui X, Zhu J, Deng J G. Asymmetric transfer hydrogenation of prochiral ketones in aqueous media with new water-soluble chiral vicinal diamine as ligand. Organic Letters, 2003, 5: 2103-2106.

[238] Zhou H F, Fan Q H, Huang Y Y, Wu L, He Y M, Tang W J, Gu L Q, Chan A S C. Mixture of poly(ethylene glycol) and water as environmentally friendly media for efficient enantioselective transfer hydrogenation and catalyst recycling. Journal of Molecular Catalysis A: Chemical, 2007, 275: 47-53.

[239] Lin G Q, Xu M H, Zhong Y W, Sun X W. An advance on exploring N-tert-butanesulfinyl imines in asymmetric synthesis of chiral amines. Accounts of Chemical Research, 2008, 41: 831-840.

[240] Zhang B, Xu M H, Lin G Q. Catalytic enantioselective synthesis of chiral phthalides by efficient reductive cyclization of 2-acylarylcarboxylates under aqueous transfer hydrogenation conditions. Organic Letters, 2009, 11: 4712-4715.

[241] Wang J B, Qin R X, Wei X, Jia Y, Liu D R, Feng J, Chen H. Asymmetric hydrogenation of benzalacetone catalyzed by RuCl$_2$(TPPTS)$_2$ -(S,S)-DPENDS in water medium. Chinese Journal of Catalysis, 2010, 31: 273-277.

[242] Zhou Z Q, Ma Q, Zhang A Q, Wu L M. Synthesis of water-soluble monotosylated ethylenediamines and their application in ruthenium and iridium-catalyzed transfer hydrogenation of aldehydes. Applied Organometallic Chemistry, 2011, 25: 856-861.

[243] Qin R X, Wang J B, Wei X, Liu D R, Jian F, Hua C. Asymmetric hydrogenation of aromatic ketones catalyzed by achiral monophosphine TPPTS-stabilized Ru in PEG-water. Chinese Journal of Catalysis, 2011, 18: 1643-1647.

[244] Huang C Y, Kuan K Y, Liu Y H, Peng S M, Liu S T. Synthesis and catalytic activity of water-soluble ruthenium(II) complexes bearing a naphthyridine-carboxylate ligand. Organometallics, 2014, 33: 2831-2836.

[245] Prakash O, Joshi H, Sharma K N, Gupta P L, Singh A K. Transfer hydrogenation (pH Independent) of ketones and aldehydes in water with glycerol: Ru, Rh, and Ir catalysts with a COOH group near the metal on a (phenylthio)methyl-2-pyridine scaffold. Organometallics, 2014, 33: 3804-3812.

[246] Soni R, Hall T H, Mitchell B P, Owen M R, Wills M. Asymmetric reduction of electron-rich ketones with tethered Ru(II)/TsDPEN catalysts using formic acid/triethylamine or aqueous sodium formate. The Journal of Organic Chemistry, 2015, 80: 6784-6793.

[247] Wu J S, Fei W, Ma Y P, Xin C, Cun L F, Jin Z, Deng J G, Yu B L. Asymmetric transfer hydrogenation of imines and iminiums catalyzed by a water-soluble catalyst in water. Chemical Communications, 2006, 37: 1766-1768.

[248] Li C J, Wang D, Chen D L. Reshuffling of functionalities catalyzed by a ruthenium complex in water. Journal of the American Chemical Society, 1995, 117: 12867-12868.

[249] Wang M, Li C J. Aldol reaction via in situ olefin migration in water. Tetrahedron Letters, 2002, 43: 3589-3591.

[250] Chen X G, Peng X, Sung H H Y, Williams I D, Peruzzini M, Bianchini C, Jia G C. Ruthenium-promoted Z-selective head-to-head dimerization of terminal alkynes in organic and aqueous media. Organometallics, 2005, 24: 4330-4332.

[251] Arockiam P B, Cédric, Christian B, Dixneuf P H. C-H bond functionalization in water catalyzed by carboxylato ruthenium(II) systems. Angewandte Chemie International Edition, 2010, 49: 6629-6632.

[252] Ackermann L, Pospech J, Potukuchi H K. Well-defined ruthenium(II) carboxylate as catalyst for direct C-H/C-O bond arylations with phenols in water. Organic Letters, 2012, 43: 2146-2149.

[253] Singh K S. Direct C-H bond arylation in water promoted by (O,O)-and (O,N)-chelate ruthenium(II) catalysts. Chemcatchem, 2013, 5: 1313-1316.

[254] Tlili A, Schranck J, Pospech J, Neumann H, Beller M. Ruthenium-catalyzed carbonylative C-C coupling in water by directed C-H bond activation. Angewandte Chemie International Edition, 2013, 52: 6293-6297.

[255] Tlili A, Schranck J, Pospech J, Neumann H, Beller M. Ruthenium-catalyzed hydroaroylation of styrenes in water through directed C-H bond activation. ChemCatChem, 2014, 45: 1562-1566.

[256] Hong Z, Tinli Z, Tao Y, Cai M Z. Recyclable and reusable[RuCl$_2$(p-cymene)]$_2$/Cu(OAc)$_2$/PEG-400/H$_2$O system for oxidative C-H bond alkenylations: Green synthesis of phthalides. Journal of Organic Chemistry, 2015, 80: 8849-8855.

[257] McGrath D V, Grubbs R H. The mechanism of aqueous ruthenium (Ⅱ)-catalyzed olefin isomerization. Organometallics, 1994, 13: 224-235.

[258] Cadierno V, García-Garrido S E, Gimeno J, Varela-Álvarez A, Sordo J A. Bis(allyl)-ruthenium(Ⅳ) complexes as highly efficient catalysts for the redox isomerization of allylic alcohols into carbonyl compounds in organic and aqueous media: Scope, limitations, and theoretical analysis of the mechanism. Journal of the American Chemical Society, 2006, 128: 1360-1370.

[259] González B, Lorenzo-Luis P, Serrano-Ruiz M, Éva P, Fekete M, Csépke K, Ősz K, Ágnes K, Joó F, Romerosa A. Catalysis of redox isomerization of allylic alcohols by[RuClCp(mPTA)$_2$] (OSO$_2$CF$_3$)$_2$ and[RuCp(mPTA)$_2$(OH^2-κO)] (OSO$_2$CF$_3$)$_3$·(H$_2$O)(C$_4$H$_{10}$O)$_{0.5}$ unusual influence of the pH and interaction of phosphate with catalyst on the reaction rat. Journal of Molecular Catalysis A: Chemical, 2010, 326: 15-20.

[260] García-Álvarez J, Gimeno J, Suárez F J. Redox isomerization of allylic alcohols into carbonyl compounds batalyzed by the ruthenium(Ⅳ) complex[Ru(η^3:η^3-C$_{10}$H$_{16}$)Cl(κ$_2$O,O-CH$_3$CO$_2$)] in water and ionic liquids: Highly efficient transformations and catalyst recycling. Organometallics, 2011, 30: 2893-2896.

[261] Lastra-Barreira B, Francos J, Crochet P, Cadierno V. Ruthenium(Ⅳ) catalysts for the selective estragole to trans-anethole isomerization in environmentally friendly media. Green Chemistry, 2011, 13: 307-313.

[262] Udvardy A, Bényei A C, Ágnes K. The dual role of cis-[RuCl$_2$(DMSO)$_4$] in the synthesis of new water-soluble Ru(Ⅱ)-phosphane complexes and in the catalysis of redox isomerization of allylic alcohols in aqueous-organic biphasic systems. Journal of Organometallic Chemistry, 2012, 717: 116-122.

[263] Serrano-Ruiz M, Lorenzo-Luis P, Romerosa A, Mena-Cruz A. Catalytic isomerization of allylic alcohols in water by [RuClCp(PTA)$_2$],[RuClCp(HPTA)$_2$]Cl$_2$·2H$_2$O,[RuCp(DMSO-γS)(PTA)$_2$]Cl,[RuCp(DMSO-γS)(PTA)$_2$](OSO$_2$CF$_3$) and[RuCp(DMSO-γS)(HPTA)$_2$]Cl$_3$·2H$_2$O. Dalton Transactions, 2013, 42: 7622-7630.

[264] Serrano-Ruiz M, Aguilera-Sáez L M, Lorenzo-Luis P, Padrón J M, Romerosa A. Synthesis and antiproliferative activity of the heterobimetallic complexes [RuClCp(PPh$_3$)-μ-dmoPTA-1κP:2κ^{2}N,N'-MCl$_2$] (M = Co, Ni, Zn; dmoPTA = 3,7-dimethyl-1,3,7-triaza-5-phosphabicyclo[3.3.1] nonane). Dalton Transactions, 2013, 42: 11212-11219.

[265] Mena-Cruz A, Serrano-Ruiz M, Lorenzo-Luis P, Romerosa A, KathóÁ, Joó F, Aguilera-Sáez L M. Evaluation of catalytic activity of[RuClCp(dmoPTA)(PPh$_3$)] (OSO$_2$CF$_3$) in the isomerization of allylic alcohols in water (dmoPTA = 3,7-dimethyl-1,3,7-triaza-5-phosphabicyclo[3.3.1] nonane). Journal of Molecular Catalysis A: Chemical, 2016, 411: 27-33.

[266] Nicolas I, Maux P L, Simonneaux G. Synthesis of chiral water-soluble metalloporphyrins (Fe, Ru,): New catalysts for asymmetric carbene transfer in water. Tetrahedron Letters, 2008, 49: 5793-5795.

[267] Lynn D M, Mohr B, Grubbs R H, And L M H, Day M W. Water-soluble ruthenium alkylidenes: Synthesis, characterization, and application to olefin metathesis in protic solvents. Journal of the American Chemical Society, 2000, 122: 6601-6609.

[268] Kirkland T A, And D M L, Grubbs R H. Ring-closing metathesis in methanol and water. The Journal of Organic Chemistry, 1998, 63: 9904-9909.

[269] Khan M M T, Halligudi S B, Shukla S. Hydration of acetylene to acetaldehyde using K[RuⅢ(EDTA-H)Cl]$_2$·H$_2$O. Journal of Molecular Catalysis A: Chemistry, 1990, 58: 299-305.

[270] Sears J M, Lee W C, Frost B J. Water soluble diphosphine ligands based on 1,3,5-triaza-7-phosphaadamantane (PTA-PR2): Synthesis, coordination chemistry, and ruthenium catalyzed nitrile hydration. Inorganica Chimica Acta, 2015, 255: 248-257.

[271] Garcia-Alvarez R, Zablocka M, Crochet P, Duhayon C, Majoral J P, Cadierno V. Thiazolyl-phosphine hydrochloride salts: effective auxiliary ligands for ruthenium-catalyzed nitrile hydration reactions and related amide bond forming processes in water. Green Chemistry, 2013, 15: 2447-2456.

[272] Tomas-Mendivil E, Suarez F J, Diez J, Cadierno V. An efficient ruthenium(IV) catalyst for the selective hydration of nitriles to amides in water under mild conditions. Chemical Communications, 2014, 50: 9661-9664.

[273] Kobayashi S, Ogawa C. New entries to water-compatible Lewis acids. Chemistry-A European Journal, 2006, 12: 5954-5960.

[274] Cornils B, Herrmann W A. Aqueous-phase organometallic catalysis. Chemie Ingenieur Technik, 1999, 71: 168-169.

[275] Tóth I, Hanson B E. Novel chiral water soluble phosphines I. Preparation and characterization of amine functionalized DIOP, Chiraphos, and BDPP derivatives and quaternization of their rhodium complexes. Tetrahedron: Asymmetry, 1990, 1: 895-912.

[276] Wu X F, Li X H, McConville M, Saidi O, Xiao J L. β-Amino alcohols as ligands for asymmetric transfer hydrogenation of ketones in water. Journal of Molecular Catalysis A: Chemical, 2006, 247: 153-158.

[277] Wu X F, Li X H, Zanotti-Gerosa A, Pettman A, Liu J K, Mills A J, Xiao J L. RhIII-and IrIII-catalyzed asymmetric transfer hydrogenation of ketones in water. Chemistry-A European Journal, 2008, 14: 2209-2222.

[278] Tang L, Lin Z C, Wang Q, Wang X B, Cun L F, Yuan W C, Zhu, Deng J G. Rh(II)-Cp*-TsDPEN catalyzed aqueous asymmetric transfer hydrogenation of chromenones into saturated alcohol: C=C and C=O reduction in one step. Tetrahedron Letters, 2012, 53: 3828-3830.

[279] Kang G, Lin S, Shiwakoti A, Ni B. Imidazolium ion tethered TsDPENs as efficient water-soluble ligands for rhodium catalyzed asymmetric transfer hydrogenation of aromatic ketones. Catalysis Communications, 2014, 57: 111-114.

[280] Li L, Wu J S, Wang F, Liao J, Zhang H, Lian C X, Zhu J, Deng J G. Asymmetric transfer hydrogenation of ketones and imines with novel water-soluble chiral diamine as ligand in neat water. Green Chemistry, 2007, 9: 23-25.

[281] Ahlford K, Lind J, Maler L, Adolfsson H. Rhodium-catalyzed asymmetric transfer hydrogenation of alkyl and aryl ketones in aqueous media. Green Chemistry, 2008, 10: 832-835.

[282] Madern N, Talbi B, Salmain M. Aqueous phase transfer hydrogenation of aryl ketones catalysed by achiral ruthenium(II) and rhodium(III) complexes and their papain conjugates. Applied Organometallic Chemistry, 2013, 27: 6-12.

[283] Paganelli S, Piccolo O, Pontini P, Tassini R, Rathod V D. Aqueous-phase hydrogenation and hydroformylation reactions catalyzed by a new water-soluble[rhodium]-thioligand complex. Catalysis Today, 2015, 247: 64-69.

[284] Lin Z C, Li J H, Huang Q F, Huang Q Y, Wang Q W, Tang L, Gong D. Y, Yang J, Zhu J, Deng J G. Chiral surfactant-type catalyst: Enantioselective reduction of long-chain aliphatic ketoesters in water. The Journal of Organic Chemistry, 2015, 80: 4419-4429.

[285] Longley C J, Goodwin T J, Wilkinson G. Hydrogenation of imines by rhodium-phosphine complexes. Polyhedron, 1986, 5: 1625-1628.

[286] Wang C, Li C Q, Wu X F, Pettman A, Xiao J L. pH-Regulated asymmetric transfer hydrogenation of quinolines in water. Angewandte Chemie International Edition, 2009, 48: 6524-6528.

[287] Guan H, Iimura M, Magee M P, Norton J R, Zhu G. Ruthenium-catalyzed ionic hydrogenation of iminium cations scope and mechanism. Journal of the American Chemical Society, 2005, 127: 7805-7814.

[288] Zhang L J, Qiu R Y, Xue X, Pan Y X, Xu C H, Li H R, Xu L J. Versatile (pentamethylcyclopentadienyl)rhodium-2,2'-bipyridine (Cp*Rh-bpy) catalyst for transfer hydrogenation of N-heterocycles in water. Advanced Synthesis & Catalysis, 2015, 357: 3529-3537.

[289] Tang Y F, Xiang J, Cun L F, Wang Y Q, Zhu J, Liao J, Deng J G. Chemoselective and enantioselective transfer hydrogenation of β,β-disubstituted nitroalkenes catalyzed by a water-insoluble chiral diamine-rhodium complex in water. Tetrahedron: Asymmetry, 2010, 21: 1900-1905.

[290] Soltani O, Ariger M A, Carreira E M. Transfer hydrogenation in water: Enantioselective, catalytic reduction of (E)-β,β-disubstituted nitroalkenes. Organic Letters, 2009, 11: 4196-4198.

[291] Sakai M, Hayashi H, Miyaura N. Rhodium-catalyzed conjugate addition of aryl-or 1-alkenylboronic acids to enones. Organometallics, 1997, 16: 4229-4231.
[292] Iguchi Y, Itooka R, Miyaura N. Asymmetric conjugate addition of arylboronic acids to enones catalyzed by rhodium-monodentate phosphoramidite complexes in the presence of bases. Synlett, 2003, 34: 1040-1042.
[293] Shi Q, Xu L J, Li X S, Jia X, Wang R H, Au-Yeung T T L, Chan A S C, Hayashi T, Cao R, Hong M C. Bipyridyl-based diphosphine as an efficient ligand in the rhodium-catalyzed asymmetric conjugate addition of arylboronic acids to α,β-unsaturated ketones. Tetrahedron Letters, 2003, 44: 6505-6508.
[294] Huang T, Venkatraman S, Meng Y, Nguyen T V, Kort D, Wang D, Ding R, Li C J. Quasi-nature catalysis. Rhodium-catalyzed C-C bond formation in air and water. Pure and Applied Chemistry, 2001, 73: 1315-1319.
[295] Venkatraman S, Li C J. Rhodium catalyzed conjugated addition of unsaturated carbonyl compounds by triphenylbismuth in aqueous media and under an air atmosphere. Tetrahedron Letters, 2001, 42: 781-784.
[296] Ding R, Chen Y J, Wang D, Li C J. Rhodium-catalyzedconjugated addition of aryllead reagents to α,β-unsaturated carbonyl compounds in air and water. Synlett, 2001, 2001: 1470-1472.
[297] Oi S C, Moro M, Ito H, Honma Y, Miyano S, Inoue Y. Rhodium-catalyzed conjugate addition of aryl-and alkenyl-stannanes to α,β-unsaturated carbonyl compounds. Tetrahedron, 2002, 58: 91-97.
[298] Sato A, Kinoshita H, Shinokubo H, Oshima K. Hydrosilylation of alkynes with a cationic rhodium species formed in an anionic micellar system. Organic Letters, 2004, 6: 2217-2220.
[299] Wu W, Li C J. A one-pot, rhodium-catalyzed hydrostannylation-conjugate addition in air and water. Letters in Organic Chemistry, 2004, 1: 122-125.
[300] Wender P A, Love J A, Williams T J. Rhodium-catalyzed[5+2] cycloaddition reactions in water. Synlett, 2003, 34(48): 1295-1298.
[301] Franke R, Selent D, Börner A. Applied hydroformylation. Chemical Reviews, 2012, 112: 5675-5732.
[302] Bohnen H. Hydroformylation of alkenes: An industrial view of the status and importance. Advances in Catalysis, 2002, 47: 1-64.
[303] Ajjou A N, Alper H. A new, efficient, and in some cases highly regioselective water-woluble polymer rhodium catalyst for olefin hydroformylation. Journal of the American Chemical Society, 1998, 120: 1466-1468.
[304] Mieczyńska E, Trzeciak A M, Ziółkowski J J. Hydrogenation and hydroformylation of C_4 unsaturated alcohols with an[Rh(acac)(CO)$_2$]/PNS catalyst in water solution (PNS-Ph$_2$PCH$_2$CH$_2$CONHC(CH$_3$)$_2$CH$_2$SO$_3$Li). Journal of Molecular Catalysis A: Chemical, 1999, 148: 59-68.
[305] Verspui G, Elbertse G, Papadogianakis G, Sheldon R A. Catalytic conversions in water: Part 19. Smooth hydroformylation of N-allylacetamide in mono-and biphasic aqueous media. Journal of Organometallic Chemistry, 2001, 621: 337-343.
[306] Monflier E, Fremy G, Castanet Y, Mortreux A. Molecular tecognition between chemically modified β-cyclodextrin and dec-1-ene: New prospects for biphasic hydroformylation of water-insoluble olefins. Angewandte Chemie International Edition in English, 1995, 34: 2269-2271.
[307] Chen H, Li Y Z, Li R X, Cheng P M, Li X J. Highly regioselective hydroformylation of 1-dodecene catalyzed by Rh-BISBIS in aqueous two-phase system. Journal of Molecular Catalysis A: Chemical, 2003, 198: 1-7.
[308] Yuan M L, Chen H, Li R X, Li Y Z, Li X J. Hydroformylation of 1-butene catalyzed by water-soluble Rh-BISBIS complex in aqueous two-phase catalytic system. Applied Catalysis A: General, 2003, 251: 181-185.
[309] Baricelli P J, López-Linares F, Bruss A, Santo R, Lujano E, Sánchez-Delgado R A. Biphasic hydroformylation of olefins by the new binuclear water soluble rhodium complex[Rh(μ-Pz)(CO)(TPPTS)]$_2$. Journal of Molecular Catalysis A: Chemical, 2005, 239: 130-137.
[310] Baricelli P J, Santos R, Lujano E, Pardey A J. Synthesis, characterization and catalytic activity in the hydroformylation of 1-hexene and styrene of water-soluble rhodium complex[Rh(μ-Pz)(CO)(TPPMS)]$_2$ (Part 1). Journal of Molecular Catalysis A: Chemical, 2004, 207: 83-89.

[311] Chen S J, Wang Y Y, Li Y Q, Zhao X L, Liu Y. Ion-pair effect of phosphino-imidazolium salts on Rh(Ⅲ)-complex catalyzed hydroformylation of 1-octene in water suspension. Catalysis Communications, 2014, 50: 5-8.

[312] Frieman R H, Kennedy E R, Lucas H J. The hydration of unsaturated compounds. V. the rate of hydration of acetylene in aqueous solution of sulfuric acid and mercuric sulfate1. Journal of the American Chemical Society, 1937, 59: 722-726.

[313] Ibrahim H, de Frémont P, Braunstein P, Théry V, Nauton L, Cisnetti F, Gautier, A. Water-soluble gold-N-heterocyclic carbene complexes for the catalytic homogeneous acid-and silver-free hydration of hydrophilic alkynes. Advanced Synthesis & Catalysis, 2015, 357: 3893-3900.

[314] Lo V K Y, Liu Y G, Wong M K, Che C M. Gold(Ⅲ) salen complex-catalyzed synthesis of propargylamines via a three-component coupling reaction. Organic Letters, 2006, 8: 1529-1532.

[315] Lo V K Y, Kung K K Y, Wong M K, Che C M. Gold(Ⅲ) (C^N) complex-catalyzed synthesis of propargylamines via a three-component coupling reaction of aldehydes, amines and alkynes. Journal of Organometallic Chemistry, 2009, 694: 583-591.

[316] Elie B T, Levine C, Ubarretxena-Belandia I, Varela-Ramírez A, Aguilera R. J, Ovalle R, Contel M. Water-soluble (phosphane)gold(I) complexes-applications as recyclable catalysts in a three-component coupling reaction and as antimicrobial and anticancer agents. European Journal of Inorganic Chemistry, 2009: 3421-3430.

[317] Katrin B, Norbert K. Ammonium-salt-tagged IMesAuCl complexes and their application in gold-catalyzed cycloisomerization reactions in water. European Journal of Organic Chemistry, 2012, 14: 1888-1895.

[318] Belger K, Krause N. Smaller, faster, better: Modular synthesis of unsymmetrical ammonium salt-tagged NHC-gold(Ⅰ) complexes and their application as recyclable catalysts in water. Organic & Biomolecular Chemistry, 2015, 13: 8556-8560.

[319] Ye D J, Wang J F, Zhang X, Zhou Y, Ding X, Feng E G, Sun H F, Liu G N, Jiang H L, Liu H. Gold-catalyzed intramolecular hydroamination of terminal alkynes in aqueous media: Efficient and regioselective synthesis of indole-1-carboxamides. Green Chemistry, 2009, 11: 1201-1208.

[320] Feng E G, Zhou Y, Zhao F, Chen X J, Zhang L, Jiang H L, Liu H. Gold-catalyzed tandem reaction in water: An efficient and convenient synthesis of fused polycyclic indoles. Green Chemistry, 2012, 14: 1888-1895.

[321] Xu S T, Zhou Y, Xu J Y, Jiang H L, Liu H. Gold-catalyzed Michael addition/intramolecular annulation cascade: An effective pathway for the chemoselective-and regioselective synthesis of tetracyclic indole derivatives in water. Green Chemistry, 2013, 15: 718-726.

[322] Shen C H, Li L L, Zhang W, Liu S, Shu C, Xie Y E, Yu Y F, Ye L W. Gold-catalyzed tandem cycloisomerization/ functionalization of in situ generated α-oxo gold carbenes in water. The Journal of Organic Chemistry, 2014, 79: 9313-9318.

[323] Arzoumanian H, Jean M, Nuel D, Cabrer A, Guiterrez J L G, Rosas N. Nickel-catalyzed carbonylation of alpha-keto alkynes under phase transfer conditions. Organometallics, 1995, 14: 5438-5441.

[324] Rosas N, Salmon M, Sharma P, Alvarez C, Ramirez R, Garcia J L, Arzoumanian H. Nickel catalyzed cascade conversion of propargyl halide and propargyl alcohol into 4,6-dimethyl-5-cyano-2-pyrone in water. Journal of the Chemical Society, Perkin Transactions 1, 2000: 1493-1494.

[325] Rosas N, Sharma P, Alvarez C, Cabrera A, Ramirez R, Delgado A, Arzoumanian H. Novel and facile catalytic synthesis of 2,4-dioxopyrido[2,3-d] pyrimidine derivatives in water. Journal of the Chemical Society, Perkin Transactions 1, 2001, 2341-2343.

[326] Suárez-Ortiz G A, Sharma P, Amézquita-Valencia M, Arellano I, Cabrera A, Rosas N. Ni(0) catalyzed one step synthesis of benzo[b] [1,8] naphthyridin-5-ones from silyl-α-ketoalkynes in water. Tetrahedron Letters, 2011, 52: 1641-1643.

[327] Alper H, Abbayes H D. Phase-transfer catalyzed organometallic chemistry. The cobalt carbonyl-catalyzed carbonylation of halides. Journal of Organometallic Chemistry, 1977, 134: C11-C14.

[328] Krafft M E, Wright J A, Boñaga L V R. Aqueous-phase, thermal Pauson-Khand reactions in the presence of surfactants. Tetrahedron Letters, 2003, 44: 3417-3422.
[329] Sun C L, Qu P P, Li F. Rearrangement of aldoximes to amides in water under air atmosphere catalyzed by water-soluble iridium complex[Cp*Ir(H$_2$O)$_3$] [OTf] $_2$. Catalysis Science & Technology, 2014, 4: 988-996.
[330] Li F, Lu L, Ma J. Acceptorless dehydrogenative condensation of o-aminobenzamides with aldehydes to quinazolinones in water catalyzed by a water-soluble iridium complex[Cp*Ir(H$_2$O)$_3$] [OTf]$_2$. Organic Chemistry Frontiers, 2015, 2: 1589-1597.
[331] Sun M X, Campbell J, Kang G W, Wang H G, Ni B K. Imidazolium ion tethered TsDPENs as efficient ligands for Iridium catalyzed asymmetric transfer hydrogenation of α-keto phosphonates in water. Journal of Organometallic Chemistry, 2016, 810: 12-14.
[332] Trost B M, Brindle C S. The direct catalytic asymmetric aldol reaction. Chemical Society Reviews, 2010, 39: 1600-1632.
[333] Dickerson T J, Lovell T, Meijler M M, Noodleman L, Janda K D. Nornicotine aqueous aldol reactions: Synthetic and theoretical investigations into the origins of catalysis. The Journal of Organic Chemistry, 2004, 69: 6603-6610.
[334] Allemann C, Gordillo R, Clemente F R, Cheong P H Y, Houk K N. Theory of asymmetric organocatalysis of aldol and related reactions: Rationalizations and predictions. Accounts of Chemical Research, 2004, 37: 558-569.
[335] Hayashi Y, Aratake S, Okano T, Takahashi J, Sumiya T, Shoji M. Combined proline-surfactant organocatalyst for the highly diastereo-and enantioselective aqueous direct cross-aldol reaction of aldehydes. Angewandte Chemie International Edition, 2006, 45: 5527-5529.
[336] Chimni S S, Mahajan D. Small organic molecule catalyzed enantioselective direct aldol reaction in water. Tetrahedron: Asymmetry, 2006, 17: 2108-2119.
[337] Raj M, Vishnumaya, Ginotra S K, Singh V K. Highly enantioselective direct aldol reaction catalyzed by organic molecules. Organic Letters, 2006, 8: 4097-4099.
[338] Maya V, Raj M, Singh V K. Highly enantioselective organocatalytic direct aldol reaction in an aqueous medium. Organic Letters, 2007, 9: 2593-2595.
[339] Vishnumaya M R, Singh V K. Highly efficient small organic molecules for enantioselective direct aldol reaction in organic and aqueous media. The Journal of Organic Chemistry, 2009, 74: 4289-4297.
[340] Gandhi S, Singh V K. Synthesis of chiral organocatalysts derived from aziridines: Application in asymmetric aldol reaction. The Journal of Organic Chemistry, 2008, 73: 9411-9416.
[341] Sathapornvajana S, Vilaivan T. Prolinamides derived from aminophenols as organocatalysts for asymmetric direct aldol reactions. Tetrahedron, 2007, 63: 10253-10259.
[342] Tzeng Z H, Chen H Y, Huang C T, Chen K. Camphor containing organocatalysts in asymmetric aldol reaction on water. Tetrahedron Letters, 2008, 49: 4134-4137.
[343] Gryko D, Saletra W J. Organocatalytic asymmetric aldol reaction in the presence of water. Organic & Biomolecular Chemistry, 2007, 5: 2148-2153.
[344] Singh Chimni S, Singh S, Mahajan D. Protonated (S)-prolinamide derivatives-water compatible organocatalysts for direct asymmetric aldol reaction. Tetrahedron: Asymmetry, 2008, 19: 2276-2284.
[345] Zu L S, Xie H X, Li H, Wang J, Wang W. Highly enantioselective aldol reactions catalyzed by a recyclable fluorous (S)-pyrrolidine sulfonamide on water. Organic Letters, 2008, 10: 1211-1214.
[346] Peng F Z, Shao Z H, Pu X W, Zhang H B. Highly diastereo-and enantioselective direct aldol reactions promoted by water-compatible organocatalysts bearing central and axial chiral elements. Advanced Synthesis & Catalysis, 2008, 350: 2199-2204.
[347] Guizzetti S, Benaglia M, Raimondi L, Celentano G. Enantioselective direct aldol reaction "on water" promoted by chiral organic catalysts. Organic Letters, 2014, 9: 1247-1250.
[348] Guillena G, Hita M D C, Nájera C. High acceleration of the direct aldol reaction cocatalyzed by BINAM-prolinamides and benzoic acid in aqueous media. Tetrahedron: Asymmetry, 2006, 17: 1493-1497.

[349] Wang C, Jiang Y, Zhang X X, Huang Y, Li B G, Zhang G L. Rationally designed organocatalyst for direct asymmetric aldol reaction in the presence of water. Tetrahedron Letters, 2007, 48: 4281-4285.

[350] Wu X Y, Jiang Z Q, Shen H M, Lu Y X. Highly efficient threonine-derived organocatalysts for direct asymmetric aldol reactions in water. Advanced Synthesis & Catalysis, 2007, 349: 812-816.

[351] Teo Y C. Direct asymmetric aldol reactions catalyzed by a siloxy serine organocatalyst in water. Tetrahedron: Asymmetry, 2007, 18: 1155-1158.

[352] Ramasastry S S V, Albertshofer K, Utsumi N, Barbas C F. Water-compatible organocatalysts for direct asymmetric syn-aldol reactions of dihydroxyacetone and aldehydes. Organic Letters, 2008, 10: 1621-1624.

[353] Zhu M K, Xu X Y, Gong L Z. Organocatalytic asymmetric syn-aldol reactions of aldehydes with long-chain aliphatic ketones on water and with dihydroxyacetone in organic solvents. Advanced Synthesis & Catalysis, 2008, 350: 1390-1396.

[354] Raj M, Parashari G S, Singh V K. Highly enantioselective organocatalytic syn-and anti-aldol reactions in aqueous medium. Advanced Synthesis & Catalysis, 2009, 351: 1284-1288.

[355] Tokuda O, Kano T, Gao W G, Ikemoto T, Maruoka K. A practical synthesis of (S)-2-cyclohexyl-2-phenylglycolic acid via organocatalytic asymmetric construction of a tetrasubstituted carbon center. Organic Letters, 2005, 7: 5103-5105.

[356] Chimni S S, Mahajan D, Suresh Babu V V. Protonated chiral prolinamide catalyzed enantioselective direct aldol reaction in water. Tetrahedron Letters, 2005, 46: 5617-5619.

[357] Wu Y S, Chen Y, Deng D S, Cai J W. Proline-catalyzed asymmetric direct aldol reaction assisted by d-camphorsulfonic acid in aqueous media. Synlett, 2005, 36: 1627-1629.

[358] Hayashi Y, Sumiya T, Takahashi J, Gotoh H, Urushima T, Shoji M. Highly diastereo-and enantioselective direct aldol reactions in water. Angewandte Chemie International Edition, 2006, 45: 958-961.

[359] Narayanaswamy A, Xu H F, Pradhan N, Peng X G. Crystalline nanoflowers with different chemical compositions and physical properties grown by limited ligand protection. Angewandte Chemie International Edition, 2006, 45: 5361-5364.

[360] Jiang Z Q, Liang Z, Wu X Y, Lu Y X. Asymmetric aldol reactions catalyzed by tryptophan in water. Chemical Communications, 2006, 37: 2801-2803.

[361] Amedjkouh M. Aqua-organocatalyzed direct asymmetric aldol reaction with acyclic amino acids and organic bases with control of diastereo-and enantioselectivity. Tetrahedron: Asymmetry, 2007, 18: 390-395.

[362] Singh N, Pandey J, Tripathi R P. d-Glucosamine, a natural aminosugar as organocatalyst for an ecofriendly direct aldol reaction of ketones with aromatic aldehydes in water. Catalysis Communications, 2008, 9: 743-746.

[363] Hayashi Y, Aratake S, Itoh T, Okano T, Sumiya T, Shoji M. Dry and wet prolines for asymmetric organic solvent-free aldehyde-aldehyde and aldehyde-ketone aldol reactions. Chemical Communications, 2007: 957-959.

[364] Deng D S, Liu P, Ji B M, Fu W J, Li L. Acyclic amino acids catalyzed direct asymmetric aldol reactions in aqueous media assisted by 2,4-dinitrophenol. Catalysis Letters, 2010, 137: 163-170.

[365] Aratake S, Itoh T, Okano T, Usui T, Shoji M, Hayashi Y. Small organic molecule in enantioselective, direct aldol reaction "in water". Chemical Communications, 2007, 24: 2524-2526.

[366] Darbre T, Machuqueiro M. Zn-Proline catalyzed direct aldol reaction in aqueous media. Chemical Communications, 2003, 9: 1090-1091.

[367] Luo S Z, Xu H, Li J Y, Zhang L, Mi X L, Zheng X X, Cheng J P. Facile evolution of asymmetric organocatalysts in water assisted by surfactant Brønsted acids. Tetrahedron, 2007, 63: 11307-11314.

[368] Giacalone F, Gruttadauria M, Meo P L, Riela S, Noto R. New simple hydrophobic proline derivatives as highly active and stereoselective catalysts for the direct asymmetric aldol reaction in aqueous medium. Advanced Synthesis & Catalysis, 2008, 350: 2747-2760.

[369] Lin J H, Zhang C P, Xiao J C. Enantioselective aldol reaction of cyclic ketones with aryl aldehydes catalyzed by a cyclohexanediamine derived salt in the presence of water. Green Chemistry, 2009, 11: 1750-1753.

[370] Emer E, Galletti P, Giacomini D. A temperature study on a stereoselective organocatalyzed aldol reaction in water. Tetrahedron, 2008, 64: 11205-11208.

[371] Fu S D, Fu X K, Zhang S P, Zou X C, Wu X. J. Highly diastereo-and enantioselective direct aldol reactions by 4-tert-butyldimethylsiloxy-substituted organocatalysts derived from N-prolylsulfonamides in water. Tetrahedron: Asymmetry, 2009, 20: 2390-2396.

[372] Yang H, Mahapatra S, Cheong P H Y, Carter R G. Highly stereoselective and scalable anti-aldol reactions using N-(p-dodecylphenylsulfonyl)-2-pyrrolidinecarboxamide: Scope and origins of stereoselectivities. The Journal of Organic Chemistry, 2010, 75: 7279-7290.

[373] Zhang S P, Fu X K, Fu S D. Rationally designed 4-phenoxy substituted prolinamide phenols organocatalyst for the direct aldol reaction in water. Tetrahedron Letters, 2009, 50: 1173-1176.

[374] Moorthy J N, Saha S. Highly diastereo-and enantioselective aldol reactions in common organic solvents using N-arylprolinamides as organocatalysts with enhanced acidity. European Journal of Organic Chemistry, 2010, 40: 739-748.

[375] Chimni S S, Singh S, Kumar A. The pH of the reaction controls the stereoselectivity of organocatalyzed direct aldol reactions in water. Tetrahedron: Asymmetry, 2009, 20: 1722-1724.

[376] Nugent T C, Umar M N, Bibi A. Picolylamine as an organocatalyst template for highly diastereo-and enantioselective aqueous aldol reactions. Organic & Biomolecular Chemistry, 2010, 8: 4085-4089.

[377] Wu C L, Fu X K, Ma X B, Li S. One-step, efficient synthesis of combined threonine-surfactant organocatalysts for the highly enantioselective direct aldol reactions of cyclic ketones with aromatic aldehydes in the presence of water. Tetrahedron: Asymmetry, 2010, 21: 2465-2470.

[378] Wu C L, Fu X K, Li S. New simple and recyclable O-acylation serine derivatives as highly enantioselective catalysts for the large-scale asymmetric direct aldol reactions in the presence of water. Tetrahedron, 2011, 67: 4283-4290.

[379] Hernández J G, Juaristi E. Efficient ball-mill procedure in the 'green' asymmetric aldol reaction organocatalyzed by (S)-proline-containing dipeptides in the presence of water. Tetrahedron, 2011, 67: 6953-6959.

[380] Tang G K, Hu X M, Altenbach H J. l-Prolinamide derivatives as efficient organocatalysts for aldol reactions on water. Tetrahedron Letters, 2011, 52: 7034-7037.

[381] Bellomo A, Daniellou R, Plusquellec D. Aqueous solutions of facial amphiphilic carbohydrates as sustainable media for organocatalyzed direct aldol reactions. Green Chemistry, 2012, 14: 281-284.

[382] Miura T, Kasuga H, Imai K, Ina M, Tada N, Imai N, Itoh A. Highly efficient asymmetric aldol reaction in brine using a fluorous sulfonamide organocatalyst. Organic & Biomolecular Chemistry, 2012, 10: 2209-2213.

[383] Zhao Q Q, Lam Y H, Kheirabadi M, Xu C, Houk, K N, Schafmeister C E. Hydrophobic substituent effects on proline catalysis of aldol reactions in water. The Journal of Organic Chemistry, 2012, 77: 4784-4792.

[384] Huang J, Chen G D, Fu X K, Li C, Wu C L, Miao Q. Highly efficient direct large-scale stoichiometric aldol reactions catalyzed by a new diamine salt in the presence of water. Catalysis Science & Technology, 2012, 2: 547-553.

[385] Kumar A, Chimni S S. Organocatalyzed direct asymmetric aldol reaction of isatins in water: low catalyst loading in command. Tetrahedron, 2013, 69: 5197-5204.

[386] He T, Li K, Wu M Y, Wu M B, Wang N, Pu L, Yu X Q. Water promoted enantioselective aldol reaction by proline-cholesterol and -diosgenin based amphiphilic organocatalysts. Tetrahedron, 2013, 69: 5136-5143.

[387] Kucherenko A S, Siyutkin D E, Dashkin R R, Zlotin S G. Organocatalysis of asymmetric aldol reaction in water: Comparison of catalytic properties of (S)-valine and (S)-proline amides. Russian Chemical Bulletin, 2013, 62: 1010-1015.

[388] Zhao H W, Sheng Z H, Meng W, Yue Y Y, Li H L, Song X Q, Yang Z. Design, synthesis and organocatalysis of 2,2'-biphenol-based prolinamide organocatalysts in the asymmetric direct aldol reaction in water. Synlett, 2013, 24: 2743-2747.

[389] Lei M, Shi L X, Li G, Chen S L, Fang W H, Ge Z, Cheng T M, Li R T. Dipeptide-catalyzed direct asymmetric aldol reactions in the presence of water. Tetrahedron, 2007, 63: 7892-7898.

[390] De Nisco M, Pedatella S, Ullah H, Zaidi J H, Naviglio D, Özdamar Ö, Caputo R. Proline-β³-amino-ester dipeptides as efficient catalysts for enantioselective direct aldol reaction in aqueous medium. The Journal of Organic Chemistry, 2009, 74: 9562-9565.

[391] Jia Y N, Wu F C, Ma X, Zhu G J, Da C S. Highly efficient prolinamide-based organocatalysts for the direct asymmetric aldol reaction in brine. Tetrahedron Letters, 2009, 50: 3059-3062.

[392] Psarra A, Kokotos C G, Moutevelis-Minakakis P. Tert-butyl esters of tripeptides based on Pro-Phe as organocatalysts for the asymmetric aldol reaction in aqueous or organic medium. Tetrahedron, 2014, 70: 608-615.

[393] Bisticha A, Triandafillidi I, Kokotos C G. Tert-butyl esters of peptides as organocatalysts for the asymmetric aldol reaction. Tetrahedron: Asymmetry, 2015, 26: 102-108.

[394] Song X X, Liu A X, Liu S S, Gao W C, Wang M C, Chang J B. Enantiopure azetidine-2-carboxamides as organocatalysts for direct asymmetric aldol reactions in aqueous and organic media. Tetrahedron, 2014, 70: 1464-1470.

[395] Bhowmick S, Kunte S S, Bhowmick K C. A new organocatalyst derived from abietic acid and 4-hydroxy-l-proline for direct asymmetric aldol reactions in aqueous media. Tetrahedron: Asymmetry, 2014, 25: 1292-1297.

[396] Zhao H W, Meng W, Yang Z, Tian T, Sheng Z H, Li H, Song X Q, Zhang Y T, Yang S, Li B. Organocatalytic stereoselective synthesis of 3-alkyl-3-hydroxy-2-oxindoles catalyzed by novel water-compatible axially unfixed biaryl-based bifunctional organocatalysts. Chinese Journal of Chemistry, 2014, 32: 417-428.

[397] Eymur S, Akceylan E, Sahin O, Uyanik A, Yilmaz M. Direct enantioselective aldol reactions catalyzed by calix[4]arene-based L-proline derivatives in the water. Tetrahedron, 2014, 70: 4471-4477.

[398] Miura D, Machinami T. Stereoselective aldol reaction in aqueous solution using prolinamido-glycosides as water-compatible organocatalyst. Modern Research in Catalysis, 2015, 4: 20-27.

[399] Fujiya A, Nobuta T, Yamaguchi E, Tada N, Miura T, Itoh A. Aerobic photooxidative direct asymmetric aldol reactions of benzyl alcohols using water as the solvent. RSC Advances, 2015, 5: 39539-39543.

[400] Valdivielso A M, Catot A, Alfonso I, Jimeno C. Intramolecular hydrogen bonding guides a cationic amphiphilic organocatalyst to highly stereoselective aldol reactions in water. RSC Advances, 2015, 5: 62331-62335.

[401] Aktas M, Uyanik A, Eymur S, Yilmaz M. L-proline derivatives based on a calix[4] arene scaffold as chiral organocatalysts for the direct asymmetric aldol reaction in water. supramolecular chemistry, 2016, 28: 351-359.

[402] Tsogoeva S B. Recent advances in asymmetric organocatalytic 1,4-conjugate additions. European Journal of Organic Chemistry, 2010, 38: 1701-1716.

[403] Christoffers J, Baro A. Construction of quaternary stereocenters: New perspectives through enantioselective Michael reactions. Angewandte Chemie International Edition, 2003, 42: 1688-1690.

[404] Berner Otto M, Tedeschi L, Enders D. Asymmetric Michael additions to nitroalkenes. European Journal of Organic Chemistry, 2002, 33: 1877-1894.

[405] Mase N, Watanabe K, Yoda H, Takabe K, Tanaka F, Barbas C F. Organocatalytic direct Michael reaction of ketones and aldehydes with β-nitrostyrene in brine. Journal of the American Chemical Society, 2006, 128: 4966-4967.

[406] Luo S Z, Mi X L, Liu S, Xu H, Cheng J P. Surfactant-type asymmetric organocatalyst: Organocatalytic asymmetric Michael addition to nitrostyrenes in water. Chemical Communications, 2006, 38: 3687-3689.

[407] Zu L S, Wang J, Li H, Wang W. A recyclable fluorous (S)-pyrrolidine sulfonamide promoted direct, highly enantioselective Michael addition of ketones and aldehydes to nitroolefins in water. Organic Letters, 2006, 8: 3077-3079.

[408] Cao Y J, Lai Y Y, Wang X, Li Y J, Xiao W J. Michael additions in water of ketones to nitroolefins catalyzed by readily tunable and bifunctional pyrrolidine-thiourea organocatalysts. Tetrahedron Letters, 2007, 48: 21-24.

[409] Vishnumaya, Singh V K. Highly enantioselective water-compatible organocatalyst for Michael reaction of ketones to nitroolefins. Organic Letters, 2007, 9: 1117-1119.

[410] Wang J, Yu F, Zhang X J, Ma D W. Enantioselective assembly of substituted dihydropyrones via organocatalytic reaction in water media. Organic Letters, 2008, 10: 2561-2564.

[411] Ma A Q, Zhu S L, Ma D W. Enantioselective organocatalytic Michael addition of malonates to α,β-unsaturated aldehydes in water. Tetrahedron Letters, 2008, 49: 3075-3077.
[412] Zhu S L, Yu S Y, Ma D W. Highly efficient catalytic system for enantioselective Michael addition of aldehydes to nitroalkenes in water. Angewandte Chemie International Edition, 2008, 47: 545-548.
[413] Carlone A, Marigo M, North C, Landa A, Jørgensen K A. A simple asymmetric organocatalytic approach to optically active cyclohexenones. Chemical Communications, 2006, 47: 4928-4930.
[414] Palomo C, Landa A, Mielgo A, Oiarbide M, Puente Á, Vera S. Water-compatible iminium activation: Organocatalytic Michael reactions of carbon-centered nucleophiles with enals. Angewandte Chemie International Edition, 2007, 46: 8431-8435.
[415] Wu J B, Ni B K, Headley A D. Di(methylimidazole)prolinol silyl ether catalyzed highly Michael addition of aldehydes to nitroolefins in water. Organic Letters, 2009, 11: 3354-3356.
[416] Chandrasekhar S, Kumar T P, Haribabu K, Reddy C R. Hydroxyphthalimide allied triazole-pyrrolidine catalyst for asymmetric Michael additions in water. Tetrahedron: Asymmetry, 2010, 21: 2372-2375.
[417] Hu F, Guo C S, Xie J, Zhu H L, Huang Z Z. Efficient and highly enantioselective Michael addition of aldehydes to nitroalkenes catalyzed by a surfactant-type organocatalyst in the presence of water. Chemistry Letters, 2010, 39: 412-414.
[418] Moon H W, Kim D Y. Enantioselective organocatalytic conjugate addition of α-nitroacetate to α,β-unsaturated ketones in water. Tetrahedron Letters, 2010, 51: 2906-2908.
[419] Xiao J, Liu Y L, Loh T P. The organocatalytic enantioselective Michael addition of aldehydes to vinyl sulfones in water. Synlett, 2010: 2029-2032.
[420] Zheng Z L, Perkins B L, Ni B K. Diarylprolinol silyl ether salts as new, efficient, water-soluble, and recyclable organocatalysts for the asymmetric Michael addition on water. Journal of the American Chemical Society, 2010, 132: 50-51.
[421] Syu S E, Kao T T, Lin W W. A new type of organocatalyst for highly stereoselective Michael addition of ketones to nitroolefins on water. Tetrahedron, 2010, 66: 891-897.
[422] Saha S, Seth S, Moorthy J N. Functionalized proline with double hydrogen bonding potential: Highly enantioselective Michael addition of carbonyl compounds to β-nitrostyrenes in brine. Tetrahedron Letters, 2010, 51: 5281-5286.
[423] Luo C H, Du D M. Highly enantioselective Michael addition of ketones and an aldehyde to nitroalkenes catalyzed by a binaphthyl sulfonimide in water. Synthesis, 2011: 1968-1973.
[424] Li J, Liu Y, Liu L. Efficient enantioselective Michael addition of nitroalkenes catalyzed by a surfactant-type bifunctional thiourea organocatalyst in the presence of water. Letters Organic Chemistry, 2012, 9: 51-55.
[425] Veverkova E, Polackova V, Liptakova L, Kazmerova E, Meciarova M, Toma, S, Sebesta R. Organocatalyst efficiency in the Michael additions of aldehydes to nitroalkenes in water and in a ball-mill. ChemCatChem, 2012, 4: 1013-1018.
[426] Qiao Y P, He J P, Ni B K, Headley A D. Asymmetric Michael reaction of acetaldehyde with nitroolefins catalyzed by highly water-compatible organocatalysts in aqueous media. Advanced Synthesis & Catalysis, 2012, 354: 2849-2853.
[427] Wei J W, Guo W G, Zhang B Y, Liu Y, Du X, Li C. An amphiphilic organic catalyst for the direct asymmetric Michael addition of cycloketone to nitroolefins in water. Chinese Journal of Catalysis, 2014, 35: 1008-1011.
[428] Kumar T P, Radhika L, Haribabu K, Kumar V N. Pyrrolidine-oxyimides: New chiral catalysts for enantioselective Michael addition of ketones to nitroolefins in water. Tetrahedron: Asymmetry, 2014, 25: 1555-1560.
[429] Zhao H W, Yang Z, Yue Y Y, Li H L, Song X Q, Sheng Z H, Meng W, Guo X Y. Asymmetric direct Michael reactions of cyclohexanone with aromatic nitroolefins in water catalyzed by novel axially unfixed biaryl-based bifunctional organocatalysts. Synlett, 2014, 25: 293-297.
[430] Mao Z F, Jia Y M, Li W Y, Wang R. Water-compatible iminium activation: Highly enantioselective organocatalytic Michael addition of malonates to α,β-unsaturated enones. The Journal of Organic Chemistry, 2010, 75: 7428-7430.
[431] Jia S J, Luo C H, Du D M. Highly enantioselective Michael addition of ketone to alkylidene malonates catalyzed by binaphthyl sulfonimides in water. Chin J Chemistry, 2012, 30: 2676-2680.

[432] Kamito Y, Masuda A, Yuasa H, Tada N, Itoh A, Nakashima K, Hirashima S, Koseki Y, Miura T. Asymmetric conjugate addition of malonate to α,β-unsaturated ketones in water using a perfluoroalkanesulfonamide organocatalyst. Tetrahedron: Asymmetry, 2014, 25: 974-979.

[433] Bae H Y, Song C E. Unprecedented hydrophobic amplification in noncovalent organocatalysis "on water": Hydrophobic chiral squaramide catalyzed Michael addition of malonates to nitroalkenes. ACS Catalysis, 2015, 5: 3613-3619.

[434] Ting A, Schaus S E. Organocatalytic asymmetric Mannich reactions: New methodology, catalyst design, and synthetic applications. European Journal of Organic Chemistry, 2007, 2007: 5797-5815.

[435] Scholz U, Winterfeldt E. Biomimetic synthesis of alkaloids. Natural Product Reports, 2000, 17: 349-366.

[436] Hayashi Y, Urushima T, Aratake S, Okano T, Obi K. Organic solvent-free, enantio-and diastereoselective, direct Mannich reaction in the presence of water. Organic Letters, 2008, 10: 21-24.

[437] Teo Y C, Lau J J, Wu M C. Direct asymmetric three-component Mannich reactions catalyzed by a siloxy serine organocatalyst in water. Tetrahedron: Asymmetry, 2008, 19: 186-190.

[438] Cheng L L, Wu X Y, Lu Y X. Direct asymmetric three-component organocatalytic anti-selective Mannich reactions in a purely aqueous system. Organic & Biomolecular Chemistry, 2007, 5: 1018-1020.

[439] Amedjkouh M, Brandberg M. Asymmetric autocatalytic Mannich reaction in the presence of water and its implication in prebiotic chemistry. Chemical Communications, 2008, 26: 3043-3045.

[440] Srinivas N, Bhandari K. Proline-catalyzed facile access to Mannich adducts using unsubstituted azoles. Tetrahedron Letters, 2008, 49: 7070-7073.

[441] An Y J, Wang C C, Liu Z P, Tao J C. Isosteviol proline conjugates as highly efficient amphiphilic organocatalysts for asymmetric three-component Mannich reactions in the presence of water. Helvetica Chimica Acta, 2012, 95: 43-51.

[442] Veverková E, Liptáková L, Veverka M, Šebesta R. Asymmetric Mannich reactions catalyzed by proline and 4-hydroxyproline derived organocatalysts in the presence of water. Tetrahedron: Asymmetry, 2013, 24: 548-552.

[443] Northrup A B, MacMillan D W C. The first general enantioselective catalytic Diels-Alder reaction with simple α,β-unsaturated ketones. Journal of the American Chemical Society, 2002, 124: 2458-2460.

[444] Lemay M, Aumand L, Ogilvie W W. Design of a conformationally rigid hydrazide organic catalyst. Advanced Synthesis & Catalysis, 2007, 349: 441-447.

[445] Bonini B F, Capitò E, Comes-Franchini M, Fochi, M, Ricci, A, Zwanenburg, B. Aziridin-2-yl methanols as organocatalysts in Diels-Alder reactions and Friedel-Crafts alkylations of N-methyl-pyrrole and N-methyl-indole. Tetrahedron: Asymmetry, 2006, 17: 3135-3143.

[446] Hayashi Y, Samanta S, Gotoh H, Ishikawa H. Asymmetric Diels-Alder reactions of α,β-unsaturated aldehydes catalyzed by a diarylprolinol silyl ether salt in the presence of water. Angewandte Chemie International Edition, 2008, 47: 6634-6637.

第三章 金属催化水介质有机合成

金属催化剂是现代化学工业中一类重要的催化剂,在医药、化工、食品以及环境治理等方面有着广泛应用,在国民生产中起着举足轻重的作用。金属催化剂可分为非负载型金属催化剂和负载型金属催化剂。本章介绍金属催化水介质有机合成,包括贵金属(如铂、钯、钌、铑、金等)及过渡元素(铁、钴、镍等)。

第一节 非负载型金属催化剂

非负载型金属催化剂不含载体,按组成可分为单金属和合金两类。由于非负载型金属催化剂分散度低,在反应过程中易团聚,相对负载型催化剂研究较少,在水介质有机反应中的应用报道甚少。

一、偶联反应

早在 1995 年,Pellon 等[1]报道在水中 Cu 可催化卤代芳烃与酚和苯胺的 Ullmann-Goldberg 缩合反应[2,3],如在碱性水溶液中,Cu 催化邻氯苯甲酸与对氯苯酚发生偶联反应,生成 2-(4-氯代酚氧基)苯甲酸[式(3-1)]。

$$\text{(Cl)C}_6\text{H}_4\text{COOH} + \text{HO-C}_6\text{H}_4\text{-Cl} \xrightarrow[\text{H}_2\text{O}]{\text{K}_2\text{CO}_3, \text{吡啶}, \text{Cu}, \text{CuI}} \text{HOOC-C}_6\text{H}_4\text{-O-C}_6\text{H}_4\text{-Cl} \quad (3\text{-}1)$$

Roucoux 等[4]合成了一种 *N*-(2-羟乙基)铵盐(247)稳定的水溶胶 Pd 纳米粒子(NPs),用其催化水介质溴苯和四苯基硼酸钠的 Suzuki 交叉偶联反应[式(3-2)]。以 Cs_2CO_3 为碱,HEA16Cl 为稳定剂,当催化剂用量为 1 mol%时,可定量地给出偶联产物,但对底物的适用范围较小,仅对溴苯催化效果较好,而对碘苯、氯苯、苯基三氟甲烷磺酸和氟苯等底物催化活性都比较差。

$$\text{Br-C}_6\text{H}_5 + \text{Ph}_4\text{BNa} \xrightarrow[\text{Cs}_2\text{CO}_3, \text{H}_2\text{O}, 60\,°\text{C}, 30\,\text{min}]{\text{HEA}n\text{X-Pd(0) NPs}} \text{C}_6\text{H}_5\text{-C}_6\text{H}_5 \quad (3\text{-}2)$$

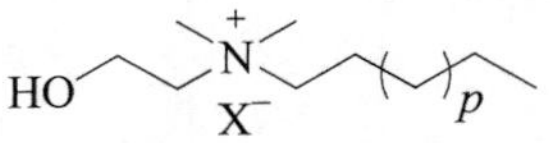

HEA16Cl, *p* = 12, *n* = 16
HEA18Cl, *p* = 14, *n* = 18
HEA16Br, *p* = 12, *n* = 16

247

2007年, Sarkar 等[5,6]在水中以钨的 Fischer 卡宾配合物作为还原剂, 合成出 PEG 包裹的胶体状 Pd 纳米粒子(图3.1), 它在空气中非常稳定, 催化水介质溴代芳烃与芳基三烷氧基硅烷的 Hiyama 交叉偶联反应时显示出高活性, 以 NaOH 为碱, 催化剂用量为 1 mol%时, 在 90℃的条件下较大范围的底物均能发生 Hiyama 交叉偶联反应[式(3-3)], 目标产物得率为 88%~98%。

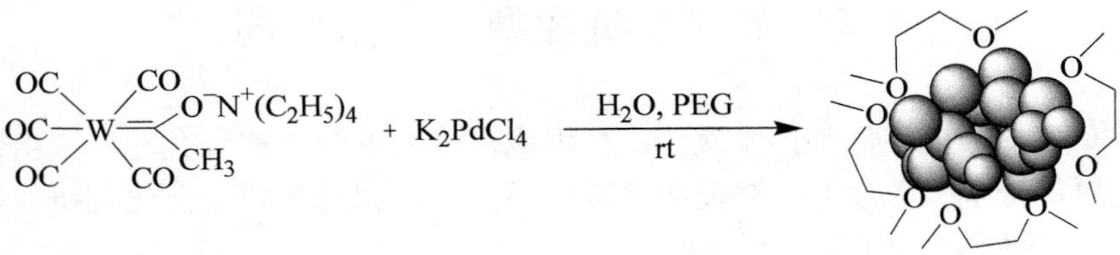

图 3.1 Pd NPs 催化剂的合成示意图

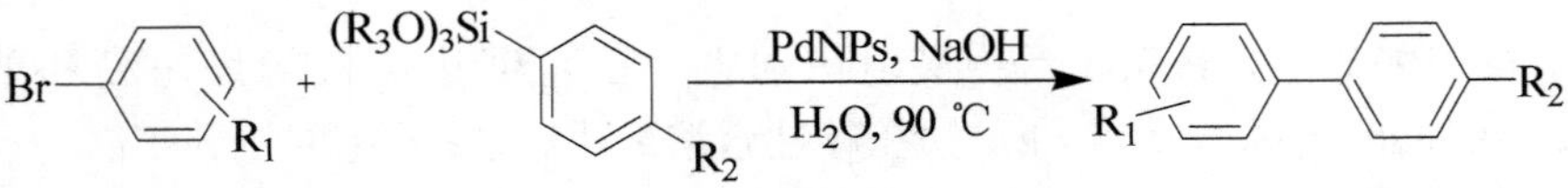

最近, Nasrollahzadeh 等[7]报道用千根草叶子与 $PdCl_2$ 反应制备 Pd 纳米粒子, 催化水中有机锡与溴代芳烃的反应[式(3-4)]以及溴代芳烃与有机硅的 Hiyama 反应[式(3-3)], 产物得率为 84%~96%, 催化剂可重复使用 5 次。

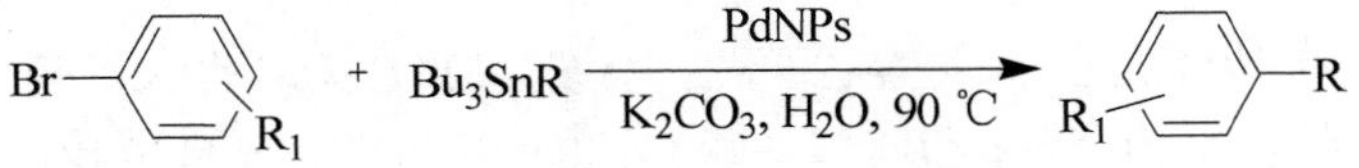

Khashab 等[8,9]合成了四种形貌(花状、球状、树枝状和分支)的 Au@Pd 核壳纳米粒子及 Au-Pd 合金纳米粒子, 并考察它们在水中 Suzuki-Miyaura 和 Heck 交叉偶联反应中的催化性能。结果表明, 无论是 Au@Pd 核壳结构催化剂还是 Au-Pd 合金催化剂, 在不同底物的 Suzuki-Miyaura [式(3-5)]和 Heck [式(3-6)]交叉偶联反应中均显示高催化效率。含有分支的 Au@Pd 核壳结构催化剂催化效率最高, 例如, 在对羧基苯硼酸与不同碘苯的反应中, 仅用 0.25 mol%的 Au@Pd 催化剂, 产物得率大于 98%。而在 Heck 反应中, 产物的得率为 90%~95%。这种 Au@Pd 核壳结构催化剂是在高温(340℃)下制得的, 所以非常稳定。

第三章 金属催化水介质有机合成

$$\text{对-B(OH)}_2\text{-苯-CO}_2\text{H} + \text{I-苯-R} \xrightarrow[\text{CTAB, H}_2\text{O, 80 ℃}]{\text{Au-Pd, NaHCO}_3} \text{HO}_2\text{C-苯-苯-R} \quad (3\text{-}5)$$

$$\text{苯-I} + \text{CH}_2\text{=CH-R} \xrightarrow[\text{CTAB, H}_2\text{O, 80 ℃}]{\text{Au-Pd, NaHCO}_3} \text{苯-CH=CH-R} \quad (3\text{-}6)$$

Nacci 等[10]报道在温和条件(40~90 ℃)下,以葡萄糖为还原剂、四甲基氢氧化铵为碱、表面活性剂为相转移催化剂,用 Pd 胶体催化水相溴代芳烃和氯代芳烃 Ullmann 反应,发现含有给电子基团的对甲氧基氯苯的自偶联反应获得产物 4,4'-二甲氧基联苯的得率在 80%以上。

有报道采用 Au/Pd 双金属活化 C—Cl,使对氯代芳烃在水相中发生 Ullmann 反应[11,12]。当用 2 atm% (原子百分数) $Au_{0.5}/Pd_{0.5}$:PVP、1.5mol KOH 和 1:1 的 H_2O/DMF 时,产物得率达到 86%~99%[11]。也有报道用莘菠合成了 Pd 纳米粒子(约 20 nm)催化剂,在无配体、胺和铜的 Sonogashira 反应[式(3-7)]中显示出非常高的催化性能,产物得率为 84%~95%,催化剂通过简单的静置、倾倒即可与反应体系分离,且重复使用 5 次[13],但该催化体系不适用于氯代芳烃的偶联反应。

$$\text{ArX} + \text{≡≡R} \xrightarrow[\text{K}_2\text{CO}_3, \text{H}_2\text{O, 80 ℃}]{\text{PdNPs}} \text{Ar≡≡R} \quad (3\text{-}7)$$
$$\text{X=Br, I}$$

Amatore 等[14]用一种新型方法制得核壳结构纳米粒子 Au@Pd 催化剂,Au 纳米核的周围包裹一层或多层 PdNPs,通过表面增强拉曼光谱和循环伏安法对其进行了表征,以 K_2CO_3 为碱,室温下可催化对甲氧基溴苯与苯硼酸的偶联反应,获得 88%得率的 4-甲氧基联苯产物。

2015 年,Nasrollahzadeh 和 Banaoei[15]报道了一种 Au/Pd 双金属纳米粒子催化剂,从 SEM 图可看出,Au 纳米粒子表面覆盖 80 nm 的 Pd 纳米粒子,EDS 分析显示,催化剂中 Pd、Au 和 O 的含量分别为 7%、16%和 77%,其中的 O 属于物理吸附态。在催化水中溴代或碘代芳烃与烯的 Heck 反应中具有较好的催化性能,在 80 ℃下用 0.3 mol%的 Au/Pd 催化剂,获得 80%~94%得率的偶联产物,催化剂可重复使用 4 次。

Lipshutz 等[16]以 $FeCl_3$ 和 CH_3MgCl 为原料制备了 Fe 纳米粒子,在制备过程中掺杂 10^{-6} 级的 $Pd(OAc)_2$,得到双金属 Fe-Pd 催化剂,可在室温下催化水相 Suzuki-Miyaura 交叉偶联反应,适应底物范围广,尤其是对于活性较低的氯代芳烃,反应 16~48 h,能获得 72%~85%的得率。

Suzuki 等[17]设计了一种微型管状反应器用于金属催化高温高压水中 Sonogashira 交叉偶联反应。当温度为 250℃，压力为 16 MPa 时，碘苯与苯乙炔在 Pd 催化下反应 1.6 s，碘苯的转化率为 47.4%，选择性达 100%。如使用 Pd-Cu 合金为催化剂，在相同的反应条件下，由于共催化作用，碘苯的转化率提高到 83.5%，选择性仍保持在 100%。然而，随着反应时间的推移，会有少许有机物覆盖在 Pd-Cu 催化剂的表面，使其失活。将此失活催化剂在 750℃下焙烧 2 h，可恢复其活性。

二、还原反应

水介质中芳环的催化加氢反应在燃料工业、造纸工业及有机合成中具有重要的应用。采用不同的 Ru 催化体系，可实现 2-甲氧基-4-丙基苯酚和磨碎的木质素的芳香加氢反应。许多过渡金属纳米粒子均显示出高催化效率。如通过 $IrCl_3$ 和 N,N-二甲基-N-十六烷基-N-(2-羟基乙基)氯化铵作用制备的 Ir 纳米粒子，在水溶液催化芳香衍生物的氢化反应中显示出高活性，在温和条件下就可顺利进行反应[式(3-8)][18]。

$$\text{PhOCH}_3 \xrightarrow[\text{H}_2\text{O, 20 ℃, 0.75 h}]{\text{Ir NPs}} \text{C}_6\text{H}_{11}\text{OCH}_3 \qquad (3\text{-}8)$$

同样地，用 $RhCl_3 \cdot H_2O$ 与 N, N-二甲基-N-十六烷基-N-(2-羟基乙基)铵盐作用得到的悬浮在水中的 Rh 纳米粒子，也能催化芳烃衍生物的加氢反应[式(3-9)]，不仅反应条件温和(1 atm① H_2，室温)，且催化剂可多次重复使用[19]。

$$\text{Ph-R} \xrightarrow[\text{20 ℃, 5~20 h}]{\text{Rh NPs}} \text{C}_6\text{H}_{11}\text{-R} \qquad (3\text{-}9)$$

$$R = OCH_3, OH, NH_2 \text{ 等}$$

在稀碱水溶液中，Ranney Ni/Al 合金[20, 21]可催化水/超临界流体体系中芳烃衍生物的加氢反应。芳香酮在水中回流反应 2 h，得到产物亚甲基化合物，得率达 89%~99.8% [式(3-10)][22]。

$$\text{PhCOCH}_3 \xrightarrow[\text{H}_2\text{O}]{\text{Ni-Al}} \text{PhCH(OH)CH}_3 \ (1\%) + \text{PhCH}_2\text{CH}_3 \ (99\%) \qquad (3\text{-}10)$$

Reinhard 报道 Pd、Pd-Cu 及 Ni 可有效催化水中 N-亚硝基二甲胺生成二甲胺和胺的加氢反应[式(3-11)][23]。Bras 等[24]以 $Pd(OAc)_2$ 为钯源，在四丁基溴化铵和三丁基胺的存在下制得 Pd 纳米粒子，催化水或甲苯溶剂中各种化合物中 C=C 的选择

---

① atm，标准大气压，1atm=1.01325×10⁵Pa。

性加氢反应[式(3-12)]，在水中获得产物的得率和选择性明显优于甲苯溶剂中。在最佳条件下(1 mmol 底物、0.01 mmol 催化剂、2 mL 水、1 atm H_2、室温)，不同底物选择性加氢反应活性均较高，且催化剂能重复使用 9 次。

$$\text{Me}_2\text{N-N=O} + H_2 \xrightarrow[H_2O]{Pd-Cu} HNMe_2 + NH_3 \qquad (3\text{-}11)$$

$$\text{R-CH=CH-FG} + H_2 \xrightarrow[H_2O]{1\% Pd(OAc)_2} \text{R-CH}_2\text{-CH}_2\text{-FG} \qquad (3\text{-}12)$$

2009 年，Wai 等[25]发展了一种在 H_2O-CO_2 微乳体系中 Pd 纳米粒子催化不饱和化合物[式(3-13)~式(3-16)]的选择性加氢反应，如硝基苯在 50℃的超临界 CO_2 体系中反应 30 min，产物苯胺的得率大于 99%。

$$\text{PhNO}_2 + H_2 \xrightarrow[H_2O]{Pd\ NPs} \text{PhNH}_2 \qquad (3\text{-}13)$$

$$\text{马来酸} + H_2 \xrightarrow[H_2O]{Pd\ NPs} \text{琥珀酸} \qquad (3\text{-}14)$$

$$\text{4-MeO-C}_6\text{H}_4\text{-CH=CH-CO}_2\text{H} + H_2 \xrightarrow[H_2O]{Pd\ NPs} \text{4-MeO-C}_6\text{H}_4\text{-CH}_2\text{CH}_2\text{-CO}_2\text{H} \qquad (3\text{-}15)$$

$$\text{PhCH=CHPh} + H_2 \xrightarrow[H_2O]{Pd\ NPs} \text{PhCH}_2\text{CH}_2\text{Ph} \qquad (3\text{-}16)$$

Roucoux 等[26]报道用含有十烷基的手性表面活性剂$(1R, 2S)$-(−)-NMeEph12X (248)稳定 Rh NPs (2.5 nm)，使用此催化体系 Rh@NMeEph12X 催化水中丙酮酸乙酯不对称还原[式(3-17)]。当使用 NMeEph12X (X = Br$^-$)(S)-(−)-乳酸-表面活性剂时，给出产物的 ee 值达 13%。如使用手性诱导剂(−)-金鸡纳啶时，ee 值可以提高到 18%。

$(1R,2S)$-(−)- NMeEph12X
248

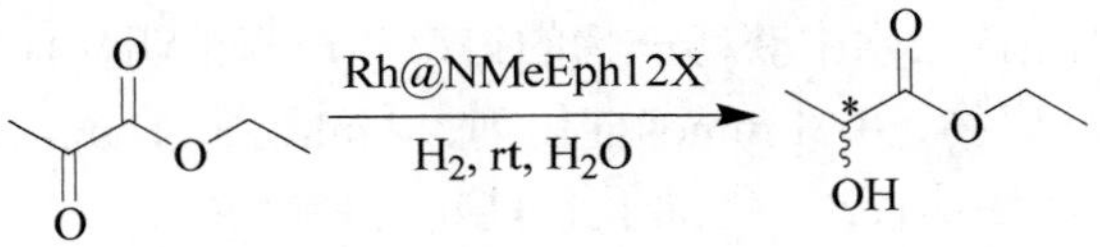

$$\text{(ethyl pyruvate)} \xrightarrow[\text{H}_2,\ \text{rt},\ \text{H}_2\text{O}]{\text{Rh@NMeEph12X}} \text{(ethyl lactate)} \tag{3-17}$$

最近，他们又发展了在手性铵盐表面活性剂作用下，通过静电作用稳定大小约为 2.5 nm 的 Rh 纳米粒子(图 3.2)，并将其用于催化水中苯及其衍生物不对称还原，如茴香醚的还原反应，用($1S$, $2S$, $4S$, $5R$)-(+)-QCl16Br (249)表面活性剂稳定 Rh 纳米粒子，在室温下反应 1 h，茴香醚的转化率达 80%，产物的立体选择性为 70% [式(3-18)]，($1S$, $2S$, $4S$, $5R$)-(+)-QCl16Br 明显优于($1R$, $2S$)-(−)-NMeEph12(S)-Lactate 稳定剂[27]。

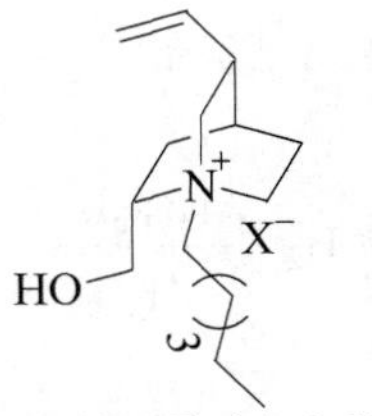

($1S$, $2S$, $4S$, $5R$)-(+)-QCl16Br

249

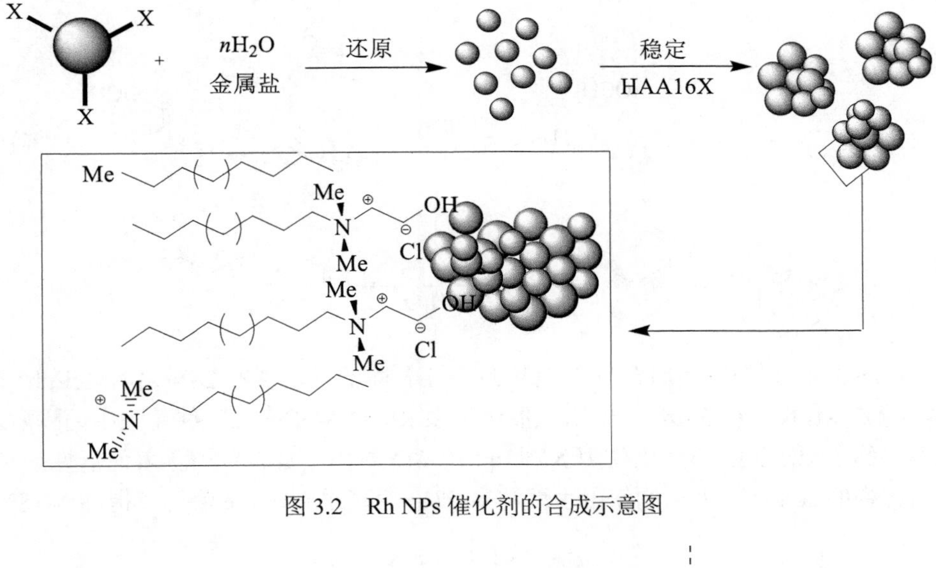

图 3.2 Rh NPs 催化剂的合成示意图

$$\text{(1,2-dimethoxybenzene)} \xrightarrow[\text{H}_2\text{O},\ \text{H}_2,\ \text{rt}]{\text{Rh NPs@手性铵盐表面活性剂}} \text{(products)} \tag{3-18}$$

三、其他反应

近来，Li 和 Wei[28]报道 Ru-In 双金属催化水中 C—H 活化实现各种醛与苯乙炔的高效加成反应[式(3-19)]，得到 Grignard 类型的亲核加成产物。

$$R'CHO + R{\equiv}H \xrightarrow[H_2O]{Ru\text{-}In} R{\equiv}\text{—}\underset{OH}{CH(R')} \qquad (3\text{-}19)$$

Chung 等[29]通过还原含有十二烷基苯磺酸钠(SDS)的乙酸钴水溶液，得到水相胶体 Co 纳米粒子，高效催化水中的分子内 Pauson–Khand 反应[式(3-20)]，催化剂可以回收重复使用 8 次。

$$\underset{Ph}{\text{烯炔底物}} \xrightarrow[H_2O, 130\ ^\circ C, 12\ h]{Co\ NPs,\ 20\ atm,\ CO} \underset{Ph}{\text{双环产物}} \qquad (3\text{-}20)$$

2012 年，Pal 等[30]以 $PdCl_2$ 为钯源，制备出 Pd 纳米粒子，在催化水相二[杂环基]甲烷衍生物的合成反应[式(3-21)~式(3-24)]中，表现出较好的活性和较长的使用寿命，对于不同底物，给出产物的得率为 86%~100%，催化剂可重复使用 4 次。

$$\text{二甲基环己二酮} + ArCHO \xrightarrow[H_2O, 回流]{Pd\ NPs} \text{产物} \qquad (3\text{-}21)$$

$$\text{吲哚} + ArCHO \xrightarrow[H_2O, 回流]{Pd\ NPs} \text{二(吲哚-3-基)甲烷} \qquad (3\text{-}22)$$

$$\text{1-苯基-3-甲基-5-吡唑酮} + ArCHO \xrightarrow[H_2O, 回流]{Pd\ NPs} \text{产物} \qquad (3\text{-}23)$$

$$\text{4-羟基香豆素} + ArCHO \xrightarrow[H_2O, 回流]{Pd\ NPs} \text{产物} \qquad (3\text{-}24)$$

2011年，Park等[31]报道了Ag纳米粒子(约80 nm)催化腈水合合成酰胺的反应[式(3-25)]，在含0.3 mol%催化剂、7 mL水和3.0 mol苯乙腈的25 mL高压釜中，于150℃下反应1 h，苯甲酰胺的得率为100%，催化剂可重复使用3次。研究表明，苯乙腈中苯环上的取代基性质对其他腈的水合反应活性没有明显的影响，但空间位阻对反应活性影响比较明显，如邻位甲基苯乙腈的水合反应活性小于对甲基苯乙腈和间甲基苯乙腈，脂肪族腈如乙腈在此体系中也能顺利地进行水合反应，但产物得率较低，180℃下反应6 h，得率仅为46%。

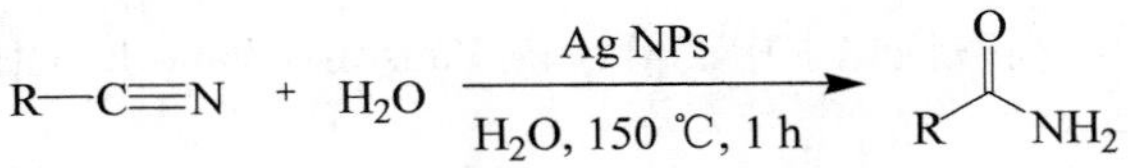

$$R-C\equiv N + H_2O \xrightarrow[H_2O,\ 150\ ℃,\ 1\ h]{Ag\ NPs} \underset{R}{\overset{O}{\|}}-NH_2 \tag{3-25}$$

第二节 负载型金属催化剂

为了提高金属催化剂的催化效率、增强催化材料抵抗机械破坏的能力以及提高分散度等，通常需要将催化活性组分负载到适宜载体上，有时还会添加共催化组分或助催化组分。载体对催化剂活性成分有重要的支撑作用，可以使催化剂活性成分的分散度大大提高，这样不仅可以抑制晶粒增长、提高选择性，还能起到导热作用，有时甚至能生成其他化合物与固溶体，调变反应微环境，改善催化性能。有些载体如碳纳米管(CNTs)，不仅具有支撑功能，还兼具助剂作用。因此，载体也是提高催化效率的关键因素之一。目前常用的载体主要有高分子、金属氧化物、活性炭、CNTs以及各种生物材料等。

一、负载型Pd催化剂

1. 偶联反应

1) 碳材料负载钯催化剂

Bumagin等首次报道[32]将Pd/C应用在水介质间溴苯甲酸与四苯基硼酸钠的Suzuki反应[式(3-26)]。如添加3 eq Na_2CO_3，10% Pd/C (0.3 mol%)在室温下能催化水中对碘苯酚和硼酸的Suzuki反应[式(3-27)]，产物得率不受硼酸种类影响，催化剂通过简单的过滤就能回收并可重复使用5次[33]。

$$\text{(3-26)}$$

X = I, Br, Cl;
R_1 = 4-OCH$_3$, 4-CHO, 2-NO$_2$, 2-COCH$_3$
R_2 = H, 4-CH$_3$, 4-F, 3-OCH$_3$

$$\text{(3-27)}$$

R = H, CH$_3$, 4-F, OCH$_3$

Xu 等[34]发展了水介质溴代芳烃与四苯基硼酸钠的偶联反应[式(3-28)]，采用 0.0025 mol% Pd/C，100 ℃下反应 1~7 h 就可使反应完成，钠盐碱比钾盐碱更有利于反应，Pd/C 催化剂可重复使用。

$$\text{(3-28)}$$

对于非水溶性卤代芳烃的偶联反应，一般是采用添加表面活性剂提高反应底物在水中的溶解度，如 Arcadi 等[35]通过添加 CTAB 提高底物的溶解度。当以 CTAB 为表面活性剂，K$_2$CO$_3$ 为碱时，5 mol% Pd/C 可成功地催化不同芳烃与硼酸的偶联反应。尤其重要的是，这种反应体系也适用于氯代芳烃的 Suzuki 反应，给出较高得率的偶联产物，不足之处是催化剂容易失活。

Kohler 和 Lysen 报道了用无配体的 Pd/C 催化水中氯代芳烃与芳基硼酸的 Suzuki 反应[36,37]，所有反应在空气氛围室温下进行，对于活泼的氯代芳烃，Pd 催化剂用量仅为 0.2 mol%~0.5 mol%，而对于不活泼的氯代芳烃，Pd 催化剂用量要提高到 2.0 mol%，反应时间也要适当延长至 6 h，同时反应需要四丁基溴化铵作为表面活性剂，KOH 为碱。对于碘代芳烃和溴代芳烃，只要稍优化反应条件，反应物即可完全转化成联苯衍生物。不仅硼酸可发生 Suzuki 反应，硼酸酯和硼酸盐同样也能发生 Suzuki 反应[式(3-29)和式(3-30)]。

Freundlich 和 Landis 将微波引入多取代的溴代芳烃如 4-羟基-3-甲氧基溴苯与硼酸的水相 Suzuki 反应中[38]，各种硼酸在 KOH 水溶液中均能发生偶联[式(3-31)和式(3-32)]，反应在 15 min 内就能完成，但不能应用于氯代芳烃的偶联反应，三氟苯硼酸钾与对溴苯酚的反应[式(3-33)]活性明显低于对溴苯酚与苯硼酸的反应。

$$\text{(3-29)}$$

$$\text{(3-30)}$$

X = I, Br, Cl

$$\text{(3-31)}$$

$$\text{(3-32)}$$

$$\text{(3-33)}$$

Arvela 和 Leadbeater[39]将 CTAB 应用到微波加热的水相氯代芳烃与硼酸的交叉偶联反应中，对于含有吸电子基团的底物，产物得率较低，这是由于氯代芳烃的偶联反应比分解反应快；而对于含有中性电子基团或给电子基团的底物，产物得率明显提高，且反应在 10 min 内就能完成。Bai[40]将相似的催化体系应用于四苯基硼钠与溴代芳烃的偶联反应，在 120 ℃下反应 20 min，获得高得率的偶联产物，但催化剂用量稍高(5mol%)，催化剂可重复使用 5 次。

Pal 等[41]用 Pd/C 分别催化水中邻碘苯酚和 4-硝基-2,4-二碘苯酚与末端炔的偶联反应合成 2-烷基/芳基取代苯并[b]呋喃/硝基苯并[b]呋喃及其衍生物[式(3-34)]，不需要相转移催化剂和水溶性膦配体，也不需要有机溶剂作共溶剂，但需要加入 PPh$_3$ 配体及 CuI 共催化剂。类似的催化体系也可用于水相邻碘苯胺与末端炔的偶联反应，不同的是，该反应中使用氨基乙醇作为碱[式(3-35)]。此催化体系还可应用在 6-溴-8-碘-1,2,3,4-四羟基喹啉与末端炔的偶联反应合成[式(3-36)]，获得高得率的目标产物[42]。

$$\text{(结构式)} \xrightarrow[\text{(S)-脯氨酸, H}_2\text{O, 80 ℃}]{10\% \text{ Pd/C, PPh}_3, \text{CuI}} \text{(产物)} \quad (3\text{-}34)$$

$$\text{(结构式)} \xrightarrow[\text{2-氨基甲醇, H}_2\text{O, 80 ℃}]{10\% \text{ Pd/C, PPh}_3, \text{CuI}} \text{(产物)} \quad (3\text{-}35)$$

R = 脂基, 芳基; X = H, COCF$_3$, SO$_2$CH$_3$

$$\text{(结构式)} \xrightarrow[\text{2-氨基甲醇, H}_2\text{O, 80 ℃}]{10\% \text{ Pd/C, PPh}_3, \text{CuI}} \text{(产物)} \quad (3\text{-}36)$$

Raju 等[43]将上述催化体系应用于水相 2-碘噻吩与不同末端炔的偶联反应，得到不同的 2-炔基噻吩产物[式(3-37)]。Reddy 等[44]用三乙胺取代上述反应体系中的氨基乙醇，将 Pd/C-CuI-PPh$_3$ 应用在水相 2-氯喹啉与末端炔的反应中，从而获得产物 2-炔基喹啉[式(3-38)]。Pd/C-CuI-PPh$_3$ 对水和空气都比较稳定，在疏水和亲水的末端炔的反应中均显示出高活性，反应中未检测到二炔和 2-氯喹啉水解等副产物。在 2,4-二氯喹啉与炔的反应中，可以得到高立体选择性的产物 2-炔基-4-氯喹啉。

$$\text{(结构式)} \xrightarrow[\text{2-氨基甲醇, H}_2\text{O, 80 ℃}]{10\% \text{ Pd/C, PPh}_3, \text{CuI}} \text{(产物)} \quad (3\text{-}37)$$

$$\text{(结构式)} \xrightarrow[(\text{C}_2\text{H}_5)_3\text{N, H}_2\text{O, 80 ℃}]{10\% \text{ Pd/C, PPh}_3, \text{CuI}} \text{(产物)} \quad (3\text{-}38)$$

R = C(CH$_3$)$_2$OH, CH$_2$OH, CH$_2$CH$_2$OH, (CH$_2$)$_3$CH$_2$OH, (CH$_2$)$_5$CH$_2$OH, CH$_2$CH(OH)CH$_3$

Layek 等[45]将该催化体系应用于水中 2-炔基-7-氮杂吲哚合成反应，80℃下用 10% Pd/C 可高效催化 4-氯-2-碘-7-氮杂吲哚与末端炔偶联，获得中等得率的目标产物 2-炔基-4-氯-7-氮杂吲哚[式(3-39)]。

$$\text{(结构式)} \xrightarrow[(\text{C}_2\text{H}_5)_3\text{N, H}_2\text{O, 80 ℃}]{10\% \text{ Pd/C, PPh}_3, \text{CuI}} \text{(产物)} \quad (3\text{-}39)$$

1999 年，Venkatraman 和 Li[46]报道了在室温下和空气中，以 Zn 为还原剂，Pd/C

为催化剂，催化水/丙酮介质中卤代芳烃的还原偶联反应，多种碘代芳烃和溴代芳烃均能有效发生反应。他们也发现，在纯水中加入表面活性剂[47-49]，如冠醚等，可大大提高产物得率。Sasson 等[50]发现，通过加入聚乙二醇(PEG-400)添加剂和提高反应温度，其他卤代芳烃也能发生还原偶联反应。在催化量的 PEG-400 和 0.4 mol%的负载在活性炭上的三金属催化剂(4%Pd，1%Pt 和 5%Bi)存在时，氯代苯在水中发生还原偶联生成联苯，产率为 93%~95%，此反应条件减少了竞争性的还原反应[51]。研究证实，H_2 是原位产生的。除 Pd/C 外，Rh/C 也是高效催化剂[50]。最近也发现，CO_2 能促进 Pd 催化 Zn 参与的卤代烃的 Ullmann 反应[式(3-40)]，在 CO_2/Zn 和 Pd/C 存在时，包括低活性的氯代芳烃的各种芳香族卤代烃都可发生 Ullmann 反应，产物得率较高[52]。

$$ArX \xrightarrow[Zn, CO_2, H_2O, 18\text{-}冠\text{-}6]{Pd/C} Ar\text{-}Ar \quad (3\text{-}40)$$

Hain 等[53]采用改性多壁碳纳米管(MWCNT)为载体负载 Pd 纳米粒子，方法是先将 $NaPdCl_2$ 和 4-二甲氨基吡啶(DMAP)混合，然后用 $NaBH_4$ 还原，再将所得的 Pd 纳米粒子在超声作用下固载在巯基功能化的 MWCNT 上，得到 MWCNT/Pd-DMAP 催化剂。在催化水中对碘代芳烃的偶联反应时，仅用 0.004 mol%Pd，在 100℃下的 N_2 氛围中反应 10 min，转化率即可达 87%，但如采用溴代芳烃和氯代芳烃，反应 6 h 转化率仅为 53%和 25%，催化剂可以重复使用 6 次。

Zhang 等[54]在 SDS 存在下，将乙酸钯还原成 4 nm 的 Pd 纳米粒子，然后固载到石墨烯上，SDS 既作表面活性剂又作还原剂。在空气氛围中和室温下，碘苯和苯硼酸的 Suzuki 反应在 5 min 内完成。但在溴苯和烯丙基碘与苯硼酸的 Suzuki 反应中活性较低，联苯和 3-苯基丙烯反应的得率分别为 29.5%和 6.5%。

Wan 等[55]通过将过渡金属氧化物纳米粒子嵌入介孔碳中，利用金属氧化物界面的相互作用，改变 Pd 纳米粒子的电荷密度及分散度，得到约三原子构成的 Pd 金属簇，有利于氯苯的活化，其中介孔 Pd/CoO-C 催化剂在无相转移试剂和配体的条件下，可以有效催化氯苯与苯硼酸的偶联，产物联苯的得率为 49%，当氯苯的苯环上含有吸电子基团时，偶联产物的得率可达 90%，催化剂可重复使用 10 次。Li 和 Huang[56]采用 Pd/C 催化水和空气中的卤代芳烃和芳基卤硅烷的偶联反应[式(3-41)]，获得高得率的联苯类产物。

$$ArX + PhRSiCl_2 \xrightarrow[KOH (或 NaF)/H_2O]{Pd\ cat.} Ar\text{-}Ph \quad (3\text{-}41)$$

Sharavath 和 Ghosh[57]将 Pd 纳米粒子通过非共价键方式固载在 1-芘甲酸功能化的石墨烯上(PCA-GNS)，得到催化剂 PCA-GNS-Pd，不需加入相转移催化剂，就能催化氯代芳烃和溴代芳烃在纯水中 Suzuki-Miyaura 和 Heck-Mizoroki 的交叉偶联反应，这可归因于 PCA-GNS-Pd 具有亲水和亲油性质，在水中非常容易分散，因而具有较高的活性。另外,由于催化剂中含有羧基,在催化剂制备过程中,Pd^{2+} 嵌入和吸附在载体 PCA-GNS 表面,这也有利于有机底物在催化剂表面的吸附。在这两个偶联反应中,催化剂可循环使用 5 次。

Kamal 等[58]设计了一种具有两亲性能的载体碳纳米球负载的 Pd 纳米粒子催化剂(Pd@CSP),并将其应用在水相 Suzuki 反应和 Ullmann 反应,可在空气氛围下进行,不需 P 配体。在 Heck 反应中,溴代芳烃能与不同的烯发生反应,而在 Ullmann 反应中,反应体系需要加入相转移试剂。实验证实,Pd@CSP 具有高活性主要是由于载体具有较强的疏水性,同时,催化剂可重复使用 4 次。

Nájera 等[59]先用乙二醇还原 Pd 和 Ni 的前驱体,然后将所得的 Pd 和 Pd-Ni 纳米粒子沉积在多壁纳米管(MWNT)上,制得催化剂 PdNi/MWNT。无配体、水溶剂等条件下,较少量的催化剂在 120℃下可以使多种 C—C 偶联反应顺利进行,如对甲氧基碘苯与苯硼酸的 Suzuki 反应[式(3-42)],催化剂可重复使用 4 次。在对甲氧基溴苯与苯乙烯的 Mizoroki-Heck 反应[式(3-43)]中,以 K_2CO_3 为碱,加入一定量的四丁基溴化铵就能给出定量的产物,但催化剂不能重复使用;而 Pd/MWCNT 在 50% NaOH 溶液中催化对甲氧基碘苯与三甲氧基苯基硅的 Hiyama 反应[式(3-44)],无需氟配体就能给出 83% 的对甲氧基联苯产物,但催化剂使用寿命差。在对甲氧基碘苯与苯乙炔的 Sonagashira 反应[式(3-45)]中,不仅可以定量给出目标产物,还可以重复使用 3 次。

$$CH_3O-\!\!\!\!\bigcirc\!\!\!\!-I + \bigcirc\!\!\!\!-B(OH)_2 \xrightarrow[120\ ℃]{PdNi/MWNT \atop K_2CO_3, TBAB} CH_3O-\!\!\!\!\bigcirc\!\!\!\!-\!\!\!\!\bigcirc \quad (3\text{-}42)$$

1st run: 98%, 4st run: 87%

$$CH_3O-\!\!\!\!\bigcirc\!\!\!\!-Br + \bigcirc\!\!\!\!=\!\!\!\!\bigcirc \xrightarrow[120\ ℃]{PdNi/MWNT \atop K_2CO_3, 添加剂, H_2O} CH_3O-\!\!\!\!\bigcirc\!\!\!\!-\!\!\!\!=\!\!\!\!-\!\!\!\!\bigcirc \quad (3\text{-}43)$$

$$CH_3O-\!\!\!\!\bigcirc\!\!\!\!-I + \bigcirc\!\!\!\!-Si(OCH_3)_3 \xrightarrow[50\% \text{ NaOH}, 120\ ℃]{PdNi/MWCNT} CH_3O-\!\!\!\!\bigcirc\!\!\!\!-\!\!\!\!\bigcirc \quad (3\text{-}44)$$

$$\underset{OH_3C}{\overset{I}{\bigcirc}} + \bigcirc\!\!\!\!-\!\!\!\!\equiv\!\!\!\!-\!\!\!\!\bigcirc \xrightarrow[120\ ℃]{PdNi/MWCNT \atop 吡咯烷, H_2O} CH_3O-\!\!\!\!\bigcirc\!\!\!\!-\!\!\!\!\equiv\!\!\!\!-\!\!\!\!\bigcirc$$

1st run: 99%, 3rd run: 64%, 4th run: 0%

(3-45)

2) 氧化物负载钯催化剂

将 NaPdCl$_2$ 与强碱性金属氧化物纳米晶 MgO (NAP-MgO)在氮气下反应 12 h，然后用水合肼还原，得 Pd NAP-MgO 催化剂[60]，应用于水相碘代和溴代芳烃与芳基硼酸的 Suzuki 偶联反应，采用 0.5 mol%催化剂，在室温下反应 5~6 h，即可完成反应，且催化剂可重复使用 5 次，优越的催化性能可归因于纳米结构的 MgO 材料具有较大的比表面积(~600 m^2/g)和较强的碱性。

Cloffi 等[61]通过电化学浸渍法制得四边形 ZrO$_2$ 负载的 Pd 纳米粒子催化剂(Pd-ZrO$_2$)，并将其应用在水相卤代芳烃的 Suzuki 反应中，展现出高活性，溴代芳烃与苯硼酸的偶联产物平均得率为 83%，但给电子的对甲氧基溴苯在此反应体系中活性较低，催化剂可循环使用 10 次。

Sardarian 等[62]以稻秆为原料制备纳米 SiO$_2$ (nSiO$_2$)，然后对其进行氨基改性后，与 Pd(OAc)$_2$ 反应后再用 NaBH$_4$ 还原，得到树枝状 nSiO$_2$ 包裹的 Pd 催化剂(图 3.3)。其在最佳条件下能成功催化水中不同卤代烃与炔的 Sonogashira-Hagihara 偶联反应[式(3-46)]。研究发现，含有给电子基团的碘苯比含有吸电子基团的碘苯需要更长的反应时间；而无论是含有吸电子基团还是给电子基团的苯乙炔与溴苯反应时均能给出高得率的偶联产物。对于位阻较大的溴代芳烃如邻溴苯甲醚、邻溴甲苯及

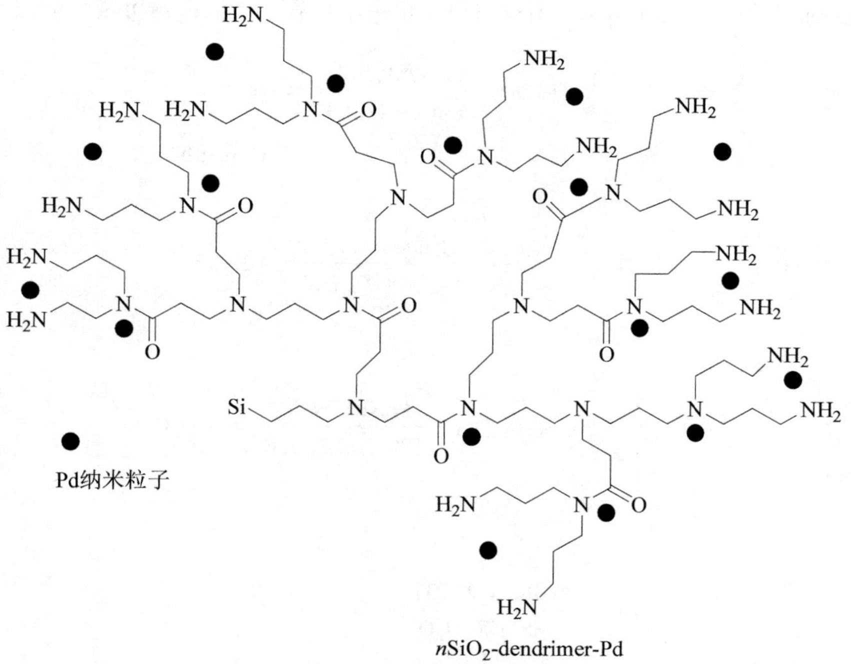

图 3.3　nSiO$_2$-dendrimer-Pd(0)催化剂结构示意图

$$\underset{X = Cl, Br, I}{R-\!\!\!\!\!\bigcirc\!\!\!\!\!-X} + \equiv\!\!-R_1 \xrightarrow[(C_2H_5)_3N, H_2O, 90\sim100\ ℃]{n\text{SiO}_2\text{-dendrimer-Pd}(0)} R-\!\!\!\!\!\bigcirc\!\!\!\!\!-\equiv\!\!-R_1 \qquad (3\text{-}46)$$

邻溴苯甲醛等均能顺利与苯乙炔反应，产物得率为 86%~90%。尽管氯代芳烃的反应活性较低，但也能给出中等到较高得率的偶联产物。此外，溴代芳烃和碘代芳烃在此催化剂存在下，也可以与脂肪炔如 1-辛炔发生偶联反应，给出较高得率的产物。而氯苯与 1-乙烯基环己醇偶联时获得目标产物的得率高达 90%，在碘苯与苯乙炔偶联反应中的 TON 值高达 1130，且催化剂可重复使用 6 次以上。

Khalafi-Nezhad 和 Panahi 采用淀粉功能化的二氧化硅载体负载 Pd 纳米粒子，在水相 Heck 反应和无铜参与的 Sonogashira 反应中，不同底物的 Heck 和 Sonogashira 反应中表现出较高的催化活性[63]，尤其是催化氯代芳烃的偶联反应，反应 8~12 h 能给出中等以上的得率，二氧化硅载体可有效抑制 Pd 纳米粒子团聚，提高了 Pd 分散度，导致高活性。Panahi 和 Khalafi-Nezhad[64]以环糊精修饰二氧化硅为载体固载 Pd 纳米粒子，使用不同环糊精功能化的二氧化硅，分别获得不同粒径的 Pd 纳米粒子，在催化水介质芳卤和烯烃的 Heck 偶联反应中显示出高效率，用含 1.1 mol% Pd 的催化剂，在 2 mL H_2O 和 2 mmol K_2CO_3 的介质中回流 5 h，溴苯与苯乙烯发生偶联反应，产物得率为 95%。在含有给电子基团和吸电子基团的溴苯与苯乙烯的反应中，产物得率均在 97%以上，对于含有取代基的氯苯，其反应活性明显较低，但延长反应时间至 12 h，产物也能有 67%以上的得率，即使是位阻较大的邻甲氧基氯苯，反应 15 h,产物的得率也可以达到 61%。同时,催化剂至少可重复使用 5 次。

3) 无机盐负载钯催化剂

Paul 等[65]先将水合肼滴加到 Pd(OAc)$_2$ 和羟基磷灰石(HAP)的乙醇混合物中，然后分别在乙醇、甲苯和乙腈溶剂中搅拌回流 6 h，得催化剂 Pd/HAP，TEM 图显示 Pd 平均粒径约为 20 nm，表明该催化剂非常稳定。其在 TBAB 表面活性剂和 K_2CO_3 存在下，可高效催化溴代芳烃和芳基硼酸的 Suzuki 反应，催化剂可重复使用 5 次以上。

Riela 等[66]用多水高岭土-三唑鎓盐作为配体负载金属 Pd 催化剂，采用含 0.1 mol% Pd 的催化剂催化水介质 Suzuki 交叉偶联反应，在 120℃微波辐射 10 min，就能获得 99%得率的偶联产物，且催化剂可重复使用 5 次。

4) 高分子材料负载钯催化剂

Wang 等[67]用离子型共聚物作为载体固载 Pd 纳米粒子催化剂，在水介质溴代芳烃的 Suzuki-Miyaura 反应中，不仅具有高活性，且至少可以重复使用 5 次。Zheng 和 Zhang 报道[68]用对 pH 响应的核壳结构聚[苯乙烯-co-2-(乙酰乙酸)甲基丙烯酸乙酯-co-丙烯酸甲酯]胶体作为载体固载 Pd 纳米粒子，这种 pH 响应的载体在碱性水溶液中高度分散，近似于均相催化剂，在 Heck 和 Suzuki 反应中显示优异催化性能，如在水介质对溴苯乙酮与苯硼酸的交叉偶联反应中，80℃下反应 8 h，产物 4-苯基苯乙酮的得率为 100%，但在对氯苯乙酮的偶联反应中，催化活性较弱。特别重要的是，反应结束后，通过调节反应体系的 pH，可使催化剂沉淀分离并重复使用。

Ohtaka 等[69]用含有聚[4-氯甲基苯乙烯-co-(4-乙烯基苯基)]三丁基氯化铵和聚丙烯酸的离子型复合物固载 Pd 纳米粒子，在 Suzuki 和 Heck 反应中均显示高活性。Nabid 和 Bide[70]用聚己内酯(PCL)和聚乙烯醇(PEG)功能化的高支链脂肪族聚酯(H40)H40-PCL-PEG 作为反应底物，原位合成负载型 Pd 纳米粒子催化剂，在室温下催化水中不同卤代芳烃与烯烃化合物的 Heck 反应，在较短的时间内就能给出较高得率的产物，且催化剂易于分离并能循环使用。也有报道称，在室温下通过 K_2PdCl_4 在水溶液中还原，再与氨硼烷原位水解得到聚(4-苯乙烯磺酸-马来酸)固载的 Pd 纳米粒子催化剂[71]，应用于水介质中各种卤代芳烃或碘代芳烃与苯硼酸进行的 Suzuki-Miyaura 交叉偶联反应，显示出优良的催化性能，在对溴苯乙酮或对碘苯乙酮与苯硼酸的反应中，TOF 值分别达到 1980 h^{-1} 和 5940 h^{-1}。另外，Dell'Anna 等[72]合成了高分子负载的 Pd 纳米粒子，在 100℃下催化水中溴代芳烃与苯硼酸的 Suzuki 反应，当在反应体系中加入相转移试剂(如 TBAB)，氯代芳烃也可以发生 Suzuki 反应。

Yokoyama 等[73]用甲醇还原 $Pd(OAc)_2$，获得小于 6 nm 的 Pd 纳米粒子并将其固载在咪唑鎓盐功能化的聚苯乙烯载体上，在不加膦配体和相转移催化剂时，就能有效地催化水相中邻位或对位取代碘苯与烯及其衍生物的 Heck 反应，催化剂活性位与载体结合牢固，可抑制活性位的流失，从而有利于催化剂的循环使用。Wu 等[74]将纳米 Ni 固载在含膦的树状聚合物载体上，在室温下催化水介质 Stille 反应中，不需要共催化剂，就能获得高得率的目标产物，催化剂可重复使用 7 次。Wan 等[75]用含氮官能团修饰的有序介孔树脂负载高分散的 Pd 纳米粒子(约 1.5 nm)，在催化水介质吲哚及 N-保护的吲哚与二芳基碘鎓盐的芳基化反应中显示高催化活性，目标产物的得率为 70%~98%，TOF 值为 82 h^{-1}，催化剂可重复使用 8 次，该催化剂也适用于其他杂环化合物的 C2 芳基化反应[式(3-47)]。

$$\underset{R_1}{\underset{X}{\text{indole}}}-H + \underset{R_2}{\underset{OTf}{\text{Ar}_2I^+}}\underset{R_2}{} \xrightarrow[H_2O, 60℃, 12 h]{Pd/11.5ODDMA-MP} \underset{R_2}{\underset{X}{\text{product}}}_{R_1} \quad (3\text{-}47)$$

X = O, S, NH, NCH_3

5) 有序介孔硅材料负载钯催化剂

介孔材料由于具有较大的比表面积和规整的孔道，在催化领域备受关注。本课题组在这方面开展了一系列研究工作。一方面，采用介孔 SiO_2 材料(如 MCM-41，SBA-15 分子筛等)负载金属 Pd 或 In 应用于水介质中的 Ullmann 反应或 Barbier 反应，由于金属纳米粒子均匀分散于介孔孔道内(图 3.4) [76-81]，其获得联苯的得率显著优

于传统 SiO_2 载体负载的 Pd 催化剂。采用共聚法合成含有苯基和氨基等有机基团的有机-无机杂化 SiO_2 载体,能显著提高催化剂外表面和孔道表面的疏水性,有利于水介质中有机反应物的扩散和吸附,从而大幅度提高催化活性和联苯的选择性,且增加了催化剂的循环使用寿命[82]。以有机基团桥联的有机硅作为硅源进行自组装,可获得有机基团均匀镶嵌于孔壁的介孔有机-无机杂化 SiO_2 载体,以此负载金属 Pd,能进一步提高水介质中 Ullmann 反应的活性和联苯的得率,达到均相催化剂的水平。将有机基团(如苯)和金属 Al 同时引入介孔材料 MCM-41 分子筛中(图 3.5),有机基团修饰可提高表面疏水性,而 Al 进入 MCM-41 分子筛孔道骨架内,能显著提高介孔结构的水热稳定性,该载体负载 Pd 获得的催化剂应用于水介质中的 Ullmann 反应,催化活性明显高于 Pd/MCM-41 催化剂,产物联苯的收率高达79%,且 Al 修饰能增加催化剂的循环使用次数。

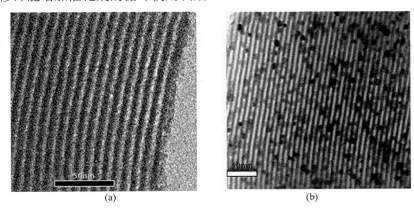

图 3.4 介孔材料 SBA-15 (a)和 Pd-SBA-15 (b)催化剂的 TEM 图

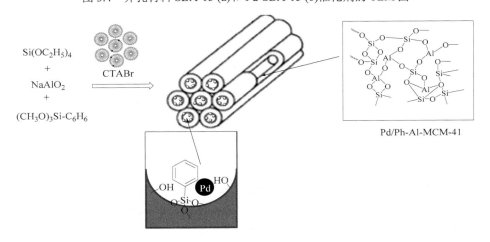

图 3.5 Pd/Ph-Al-MCM-41 结构示意图

6) 天然物质负载钯催化剂

D'Alessandro 等[83]在室温下的空气氛围中用水溶性的木质素能稳定球状 Pd 纳米粒子，其在催化水介质 Heck 和 Suzuki 反应中表现出优良的催化性能。采用不同的底物和碱时，产物得率和选择性都较高，在水介质 Sonogashira 和 Stille 反应中，碘代的底物均显示较高的反应活性，而氯代底物不能进行上述反应。Paul 等[84]用氨基功能化的二氧化硅-木质素负载 Pd 纳米粒子，在碱性水溶液中可高效催化 C—C 键[式(3-48)~式(3-51)]和 C—S 键[式(3-52)]的偶联反应，催化剂可重复使用 5 次。

$$R\text{-}C_6H_4\text{-}Br + ArB(OH)_2 \xrightarrow[K_2CO_3, TBAB, H_2O, 100\ ℃]{Pd(0)\text{-}EDA/SCs} R\text{-}C_6H_4\text{-}Ar \quad (3\text{-}48)$$

$$RCHO + \text{dimedone} \xrightarrow[H_2O,\ 100\ ℃]{Pd(0)\text{-}EDA/SC\text{-}2} \text{产物} \quad (3\text{-}49)$$

$$RCHO + \text{1-phenyl-3-methyl-pyrazolone} \xrightarrow[H_2O,\ 100\ ℃]{Pd(0)\text{-}EDA/SC\text{-}2} \text{产物} \quad (3\text{-}50)$$

$$RCHO + \text{indole} \xrightarrow[H_2O,\ 100\ ℃]{Pd(0)\text{-}EDA/SC\text{-}2} \text{产物} \quad (3\text{-}51)$$

$$R\text{-}C_6H_4\text{-}X + H_2N\text{-}C(=S)\text{-}NH_2 + PhCH_2Br \xrightarrow[TBAB,\ H_2O,\ 100\ ℃]{Pd(0)\text{-}EDA/SCs,\ K_2CO_3} \text{产物} \quad (3\text{-}52)$$

Lee 等[85]用壳聚糖-g-甲氧基聚乙烯乙二醇负载金属 Pd 催化剂，用于水中对溴苯乙酮与苯硼酸的 Suzuki 交叉偶联反应[式(3-53)]，不需额外加入相转移催化剂，产物 4-乙酰基联苯的得率就能达到 92%，这可能是由于嫁接在壳聚糖中的聚乙二醇能促进有机底物在水中的溶解能力。如采用未经修饰的壳聚糖负载的 Pd 催化剂，产物的得率仅为28%，但加入 1 eq 四丁基溴化铵后，得率提高到95%。这表明，四丁基溴化铵对此反应具有明显的促进作用。此外，催化剂可重复使用 5 次。

$$\text{Br-C}_6\text{H}_4\text{-COCH}_3 + \text{C}_6\text{H}_5\text{-B(OH)}_2 \xrightarrow[\text{H}_2\text{O, K}_3\text{PO}_4]{0.5 \text{ mol}\% \text{ Pd (0)}} \text{biphenyl-COCH}_3 \quad (3\text{-}53)$$

有报道将生物纤维素纳米材料与硝酸钯或氯化钯混合在一起，经一步法制备生物材料负载的 Pd 纳米催化剂(20 nm)[86]，Pd 的负载量高达 0.31 mmol/g，其在催化 DMF 介质中碘苯与苯硼酸的 Suzuki 反应时，85℃下反应 3.5 h，联苯得率不到 80%，这可能是由于生物材料在此介质中分散度较低；而水介质中，65℃下反应 3.5 h，联苯的得率达 79%，当提高到 85℃时，联苯得率高达 98%。该催化剂在氯苯与含有不同取代基的苯硼酸反应中也具有较高的催化性能，如在催化具有空间位阻的富电子的邻甲氧基苯硼酸与氯苯的偶联反应中，产物 2-甲氧基联苯的得率也能达到 94%。在对氰基苯硼酸与碘苯的反应中，催化剂用量很少，仅 0.05 mol% Pd。此外，催化剂可重复使用 5 次以上。

7) 有机金属骨架材料负载钯催化剂

有机金属骨架材料(MOFs)是一类由金属离子或金属簇与有机配体通过自组装形成的多孔晶态材料，具有高比表面积和孔容积、孔结构可调、功能可调等多种优点[87]，近年来在催化等领域获得了广泛应用[88]。MOFs 材料的金属位点及有机配体都可作为潜在的催化活性中心[89]。此外，MOFs 材料孔结构比较丰富，可将纳米金属分散于孔道内而引入新的催化活性中心[90]。MIL-101(Cr)是一类具有良好热稳定性和水稳定性的 MOFs 材料。有报道以 MIL-101(Cr)为载体、硝酸钯为钯源，通过浸渍法制得催化剂 Pd/MIL-101，Pd 的粒径为(1.9±0.7) nm。其在催化水相对甲氧基氯苯和苯硼酸的 Suzuki 反应[式(3-54)]中表现出高活性。在四丁基溴化铵和 NaOCH$_3$ 存在下反应 6 h，给出 82%得率的产物 4-甲氧基联苯，而用 1 wt% Pd/C 催化剂，4-甲氧基联苯的得率仅为 35%。无论是含有吸电子基团的氯苯还是给电子基团的氯苯，在与苯硼酸的偶联反应中均展现出高效率，即使采用位阻较大的邻甲氧基氯苯为底物，2-甲氧基联苯的得率也能达到 81%。Pd/MIL-101 催化剂同样也可用于水介质对甲氧基氯苯的 Ullmann 偶联反应[式(3-55)]，在 N$_2$ 和空气氛围下，联苯的得率分别为 97%和 96%。Pd/MIL-101 催化剂非常稳定，能够回收并重复使用多次[91]。

$$\text{Cl-C}_6\text{H}_4\text{-COCH}_3 + \text{C}_6\text{H}_5\text{-B(OH)}_2 \xrightarrow[\text{H}_2\text{O}]{\text{Pd/MIL-101}} \text{biphenyl-COCH}_3 \quad (3\text{-}54)$$

$$2 \quad \text{Cl-C}_6\text{H}_4\text{-COCH}_3 \xrightarrow[\text{H}_2\text{O}]{\text{Pd/MIL-101}} \text{H}_3\text{COC-C}_6\text{H}_4\text{-C}_6\text{H}_4\text{-COCH}_3 \qquad (3\text{-}55)$$

Martín-Matute 等[92]将 Pd 纳米粒子固载于氨基功能化的 MOF 材料[MIL-101(Cr)-NH$_2$]中，分别得到负载量为 4wt%、8wt%、12wt%和 16wt%的 Pd@MIL-101(Cr)-NH$_2$ 催化剂，XRD 图谱表明 Pd 已嵌入 MIL-101(Cr)孔道中。在水相甲氧基溴苯与苯硼酸的 Suzuki 偶联反应中，负载量为 8wt%的 Pd@MIL-101(Cr)-NH$_2$ 在 Cs$_2$CO$_3$ 存在下显示出高活性，在室温下反应 6 h，能定量地给出 4-甲氧基联苯产物。同时，Pd@MIL-101(Cr)-NH$_2$ 在 NaOH 的水溶液中非常不稳定，而在有机胺，如 iPr$_2$NH、iPr$_2$EtN 和 Et$_3$N 中却很稳定，当反应温度在 80℃时，对氯苯甲醛或对氯甲基苯甲酸与苯基硼酸频哪醇反应，产物得率分别为 99%和 77%。负载量为 8wt%的 Pd@MIL-101(Cr)-NH$_2$ 催化剂在对溴甲苯和苯基硼酸频哪醇的反应中可重复使用 10 次。

Islam 等[93]采用嫁接法，将 Pd 纳米粒子固载在含 Co 的 MOF 材料(MCoS-1)中，获得高活性催化剂 Pd(0)/MCoS-1，Pd 的负载量为 1.0 wt%，Pd 纳米粒子均匀分散，粒径为 5~10 nm。XPS 结果证实，载体中的 Co 以+2 价存在，在不需要添加剂和 Cu 作共催化剂时，可成功地催化水中末端炔和卤代芳烃的 Sonagashira 反应。以三乙胺为碱，在 80℃下反应 6 h，目标产物得率达 94%，采用不同碘代芳烃和溴代芳烃与炔作底物，目标产物得率为 70%~95%。但在此反应条件下，氯苯和苯乙炔不发生反应，如果提高三乙胺和催化剂的用量以及反应温度，则氯苯和苯乙炔也能发生反应，给出 44%的联苯产物。Pd(0)/MCoS-1 也可催化水中对溴苯甲醛与苯硼酸的 Suzuki 偶联反应，以 K$_2$CO$_3$ 为碱，在 70℃的空气氛围下，产物 4-苯甲醛的得率达 96%。在上述两个反应中，催化剂均能重复使用 6 次。

Li 等[94]用 Pd/MIL-101 催化水中苯乙炔和 2-碘苯胺的反应，合成 2-芳基吲哚，Pd/MIL-101 的催化性能明显优于相似负载量和粒径大小相近的 Pd/MCM-41，2-碘苯胺在 2 h 内完全转化(后者需要反应 4 h)，这归因于 Pd 在 MIL-101 载体中的高度分散。此外，由于 Pd 被限制在 MIL-101 孔中不易团聚、失活，催化剂显示出较长的寿命，可重复使用 10 次，而 Pd/MCM-41 催化剂使用 6 次即失活。

2. 还原反应

在常温常压下，以分子氢为还原剂，Pd/C 催化水溶液中芳香卤化物的加氢脱卤反应，获得芳烃化合物。研究发现，反应介质对 C—Cl 键和 C═O 键的活性有明显的影响，如在水介质对氯苯乙酮的加氢脱卤反应[式(3-56)]中，C—Cl 键比 C═O 键更容易被还原，而在异丙醇中，它们的活性刚好相反[95]。

第三章 金属催化水介质有机合成

$$\text{4-Cl-C}_6\text{H}_4\text{-COCH}_3 \xrightarrow[\text{NaOH}]{\text{Pd/C}} \text{C}_6\text{H}_5\text{-COCH}_3 \qquad (3\text{-}56)$$

Zhang 等[96]通过包裹、刻蚀及修饰等方法合成了氟功能化的核壳结构催化剂 yPd@mSiO$_2$-F (图 3.6), 在催化水中烯烃加氢反应中, 其催化活性高于未用氟修饰的 Pd NPs 催化剂(yPd@mSiO$_2$), 且稳定性高, 可重复使用 5 次以上。

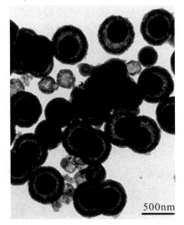

图 3.6　yPd@mSiO$_2$-F 催化剂的 SEM 图

Dell'Anna 等[97]用聚合物原位还原并负载 6~10 nm Pd 纳米晶催化剂, 在催化水中芳香族硝基化合物的还原反应[式(3-57)]中显示出优异的催化性能。在 1 atm H$_2$ 及 NaBH$_4$ 存在下, 催化活性明显高于非负载型 Pd 纳米催化剂, 且催化剂可循环使用 12 次。

$$\text{R-C}_6\text{H}_4\text{-NO}_2 \xrightarrow[\text{H}_2\text{O}]{\text{Pd-pol, NaBH}_4} \text{R-C}_6\text{H}_4\text{-NH}_2 \qquad (3\text{-}57)$$

Uozumi 等[98]用具有两亲性质的聚苯乙烯-聚乙烯乙二醇树脂负载 Pd 纳米粒子, 并催化水介质烯烃加氢反应[式(3-58)], 显示高活性和使用寿命(可重复使用 10 次)。

$$\text{Ar-CH=CH}_2 \xrightarrow[\text{H}_2]{\text{ARP-Pd}} \text{Ar-CH}_2\text{-CH}_3 \qquad (3\text{-}58)$$

Neumann 和 Levi[99]用烷基化的聚苯乙烯亚胺固载 Pd 纳米粒子,并催化水中酮和手性胺的还原氨化反应,加入一定量的 $MgCl_2$ 可促进胺析出进入烷基化聚苯乙烯亚胺的疏水区域,从而稳定了 Pd 纳米粒子,有利于提高产物得率,通过与手性聚苯乙烯亚胺的加成将 Pd 纳米粒子固载在载体上,得到一定的对映选择性产物。Cao 等[100]将高分散的 Pd 纳米粒子包裹在氨基功能化的 MOF [NH_2-MIL-101(Cr)]介孔笼中,得到催化剂 Pd/NH_2-MIL-101(Cr),Pd 的粒径约为 2.49 nm,与 MOF 的笼大小(2.9 nm,3.4 nm)基本一致。以室温下水相氯代芳烃脱氯还原反应为探针,考察了其催化性能。在以甲酸铵为还原剂时,该催化剂可以使对氯苯酚有效地脱去氯,给出得率为 98%的苯酚。该催化体系也适用于一系列底物的脱氯反应,如邻氯或间氯苯酚、含有给电子基团或吸电子基团的氯苯衍生物,产物得率均在 90%以上。有趣的是,对氯硝基苯在此反应中同时发生硝基还原和脱氯反应,给出 96%得率的苯胺。α-氯萘和 3-氯吡啶在该反应体系中反应 6 h,分别给出得率为 91%和 95%的脱氯产物。将该催化剂应用于 1,2-二氯苯、1,3-二氯苯、1,4-二氯苯、2,4-二氯苯酚、1,2,4-三氯苯和 2,4,6-三氯苯酚等的脱氯反应,反应 8 h 产物得率也能在 95%以上。Pd/NH_2-MIL-101(Cr)催化剂也可重复使用 6 次。另外,以 MIL-101 为载体负载的 Pd 纳米粒子催化水相苯酚加氢反应,在 1atm H_2 存在时,30℃下反应 10 h,苯酚完全反应,产物环己酮选择性大于 99.9%。如果提高反应温度至 40℃,则反应时间分别缩短到 7 h 和 3 h,苯酚的转化率和环己酮的选择性均为 99.9%,催化剂可重复使用 5 次。

Zhang 等[101]报道水分散的磁性 Fe_3O_4 负载 Pd 催化剂用于对硝基苯酚加氢反应和还原反应,催化剂不仅显示高活性,且易分离和重复使用。Han 等[102]采用 Pd/TiO_2 为催化剂,甘油为还原剂,水中的硝基苯在紫外光照射下被还原为苯胺,硝基苯的转化率达 100%,苯胺选择性为 95%,同时,甘油被氧化成精细化工原料,如 1,2-二羟基丙酮、甲酸、羟基乙酸和甘油醛等。Oberhauser 等[103]用 2,2′-联吡啶功能化的聚乙二醇单甲醚为载体固载 2 nm Pd 纳米粒子,并将其用于水介质腈还原合成胺的反应中,显示了高 TON 值。Yang 等[104]设计了一种"类似胶束"的非均相 Pd 催化剂,它由单分散的硅球构成,每一个硅球都含有高分散的催化活性中心,形成了疏水性核和亲水性壳,且这种核和壳与孔相通。亲水性壳可以使固体催化剂在水中高度分散,而疏水性核使有机物分子从水中吸附到催化剂表面,这种特殊结构使其在水中苯乙烯、肉桂醇等的加氢反应中表现出独特的催化活性和化学选择性。

3. 氧化反应

早在 2005 年,Okuhara 等[105]将(NH_4)$_{10}W_{12}O_{41}$·$5H_2O$ 和 Zr(OH)$_4$ 混合溶解在水中,然后在 100℃干燥 24 h,再在 800℃焙烧 5 h,获得 W/Zr 摩尔比为 0.2 的 WO_3-ZrO_2 载体,通过浸渍法负载 $PdCl_2$,经 350℃焙烧获得催化剂 Pd/WO_3-ZrO_2,该催化剂

能高效地催化水中乙烯氧化生成乙酸的 Wacker 反应，乙酸选择性高达 73%。

2010 年 Hou 等[106]报道了功能化聚乙二醇负载的 Pd 纳米粒子催化水中醇氧化生成醛或酮的反应。他们研究了还原剂种类和保护配体的用量对 Pd 纳米粒子大小及表面性质的影响，结果显示 Pd 粒径大小和表面性质对催化剂性能影响较大。研究发现，以 $NaBH_4$ 为还原剂获得的催化剂中，Pd 粒径为(2.6±0.4) nm，在 K_2CO_3 存在下，80℃催化水中苯甲醇氧化反应，转化率达 96.4%，苯甲醛选择性高达 98.8%，TOF 值为 56 h^{-1}，高于以甲醇、H_2 等作为还原剂的反应，且催化剂可重复使用 4 次以上，该体系也适用于其他醇的氧化，如肉桂醇、3,7-辛二烯-1-醇和 3-甲基-2-丁烯-1-醇等氧化生成肉桂醛，3,7-辛二烯醛、3-甲基-2-丁烯醛、2-辛醇和环己醇等氧化生成酮的反应。

Xie 等[107]设计了一种 Pd 和 Ni 纳米粒子分散在多壁碳纳米管上的催化剂(Pd@Ni/MWCNT)。Pd 和 Ni 粒径约为 4.0 nm。在 80℃下，以 H_2O_2 为氧化剂，含 0.2 mmol Pd 的催化剂 Pd@Ni/MWCNT 在水介质苯甲醇氧化反应[式(3-59)]中，苯甲醇转化率为 99%，苯甲醛的选择性为 98%，明显优于同样负载量的 Pd/MWCNT，且催化剂可循环使用 6 次[108]。

$$\underset{}{\text{PhCH}_2\text{OH}} \xrightarrow[\text{H}_2\text{O, O}_2\text{或空气}]{\text{Pd@Ni/MWCNT, K}_2\text{CO}_3} \underset{}{\text{PhCHO}} \tag{3-59}$$

Yang 等[109]先通过吸附将 $Pd(AcO)_2$ 引入笼状介孔材料 SBA-16 中，然后用 $NaBH_4$ 还原，得到 Pd/SBA-16 催化剂，Pd 纳米粒子均匀地限制在笼腔中。室温下，在没有碱的空气或氧气中，可催化水中的苄醇、1-苯基乙醇和烯丙醇氧化成相应的醛。与 Pd/SiO_2 催化剂相比，Pd/SBA-16 催化剂不仅显示出高活性，且使用寿命明显优于前者，Pd/SiO_2 使用到第 6 次就明显失活，而 Pd/SBA-16 可循环使用 12 次。

2013 年，Oberhauser 等[110]以吡啶功能化的聚乙二醇为载体固载 Pd 纳米粒子，在空气中催化水中 α,β-不饱和醇(如肉桂醇、2-甲基-4-苯基-3-丙烯醇等)的氧化生成相应的酮，TOF 值为 200 h^{-1}，催化剂可连续使用 4 次。

Hegyessy 等[111]以层状结构蒙脱石、锂皂石及双层氢氧化物等为原料，合成出大比表面积载体并以此负载 Pd 催化剂，其中，锂皂石负载的 Pd 催化剂在水中乙烯氧化成乙酸反应中展示出最佳活性，这可归因于载体中 V^{3+} 能将 O_2 还原成 O^{2-}，而由氧化得到的 V^{5+} 将 Pd 氧化成 PdO，促进了乙烯氧化获得乙酸产物。

最近，Wang 等[112]采用二乙烯基苯和咪唑盐的自由基聚合反应合成了一系列功能化的交联介孔聚离子液体(MPILs)，再由离子交换固载 Na_2PdCl_4，通过 $NaBH_4$ 还原，得到负载型 Pd 催化剂。Pd 纳米粒子具有高分散度，载体具有大比表面积(259 m^2/g)和离子密度，在水中苯甲醇氧化到苯甲醛的反应中，表现出较高的催化活性和寿命,这可能是由于含有羧基的 MPILs 载体有利于超细 Pd 纳米粒子的形成和稳定。而用 Pd(OAc)$_2$ 和 PdCl$_2$ 为钯源制得的负载型催化剂的活性明显下降,原因可能是 Pd(Ⅱ)与载体的键合方式不同,用 Na$_2$PdCl$_4$ 为 Pd 源时,PdCl$_4^{2-}$ 易于与介孔结构离子型聚合物中 Br$^-$ 进行离子交换,从而牢固地固载在载体上,这种主-客体之间的相互作用有利于 Pd(Ⅱ)的高度分散,从而得到高分散 Pd 纳米粒子,且不易团聚。而 Pd(OAc)$_2$ 和 PdCl$_2$ 与载体相互作用相对较弱,不利于获得高分散 Pd 纳米粒子。

Qiu 等[113]首次合成了疏水性介孔壳材料包裹的 Au@Pd 双金属催化剂 Au@Pd@Ph-PMO。先通过刻蚀方法,制得含有苯基桥联的介孔有机硅材料(Ph-PMO)包裹的 Au 纳米粒子(Au@Ph-PMO),由于疏水性的-O$_{1.5}$Si-Ph-SiO$_{1.5}$-单元和亲水性的-SiO$_2$-共存于一个载体中,因此,Au@Ph-PMO 具有非常好的亲-疏水性两亲性能,可作为一种优良的乳化剂。然后 Pd 纳米粒子在 Au 纳米球表面生成,得到高活性的双亲纳米催化剂(Au@Pd@Ph-PMO)。催化剂在水中高度分散,且疏水性底物易于在表面吸附,因此,Au@Pd@Ph-PMO 在水介质伯醇和仲醇的选择氧化反应中展示出优良的催化活性和稳定性;在对甲氧基苯甲醇的氧化反应中,催化剂能重复使用 10 次,这归因于双金属纳米粒子的抗团聚和抗流失特性[114]。

4. 其他反应

Shimizu 等[115]考察了商业用的 Pd/C 在水中腈水合生成酰胺[式(3-60)]的催化性能,发现 Pd 纳米粒子表面覆盖的吸附氧可提高催化剂在水中的分散度,从而提高催化效率。研究表明,这种吸附氧作为 Brønsted 碱性位与金属 Pd 在反应中存在协同效应,不仅可显著提高催化效率,且可以改善催化剂使用寿命。

$$\begin{matrix} R_1 \\ R_2 \end{matrix}\!\!-CN + H_2O \xrightarrow[H_2O]{Pd/C} \begin{matrix} R_1 \\ R_2 \end{matrix}\!\!-CONH_2 \qquad (3\text{-}60)$$

Liu 等[116]采用含有离子液体的交联聚合物为载体固载 Pd 纳米粒子,得到非均相催化剂,无需加入含碘促进剂就能有效催化水中 2-氨基-烷醇的氧化羰基化反应获得 2-噁唑烷酮[式(3-61)],催化剂可循环使用 5 次。

$$\begin{matrix} R_1 & R_2 \\ H_2N & OH \end{matrix} + CO + O_2 \xrightarrow[H_2O]{P(DVB\text{-}IL)\text{-}Pd} \underset{R_1 \ R_2}{HN\!\!\diagup\!\!\!\diagdown\!\!O} \qquad (3\text{-}61)$$

二、负载型金催化剂

1. 偶联反应

早在 2004 年,Tsukuda 等[117]报道了在聚 N-乙烯基-2-吡咯烷酮(PVP)存在下,用 $NaBH_4$ 还原氯金酸酸根离子($AuCl_4^-$)得到 PVP 稳定的 Au 纳米粒子催化剂,可有效地催化苯硼酸偶联生成联苯[式(3-62)]。反应在空气氛围下的水中进行,联苯得率大于 99%。

$$(HO)_2B-C_6H_5 \xrightarrow[K_2CO_3, H_2O, 空气, rt]{1 \text{ mol}\% \text{ Au : PVP}} C_6H_5-C_6H_5 + C_6H_5-OH \quad (3-62)$$

Chen 等[118]采用含有金属氧化物(CeO_2、TiO_2 或 ZrO_2)的 SBA-15 为载体,通过沉积法制备了负载型金催化剂 Ce-Au、Ti-Au、Zr-Au,并考察了它们在水中无碱存在下苯硼酸自偶联生成联苯的催化性能。结果表明,粒径为 1.3~2.3 nm 的 Au 纳米粒子显示出较高的活性,且可以重复使用。XPS 表征结果也证实 Au 纳米粒子在载体中呈金属态。

Guo 等[119]报道以聚(2-氨基苯硫酚)固载 Au 纳米粒子并催化 Suzuki 反应。以 NaOH 为碱,在 80℃下苯硼酸可与活性较差的氯苯顺利地发生偶联反应[式(3-63)],反应 4 h 后联苯的得率为 87%,且催化剂可重复使用 6 次。如增加 NaOH 的用量,即使空间位阻大以及含给电子基团的氯苯,也可顺利地发生偶联反应。

$$C_6H_5-Cl + C_6H_5-B(OH)_2 \xrightarrow[4 \text{ eq NaOH, } H_2O, 80℃, 4 \text{ h}]{0.05 \text{ mol}\% \text{ Au}} C_6H_5-C_6H_5 \quad (3-63)$$

Ying 等[120]报道了半导体材料负载的金催化剂(PbS-Au)并应用于水中醛、胺及炔的偶联反应[式(3-64)]。虽然含吸电子基团的对氰基苯甲醛、四氢吡啶及苯乙炔等不发生反应,但脂肪醛可顺利地反应,催化剂不仅对水和空气稳定,且易回收和重复使用。

$$R_1CHO + R_2R_3NH + R_4-\equiv-H \xrightarrow[H_2O, 100℃]{PbS-Au} R_2-\underset{R_1}{\overset{R_3}{N}}-\equiv-R_4 \quad (3-64)$$

Monopoli 等[121]以葡萄糖为还原剂,在不同介质中制备了 Au 纳米粒子催化剂并应用于水相 Ullmann 还原偶联反应。在水介质中制备的 Au 纳米粒子因粒径较小(约 1 nm)而具有大比表面积和高催化活性;而在熔融盐介质中制备的 Au 纳米粒子较大(约 20 nm)，其活性也随之降低。Nasrollahzadeh 和 Banaei[122]采用双金属 Au/Pd 纳米粒子催化水介质中不同溴代或碘代芳烃与烯及其衍生物的 Heck 反应[式 (3-65)]，在无配体存在时，反应顺利地进行，目标产物得率均在 89%以上，催化剂可循环使用 4 次。

$$\text{R}_1\text{-C}_6\text{H}_4\text{-X} + \text{CH}_2=\text{CHZ} \xrightarrow[\text{K}_2\text{CO}_3, \text{CTAB}, 80\ ℃]{\text{Au/Pd NPs}} \text{R}_1\text{-C}_6\text{H}_4\text{-CH=CH-Z} \quad (3\text{-}65)$$

Shim 等[123]采用微管 Cu_2O 为载体，在微波辅助加热下原位还原 $PdCl_4^-$ 得负载型 Pd 催化剂(Cu_2O/Pd)，并将其应用于水相芳基硼酸与碘代芳烃的 Suzuki 反应[式 (3-66)]中，偶联产物的得率均大于 90%，催化剂可重复使用。

$$\text{R-C}_6\text{H}_4\text{-B(OH)}_2 + \text{R}_1\text{-C}_6\text{H}_4\text{-I} \xrightarrow[\text{H}_2\text{O}, \text{K}_2\text{CO}_3, 80\ ℃, 12\ \text{h}]{\text{Cu}_2\text{O/Pd}} \text{R-C}_6\text{H}_4\text{-C}_6\text{H}_4\text{-R}_1 \quad (3\text{-}66)$$

Muthusubramanian 等[124]先采用溶剂热法合成粒径为 150 nm 的纳米多孔 TiO_2 载体，然后通过沉积-浸渍法获得负载型 Au 纳米粒子催化剂(Au/TiO_2)，在水相有机叠氮化合物与末端炔的 1,3-偶极环加成反应[式(3-67)]中显示出高活性，反应 20~30 min 就能给出 100%选择性的产物 1,4-二取代-1,2,3-三唑衍生物。同时，由于 Au 纳米粒子同时具有亲氧和疏水性质，Au/TiO_2 也能催化非末端炔与叠氮物的反应[式(3-68)]，但需要适当地延长反应时间(45 min)，可给出 86%~97%得率的目标产物。

$$\text{R}_1\text{-C≡CH} + \text{R}_2\text{-C}_6\text{H}_4\text{-COCH}_2\text{N}_3 \xrightarrow[\text{H}_2\text{O, 20~30 min}]{\text{Au/TiO}_2} \text{产物} \quad (3\text{-}67)$$

R_1 = Ph, 2-CH_3OPh, $CO_2C_2H_5$, 环己烯基

$$\text{R}_1\text{-C≡C-R} + \text{R}_2\text{-C}_6\text{H}_4\text{-COCH}_2\text{N}_3 \xrightarrow[\text{H}_2\text{O, 45 min}]{\text{Au/TiO}_2} \text{产物} \quad (3\text{-}68)$$

R_1 = CO_2CH_3, $CO_2C_2H_5$

2. 氧化反应

自从 Haruta 等[125]研究发现粒径小于 5.0 nm 的金纳米粒子，在催化一氧化碳低温氧化中表现出较高的催化活性以来，金纳米粒子在传感器、基因测定和催化等

领域,尤其在催化领域中引起了科学家的广泛关注[126-128]。大量的理论和实验研究表明,金纳米粒子的大小决定金催化剂的催化性能。一般来说,非常小的金纳米粒子由于量子尺寸效应,在催化反应中通常表现出较高的催化活性和选择性。然而,较小的金纳米粒子不仅难以从反应体系中分离,而且由于其具有较高的表面能而容易发生烧结。为了克服金纳米粒子本身的限制,人们研究出负载型的金纳米粒子。

1) 碳材料负载金催化剂

Wang 等[129]报道了用碳纳米管负载的 Pt-Au 纳米合金催化剂,并应用在水中5-羟基呋喃(HMF)生成 1,5-呋喃二甲酸(FDCA) [式(3-69)]的氧化反应,碳纳米管中丰富的羰基/醌有利于增强反应物和反应中间体在催化剂表面的吸附,有利于反应的进行。

$$\text{HOCH}_2\text{-furan-CHO} \xrightarrow[\text{O}_2, \text{H}_2\text{O}]{\text{Pt-Au/CTN}} \text{HOOC-furan-COOH} \quad (3\text{-}69)$$

Shaabani 等[130]以巯基功能化的果糖为原料制备介孔碳,并以此负载 Au 纳米粒子,所得的 Au 催化剂能够高效地催化水中苯甲醇氧化反应。以空气为氧化剂,K_2CO_3 为碱,反应 1 h 后,苯甲醛得率超过 99%。作者也考察了其在其他醇如对硝基苯甲醇、对溴苯甲醇、对甲基苯甲醇、1-(4-硝基苯基)乙醇等中的氧化反应性能,结果显示,延长反应时间,目标产物的得率均可达到 99%。除水外,该 Au 催化剂在其他有机溶剂如乙腈、甲苯、二甲苯、THF 等介质中,也能给出较高得率的苯甲醛。此外,催化剂能重复使用 6 次。Wang 等[131]用共组装方法一步合成了高稳定和单分散的负载于介孔碳上的 Au 纳米粒子催化剂,其中 3-巯丙基三甲氧基硅烷偶联剂可有效分散 Au 纳米粒子,同时贡献了大量的二级介孔孔道,Au 纳米粒子被锚定在介孔孔道内,具有高的稳定性,即使经过 900℃焙烧,仍可保持高分散性。在水相苯甲醇的选择性氧化中显示出高活性,在 90℃和 1 MPa O_2 的条件下,反应 50 min 转化率和选择性均为 100%;60℃和常压条件下,反应 12 h,转化率和选择性均为 100%;催化剂可重复使用。在 25℃时,虽然催化剂具有较高的 TOF 值(478 h^{-1}),但苯甲醛选择性降低,主要生成苯甲酸。后来他们又报道了以嫁接超分子离子液体的石墨烯为载体,通过共价键固载 Au 纳米粒子,在催化苯甲醇氧化反应中展现了高催化活性。即在室温下,无需添加无机碱等活化剂,反应 1 h 后,苯甲醇转化率和苯甲醛的选择性均达到 99%;采用其他醇包括脂肪族伯醇、仲醇以及芳香醇为底物时,通过调节反应时间,相应氧化产物的得率和选择性也均可达 99%以上,如对硝基苯甲醇,反应 1.7 h 后,其氧化产物的得率和选择性均为 99%[132]。

最近,Guo 等[133]报道了用石墨烯量子点为载体,$HAuCl_4$ 为金源,柠檬酸三钠为还原剂,通过"一锅"反应制得核壳结构负载型 Au 催化剂,在还原金前驱体过程中，石墨烯量子点中的含氧官能团也随之被还原，催化剂在碱性水溶液中高度分散，以 H_2O_2 为氧化剂，在 3,4-二甲氧苯甲醇氧化为 3,4-二甲氧苯甲醛的反应中表现出高活性，转化率超过 99%，选择性达到 99%。Doris 等[134]以双亲的氨三乙酸二炔酯功能化的碳纳米管为载体固载 Au 纳米粒子，这种催化剂在室温下的敞口反应器中能直接催化醇氧化。随后他们又报道了碳纳米管负载的两种不同尺寸 Au 纳米粒子催化剂，当粒径为 3 nm 时，催化剂的 TOF 值达 310 h^{-1}，远高于粒径为 20 nm 的催化剂(45 h^{-1})。催化剂可重复使用，其缺陷是对苯甲醛的选择性较低[135]。

Li 等[136]报道了以氧化的多壁碳纳米管为载体，采用气-液等离子"一锅"法，获得负载型 Au 纳米粒子催化剂，并将其用于水中硅烷水解制备硅醇[式(3-70)]。在各种硅烷的氧化反应中，催化剂在水介质中的性能不仅高于有机溶剂介质(如乙酸乙酯、丙酮，THF)，且易回收和重复使用，说明氧化型的碳纳米管和稳定剂油胺是提高催化剂性能的关键组分，因为氧化型的碳纳米管使催化剂具有较强的亲水性，而油胺稳定剂使催化剂具有强亲脂性，有利于提高水介质中的催化效率。

$$R_1R_2R_3Si-H + H_2O \xrightarrow[H_2O, rt]{Au/o\text{-}CNT} R_1R_2R_3Si-OH \qquad (3\text{-}70)$$

2) 高分子聚合物负载金催化剂

2005 年，Tsukuda 等[137]报道了聚(N-乙烯基-2-吡咯烷酮)负载的 Au 纳米簇(1.3 nm)催化剂，室温下催化水中苄醇氧化到相应的醛和/或羧酸[式(3-71)]。动力学研究揭示，Au 纳米粒子尺寸减小有利于提高催化活性[137]。与相似尺寸的 Pd 纳米粒子催化剂相比，Au 纳米粒子具有更高的活性[138]。在 0.5 MPa 的氧气氛围下，聚乙烯基醇负载的 Au 纳米粒子能有效催化不同链长的 1,2-二醇[式(3-72)]，如 1,2-丙二醇或 1,2-辛二醇的氧化反应，给出相应的 α-羟基羧酸，获得高得率的产物[139]。随后，Prati 等[140]用微溶胶稳定 Au 纳米簇(2.5 nm)催化剂并应用于水中各种醇(苄醇、脂肪醇和多元醇)的氧化反应中，在 50~70℃和 1~3 atm O_2 的条件下，催化剂 TOF 值超过 960 h^{-1}。

$$\text{R}\underset{}{\bigcirc}\text{CH}_2\text{OH} \xrightarrow[\substack{300 \text{ mol\% K}_2\text{CO}_3 \\ H_2O, \text{空气}, 27\ ℃}]{2 \text{ atom\% Au : PVP}} \text{R}\underset{}{\bigcirc}\text{CHO} + \text{R}\underset{}{\bigcirc}\text{COOH} \qquad (3\text{-}71)$$

$$\text{HO-CH}_2\text{-CH(OH)-R} \xrightarrow[H_2O]{\text{Au 纳米胶体，0.5 MPa}} \text{HOOC-CH(OH)-R} \qquad (3\text{-}72)$$

Hirao 等[141]采用聚苯胺磺酸负载 Au 纳米粒子，这种催化剂可作为转移质子和电子的氧化还原介质应用在各种醇的氧化反应中。对于体积较大的醇如苄醇，在较短时间内能给出以酮为主的产物，而对于链状的脂肪族伯醇，如正癸醇，经过长时间(如 72 h)的反应才能给出主要产物羧酸。Li 等[142]采用两亲的树脂负载 Au 纳米粒子，在催化水相醇的氧化反应中，不需要加入碱就能给出高得率和高选择性的产物，催化剂可重复使用。Tsukuda 等[143]采用一种星状的热敏性高分子材料聚乙烯乙烯醚为载体，负载 Au 纳米簇(约 4 nm) (见彩图 1)。这种催化剂由于具有较强的疏水性，能够高效催化水中苯甲醇氧化成苯甲醛，转化率超过 99%，选择性为 99%。在不同醇的氧化反应中，除了 3-苯基丙醇主要被氧化成 3-苯基丙酸外，其他芳香醇均被氧化成相应的羧酸，得率在 89%以上。

Tsukuda 等[144]采用 PVP 稳定 Au 纳米簇催化剂，催化苯甲醇氧化生成苯甲酸。在 pH >2.5 时，催化剂与反应体系成均相，而 pH <2.4 时，催化剂沉淀析出，分离回收后可重复使用。在室温下苯甲醇氧化产物更多的是苯甲酸而不是苯甲醛[145]。Tsukuda 等发现在室温下以分子氧为氧化剂，可有效催化水介质苯甲醇氧化成醛，动力学研究表明，Au 纳米粒子尺寸对催化活性具有重要影响[146]。Li 等[147]将小于 10 nm 的 Au 纳米粒子负载在聚苯乙烯(PS)小球上，并应用在苯甲醇氧化生成苯甲酸的反应中，在温和反应条件(30℃，1 atm 空气为氧化剂，K_2CO_3 为碱)下就能顺利完成反应，催化剂可重复使用 5 次，显示出较好的实际应用价值。Wang 等[148]采用一步自组装法，得到了有序介孔 Au-碳催化剂，Au 纳米粒子尺寸可从 3 nm 调节到 18 nm，催化剂都具有大的比表面积和大的孔体积，以及分布均一的孔径，其中负载 3 nm 的 Au 纳米粒子的催化剂比表面积为 1211 m^2/g，孔体积为 0.7 cm^3/g。经 600℃焙烧后，Au 纳米粒子仍高度分散在介孔孔道内。催化剂在水相对苯甲醇选择性氧化反应中显示出高活性，常压室温条件下，选择性和转化率都可达到 100%，TOF 值可从 132 h^{-1} 提高到 320 h^{-1}。Li 等[149]将 Au-Pt 合金纳米粒子负载于聚苯乙烯小球上，其优点是载体表面不需处理。考察了催化剂中 Au 与 Pt 的含量对催化活性的影响，实验结果显示，$Au_{1.25}Pt_1$ 活性最佳，在 1 atm 空气中 40℃下反应 24 h，产物苯乙酮得率为 100%；环己醇氧化为环己酮得率达 80%；在肉桂醇的氧化反应中，得到 99%的肉桂醛，催化剂可重复使用 7 次。

Ebitani 等[150]还设计了一种 N,N-二甲基十烷基氨氮氧化物、PVP 和 PVA 包裹的 AuPd 双金属催化剂，在 1,6-己二醇(HDO)氧化成 6-羟基羧酸[式(3-73)]反应中，于最佳反应条件下，第一种催化剂活性明显优于后两种，这主要与金属表面电子密度相关，此催化剂也适用于 1,7-庚二醇、1,8-辛二醇和 1,2-己二醇等的氧化反应，给出得率为 76%~84%的产物 ω-羟基酸。

$$HO\diagdown\diagup\diagdown\diagup\diagdown\diagup OH \xrightarrow[O_2,\ 70\ ℃,\ 12\ h]{AuPd\text{-}DDAO/HT} HO\diagdown\diagup\diagdown\diagup\diagdown COOH \quad (3\text{-}73)$$

3) 氧化物负载型金催化剂

Riisager 等[151]报道了 Au/TiO_2 催化剂在室温下催化空气氧化 5-羟甲基呋喃甲醛

成 2,5-呋喃二甲酸[式(3-69)]的反应,中间体氧化产物 5-羟甲基-2-呋喃甲酸的生成取决于反应体系中加入的碱量以及 O_2 的压力,在最佳条件下,转化率达 100%,产物 2,5-呋喃二酸的得率达 71%。Ebitani 等[152]采用水滑石负载 Au 纳米粒子,在 90℃下,可高效地催化水介质中空气氧化 5-羟甲基呋喃甲醛转变成 2,5-呋喃二酸,反应体系中不需均相碱,产物也比较容易纯化,催化剂可重复使用 3 次。Kaneda 等[153]首次制备羟基磷灰石负载的 Au 纳米粒子催化剂,在脂肪族硅烷选择性氧化生成硅醇的反应[式(3-70)]中,显示出高催化效率,产物硅醇选择性和得率达 99%,催化剂可重复使用。除了三苯基硅烷外,在其他硅烷氧化反应中,相应硅醇的得率均在 92%以上,且选择性高于 99%。

Wang 等[154]采用聚多金属氧酸盐负载 Au 纳米粒子,催化水中纤维二糖和纤维素选择性地转化成葡萄糖酸,催化性能明显优于典型金属氧化物(SiO_2、Al_2O_3、TiO_2)、碳纳米管以及分子筛负载的 Au 纳米粒子,这可能是由于载体具有较强的酸性及高分散 Au 纳米粒子。一方面,载体的强酸性不仅有利于纤维二糖的转化,而且使葡萄糖酸容易解析阻止其降解从而提高其选择性;另一方面,小尺寸 Au 纳米粒子可加速葡萄糖中间体的氧化,从而提高纤维二糖转化和葡萄糖酸的生成。缺点是在长时间的水热反应中 H^+ 流失,导致催化活性下降。Samanta 等[155]以 4-氨基巯基苯酚自组装的高分子材料负载 Au 纳米粒子,首次开发了一种高效、简单、绿色方法用于水中芳基取代的氧化合成 1,2-二酮[式(3-74)]。在最佳反应条件下,苯环上含有不同取代基的 α-羟基酮均能发生氧化反应,如含有吸电子基团如氟,氧化产物的得率可高达 95%。对于含有杂芳环的 α-羟基酮,在较低温度下就能给出较高得率的氧化产物(90%~93%),而且,α-羟基酮通过原位氧化,随后再与芳基-1,2-二胺缩合可以成功地给出产物喹喔啉及其衍生物[式(3-75)],无论 α-羟基酮是否含有取代基,"一锅"法反应均能顺利地进行,反应 2~5 h,给出目标产物得率为 82%~93%,但由于在反应过程中 Au 纳米粒子团聚,催化活性下降。

$$Ar\underset{OH}{\overset{O}{\underset{|}{C}}}Ar \xrightarrow[H_2O, 空气, 80\ ℃]{4\ mol\%\ 催化剂\ Au\ NPs \atop 2\ eq\ K_2CO_3} Ar\underset{O}{\overset{O}{C}}Ar \qquad (3-74)$$

$$R\!-\!\!\underset{N}{\overset{N}{\bigcirc}}\!\!\underset{Ar}{\overset{Ar}{}} \xleftarrow[碱, H_2O, 空气, \triangle]{4\ mol\%\ Au\ NPs} Ar\underset{OH}{\overset{O}{C}}Ar \xrightarrow[H_2O, 空气]{Au\ NPs, 碱} Ar\underset{O}{\overset{O}{C}}Ar \qquad (3-75)$$

Shim 等[123]采用微管 Cu_2O 为载体,在微波辅助加热下原位还原 $AuCl_4^-$ 得负载型 Au 纳米粒子催化剂,以分子氧为氧化剂,催化苄醇氧化反应[式(3-76)],无论是对氯苯甲醇还是对甲基苯甲醇、对甲氧基苯甲醇,氧化产物的选择性均大于99%,得率也在90%以上。

$$R-\!\!\!\!\bigcirc\!\!\!\!-CH_2OH \xrightarrow[H_2O, NaOH, O_2, 80\ ℃, 5\ h]{Cu_2O/Au} R-\!\!\!\!\bigcirc\!\!\!\!-CHO \quad (3\text{-}76)$$

Yang 等[156]采用沉积-浸渍法将 Au 负载在 Y 型分子筛上并用于伯醇直接氧化制备羧酸的反应中。在最佳反应条件下,对甲基苯甲醇的氧化产物的得率高达 99%,如果适当提高反应温度(150℃),脂肪族伯醇也能发生氧化反应,但产率不高。他们还采用水热法合成介孔结构球状 TiO_2,然后在 $HAuCl_4$ 溶液中,通过尿素沉积-浸渍法制备出新型负载型 Au 催化剂[157]。以分子氧为氧化剂,在碱性水溶液中该催化剂能高效地催化 1-戊醇氧化成正戊酸,活性高于分子筛、水滑石和纳米 TiO_2 负载的 Au 催化剂。该催化剂也可催化 C_4~C_{10} 的脂肪族醇和芳香醇的氧化。在正辛醇氧化反应中,TOF 值达 312 h^{-1},远高于活性炭负载的 Au 和 Pt 纳米粒子催化剂以及 Pd-Au 合金催化剂[158]。在正癸醇的氧化反应中,由于癸醇在水中溶解性差,阻碍了癸醇转化,因此,120℃下反应 6 h,仅得到 39.6%的正癸酸。如将反应温度提高 150℃,产物得率可提高到 80.8%[159]。介孔 TiO_2 负载的 Au 催化剂在芳香醇氧化反应中也表现出高活性,如苄醇在 80℃下反应 6 h,就可以定量地转化成苯甲酸,显著优于活性炭负载的 Au 纳米粒子催化剂[151],苯环上含有强给电子基团如—OCH_3 的苄醇比含有弱给电子基团或吸电子基团如—Cl,—Br 和—NO_2 的苄醇活泼,对甲氧基苯甲醇比较容易转化成对甲氧基苯甲酸,得率可达 100%,对羟基苯甲醇的活性也比较高,但产物得率仅为 55.5%,肉桂醇的氧化产物得率为 55.7%。此外,取代基位置对活性也有非常明显的影响,如 3,4-二氯苯甲醇反应活性低于 3,4-二氯苯甲醇,而大于 2,6-二氯苯甲醇,表明羟甲基的邻位基团不利于反应进行[156]。Hao 等[160]通过浸渍法制得催化剂 Au/HMS,并将其应用在 80℃和 1 atm O_2 下,Na_2CO_3 水溶液中的苯甲醇选择性氧化生成苯甲酸,转化率可达 42.9%,产物苯甲酸的选择性达 95%。

Xu 等[161]采用不同金属氧化物(Al_2O_3,TiO_2,ZrO_2,NiO 和 CuO)负载 Au 纳米粒子催化剂,首次报道在弱酸性和温和条件下催化由甘油醇氧化制备二羟基丙酮的反应[式(3-77)],其中以 CuO 为载体负载的催化剂 Au/CuO 具有高活性和选择性,优于 Au/C[162]、Au/CeO_2 [163]及 Bi 和 Au 修饰的 Pt 和 Pd-Ag 催化剂[164],且反应中不需要碱活化。

$$HO\!\!-\!\!\overset{OH}{\underset{}{\mid}}\!\!-\!\!OH \xrightarrow[O_2, H_2O]{Au/MO_x} HO\!\!-\!\!\overset{O}{\underset{\parallel}{\mid}}\!\!-\!\!OH \quad (3\text{-}77)$$

MO_x = CuO, NiO, ZrO_2, Al_2O_3

4) 无机盐负载金催化剂

Ebitani 等[165]用水滑石负载的 Au 催化分子氧氧化葡萄糖胺衍生物生成相应的 α-氨基酸的反应[式(3-78)]，因水滑石载体具有一定的碱性，所以反应中不需额外加入碱，反应在温和条件下就能顺利进行，催化剂可重复使用。

$$\text{葡萄糖胺盐酸盐} \xrightarrow[\text{1 bar}^① \text{ O}_2, \text{H}_2\text{O, 40 ℃, 3 h}]{\text{Au/HT}} \text{葡萄糖胺酸} \quad (3\text{-}78)$$

2015 年，Ebitani 报道[166]在 70℃和 1 atm O_2 条件下，负载型 AuPd 合金催化水中 1,3-丙二醇氧化成 3-羟基丙酸[式(3-79)]，得率和选择性分别为 30%和 67%，Au 和 Pd 具有协同促进作用，催化剂可重复使用。

$$\text{HO}\text{~}\text{OH} \xrightarrow[\text{H}_2\text{O, O}_2, \text{70 ℃, 12 h}]{\text{AuPd-PVP/HT}} \text{HO}\text{~}\text{COOH} \quad (3\text{-}79)$$

3. 还原反应

Huang 等[167]采用两种方法制得了负载型 Au 催化剂，一种是将柱[5]芳烃自组装形成的微米管浸没在稳定化的金纳米粒子溶液中，通过模板法制备了微米管杂化材料负载的 Au 纳米粒子，另一种是利用修饰的 Au 纳米粒子具有亲水和疏水两部分，在水中自组装得到微米管杂化材料负载的 Au 纳米粒子催化剂。并考察了这两种杂化材料作为催化剂，在 $NaBH_4$ 还原对硝基苯胺成对苯二胺[式(3-80)]反应中的性能。一般情况下，由于 Au 纳米粒子的流失，催化剂只能循环使用几次就会失活。但通过自组装合成的催化剂可循环使用 20 次，说明其高稳定性和优异的催化性能。

$$\text{O}_2\text{N-C}_6\text{H}_4\text{-NH}_2 \xrightarrow[\text{NaBH}_4, \text{H}_2\text{O, rt, 30 min}]{\text{TCMTs 或 SCMTs}} \text{H}_2\text{N-C}_6\text{H}_4\text{-NH}_2 \quad (3\text{-}80)$$

2014 年，Dai 等[168]首次报道了 Au/TiO_2 催化水中硝基苯加氢生成对氨基苯酚，反应中不用加入相转移催化剂，在低浓度酸中，给出产物得率为 63%，选择性为 81%，催化剂可以重复使用 4 次。Ebitani 等[169]报道了以甲酸为氢源，Au/ZrO_2 催化水中生物质衍生物分解合成 γ-内酯[式(3-81)]。与 Ru/C、Ru/SBA、Au/ZrC 催化剂相比，Au/ZrO_2 显示出高活性和良好的使用寿命，产物产率达到 97%。

---

① bar，压强单位，1bar=10^5Pa。

$$\text{(结构式)} + \text{HCOOH} \xrightarrow[\text{H}_2\text{O, 150 °C}]{\text{Au/ZrO}_2} \text{(内酯结构)} \quad (3\text{-}81)$$

Mondal 等[170]设计了一种新型碳量子点复合材料负载的 Au 纳米粒子催化剂,在不同芳烃硝基化合物加氢反应[式(3-82)]中显示出高活性,这主要归因于碳量子点载体加快了电子转移速度,并保护在催化剂表面上的还原分子,从而提高了催化活性。该催化剂也具有比较好的水热稳定性和使用寿命,可重复使用 4 次,有趣的是,该催化剂在不溶于水的反应底物间二硝基苯反应中也表现出良好的催化性能。

$$O_2N\text{-Ar(R)-CHO} \xrightarrow[\text{吡啶, 80 °C, 12 h}]{\text{Au-CD}} O_2N\text{-Ar(R)-CH}_2\text{OH} \quad (3\text{-}82)$$

Wang 等[148]用一步法,通过自组装得到了有序介孔 Au-碳催化剂,通过改变合成过程中添加 3-巯丙基三甲氧基硅烷的量来控制 Au 纳米粒子的尺寸,约为 3 nm,且 Au 与巯基具有强相互作用,可在 0 ℃下催化对硝基苯酚还原成对氨基苯酚的反应。Jin 等[171]报道了球形和纳米棒状 Au 纳米粒子催化 80 ℃水中对硝基苯甲醛选择性加氢生成对硝基苯甲醇的反应,纳米棒状 Au 纳米粒子催化剂比球状 Au 纳米粒子催化剂显示出更优异的催化性能和使用寿命,转化率大于 99%,选择性为 100%,催化剂可重复使用 5 次。他们还设计了一种热稳定的 Au 纳米簇催化剂并负载在 CeO_2 载体上,可催化水中硝基芳香醛选择性加氢生成醇,其催化性能明显高于非负载催化剂[172]。有趣的是,传统的硝基醛在加氢反应中硝基常被还原,但在此反应中只有甲酰基被还原,相应产物的选择性达到 100%,该催化剂也适用于不同硝基化合物的加氢反应,除邻硝基苯甲醛加氢产物得率较低(73%)外,其他硝基醛加氢产物的得率均在 90%以上。

三、其他负载型金属催化剂

1. 负载型镍催化剂

Wang 等[173]采用浸渍法制得 Ni/HZSM-5 催化剂,并将其应用于水中山梨醇转化成液体烷烃如戊烷和己烷[式(3-83)]。研究证实,通过调节催化剂的酸性可以控制山梨醇加氢产物的组成。在 Brønsted 强酸性溶液中,Ni/HZSM-5 催化剂可直接催化 C—C 的断裂;如在反应体系中加入含有 Lewis 酸性位的 MCM-41,可使 C—O 键断裂,山梨醇的转化率和液体烷烃的选择性分别达 67.1%和 98.7%,MCM-41 不仅可以提高 Ni/HZSM-5 催化剂的活性,而且可有效地防止有机碳在催化剂表面的沉积。

$$\text{HOCH}_2(\text{CHOH})_4\text{CH}_2\text{OH} \xrightarrow[\text{H}_2\text{O, 240 °C}]{\text{Ni/HZSM-5}, 4.0 \text{ MPa H}_2,} \text{C}_5\text{H}_{12} + \text{C}_6\text{H}_{14} \quad (3\text{-}83)$$

Zhao 等[174]研究了水介质中 Ni/TiO$_2$ 催化硝基苯的加氢反应,结果发现,Ni/TiO$_2$ 催化剂在反应过程中出现明显的失活现象。通过对催化剂反应前后的 XRD、XPS、ICP 的对比和分析,催化剂失活主要是由于金属镍和水发生反应生成氢氧化镍。为了解决催化剂失活的问题,他们以葡萄糖为碳源,通过在 180 ℃下焙烧,使 Ni/TiO$_2$ 催化剂表面包覆一层疏水性的碳,抑制了水和镍的接触,反应后催化剂表面没有氢氧化镍的生成,催化剂的活性和稳定性都得到了提高。Zhang 等[175]考察了在 N$_2$ 保护下,以水为溶剂,甲醇为氢供体,Raney Ni 催化呋喃甲醛加氢反应中的催化效率。结果表明,添加剂的加入导致呋喃甲醛的转化率降低,产物得率也随之降低。当用呋喃甲醛为单一反应物时,未发现呋喃甲醇的生成[式(3-84)],但在添加剂存在下,有呋喃甲醇的生成。以冰醋酸为添加剂,可以检测到 19.1%得率的呋喃甲醇,丙酮为添加剂时,可使呋喃甲醛发生脱羧反应[式(3-85)],THF 的选择性由 24%增加到 38%。苯酚添加剂可使呋喃甲醛发生重排反应[式(3-86)],五元环产物的选择性由 48%提高到 89%。与单一反应物原位加氢反应相比,在呋喃甲醛存在下,丙酮转化率增加,而冰醋酸转化率降低,苯酚转化率降低。

$$\text{furfural} \xrightarrow[\text{H}_2\text{O}]{\text{Raney Ni}} \text{furfuryl alcohol} \xrightarrow[\text{H}_2\text{O}]{\text{Raney Ni}} \text{tetrahydrofurfuryl alcohol} \quad (3\text{-}84)$$

$$\text{furfural} \xrightarrow[\text{或 HAc, H}_2\text{O}]{\text{Raney Ni, CH}_3\text{COCH}_3,} \text{furan} \xrightarrow[\text{或 HAc, H}_2\text{O}]{\text{Raney Ni, CH}_3\text{COCH}_3,} \text{THF} \quad (3\text{-}85)$$

$$\text{furfural} \xrightarrow[\text{PhOH, H}_2\text{O}]{\text{Raney Ni}} \text{cyclopentanol} \xrightarrow[\text{PhOH, H}_2\text{O}]{\text{Raney Ni}} \text{cyclopentanol} \quad (3\text{-}86)$$

Kalbasi 和 Mazaheri[176]首次报道了用 KIT-6 为硅模板合成多级孔结构的 Ni/H-mZSM-5 催化剂。与纯微孔分子筛相比,这种材料具有较强的扩散性质。这种多级分子筛可以作为一种酸羟金属双功能非均相催化剂应用在芳香族硝基化合物的还原反应中。结果显示,以 NaBH$_4$ 为还原剂,在室温下短时间内就能给出优良得率的产物。催化剂可重复使用 7 次。Yamato 等[177]报道 Raney Ni-Al 合金可以有效地催化水中二苯甲酮加氢还原生成二苯甲烷[式(3-87)]，在密闭反应器中 60℃ 下反应 3 h，产物得率达到 93%。如在 Pt/C 存在下，用 Raney Ni-Al 和 Al 粉催化剂可使二苯甲酮还原，给出二环己基甲烷产物。研究发现，反应温度、时间、水的用量以及 Raney Ni-Al 合金的用量对加氢产物的种类具有非常重要的影响。

$$\text{(3-87)}$$

最近，Cheng 等[178]报道了一种疏水性且对空气和水比较稳定的高活性 Ni 催化剂，其特点是介孔结构中嵌入乙基桥联有机硅组分，提高了载体的疏水性，此外活性物种 Ni 在载体中高度分散，在水中硝基苯加氢反应中显示出高活性，且催化剂能重复使用 5 次。研究证实，嵌入介孔结构中的与乙基桥联的有机硅组分可有效保护 Ni 活性物种不与溶剂水接触，使其不会产生惰性 Ni 物种，从而保持较高的活性。

2. 负载型钌催化剂

Liu 和 Kou 等[179]报道了 Ru 纳米簇分散在 pH=2 的水中催化纤维二糖转化产生果糖。Liu 等[180]进一步研究了 Ru/C 在热水中催化可逆产生 H^+，使纤维素发生加氢反应获得 40%的己糖醇。Wang 等[180,181]也证实负载在碳纳米管上的 Ru 纳米粒子因含有酸性基团，可有效催化纤维素还原产生纤维二糖，其他 Ni 基催化剂也可用于该反应[182,183]。Zhang 等[184]发现 Ni 促进 W_2C/AC 催化热水中纤维素加氢水解生成乙二醇，但催化剂不稳定。Wang 等[185-187]研究 Ru/碳纳米管催化纤维素转化的反应发现，反应过程中加氢反应速率比水解快。研究也显示，含有强酸性的矿物酸和液体杂多酸与 Ru/C 结合，可高效催化纤维素转化成相应的糖醇。Xue 等[188]报道用水合肼还原制得 Ru-Zn/SiO_2 催化剂并用于水中苯部分加氢反应。研究发现，水合肼的用量对 Ru 粒子大小没有明显影响，但对催化活性有明显影响。Hardacre 等[189]报道 Ru/SiO_2 催化水中 2-丁酮液相加氢生成 2-丁醇显示出高活性，但使用寿命较差，归因于部分金属 Ru 被氧化成 $Ru(OH)_x$。Wu 等[190]通过浸渍及氢气还原，制得包裹在 FUD-15 介孔高分子材料中的 Ru 纳米粒子催化剂，可催化含有不同取代基团的苯环甚至羰基化合物[式(3-88)]的加氢反应，在温和的条件下，显示高活性和选择性，如在环己酮的反应中 TOF 值达 4756 h^{-1}，同时，催化剂可以重复使用。

$$\underset{R}{\text{C}_6\text{H}_4}\text{-CHO} \xrightarrow[\text{H}_2\text{O, 4.0 MPa H}_2]{\text{Ru/FDU-15-H}_2} \underset{R}{\text{C}_6\text{H}_4}\text{-CH}_2\text{OH} \qquad (3\text{-}88)$$

Miyazawa 等[191]报道以 Ru/C 和 Al 粉催化水中 α,ω-烷基二胺、1,4-丁二胺、1,5-戊二胺以及 1,6-己二胺反应生成环状胺[式(3-89)], 得率为 42%~99%, 产物选择性为 86%~98%, 明显高于 Pd/C。

$$\text{H}_2\text{NCH}_2\text{CHR(CH}_2)_n\text{NH}_2 \xrightarrow[\text{H}_2\text{O, 158 °C}]{\text{Ru/C, Al 粉}} \underset{}{\text{环状胺}(CH_2)_n} \qquad (3\text{-}89)$$

Fukuoka 等[192, 193]以介孔碳材料负载 Ru 纳米粒子, 催化水中纤维素水解, 在温和条件下, 给出高 TON 值和高得率的葡萄糖产物, 这归因于载体和 Ru 之间存在协同作用, 其中介孔碳有利于纤维素水解成低聚糖, 而 Ru 催化低聚糖转变成葡萄糖。Tan 等[194]报道在超临界 CO_2 中将约 3.4 nm 的 Ru 纳米粒子负载在介孔二氧化硅 MCM-41 载体上, 在 75~85℃以及 50 bar H_2 条件下, Ru/MCM-41 催化水中双酚 A 的加氢反应[式(3-90)], 效率明显高于其他有机溶剂如异丙醇, 乙酸乙酯作为反应介质。与 Ru/C 相比, Ru/MCM-41 不仅具有高催化活性, 且可重复使用。

$$\text{双酚A} \xrightarrow[\substack{85\,°C,\ 50\ \text{bar}\ H_2,\ 3\ h \\ H_2O}]{\text{Ru/MCM-41}} \text{产物} \qquad (3\text{-}90)$$

Ru/MCM-41 TON值: 3233
Ru/C TON值: 2495

Li 等[195]采用共浸渍法制得 Ru/AlO(OH)催化剂, 在水介质柠檬醛加氢生成香叶醇和橙花醇反应中显示出优良催化效率, 柠檬醛转化率达 61%, 产物选择性达 74%, 这归因于载体具有高结晶度及存在大量的羟基, 水和载体表面羟基相互作用形成一层水膜, 从而有利于在催化剂表面吸附柠檬醛的 C=O 键而不是 C=C 键, 如加入有机添加剂, 则可进一步提高金属粒子的疏水性和电子密度, 提高香叶醇和橙花醇的选择性。另外, 催化剂可以多次重复使用。Iwamoto 等[196]在 250~270℃和 5 MPa H_2 条件下, 用 Ru-PVP 催化亚临界水中纤维素和木聚糖加氢获得甲烷和低级烷烃, 如采用 Rh、Ir、Pd、Pt 和 Au 催化剂则获得二醇, 证明 Ru 催化剂有利于 C—C 键断裂和加氢。2013 年，Lu 等[197]首先将 RuCl$_3$ 浸渍在介孔锆硅载体中，烘干并用氢气还原，获得 Ru/MSN-Zr 催化剂，其中 Ru 粒子均匀分散在介孔孔道中，粒径为 1 nm 左右，且在 500℃具有高热稳定性。应用于水中呋喃衍生物的加氢反应[式(3-91)]，在室温和 5 bar H$_2$ 下反应 4 h，转化率达 87.4%~100%，选择性达到 68.9%~92.1%。

$$R-\text{furan}-CHO \xrightarrow[\text{H}_2\text{O}]{\text{Ru/MSN-Zr}} R-\text{furan}-CH_2OH + R-\text{THF}-CH_2OH \quad (3\text{-}91)$$

Ebitani 和 Patankar[169, 198]报道了 Ru/C、Ru/SBA、Au/ZrC 和 Au/ZrO$_2$ 催化水中 γ-戊内酯的合成[式(3-92)]。以甲酸为氢源，乙酰丙酸为底物，获得产物 γ-戊内酯的得率均超过 90%，Au/ZrO$_2$ 催化活性和稳定性最佳，能使甲酸完全降解成 CO$_2$ 和 H$_2$，产物 γ-戊内酯的得率为 97%，另外，Pd-Cu/ZrO$_2$ 双金属催化剂也表现出较高活性和较长使用寿命。

$$\text{levulinic acid} + \text{HCOOH} \xrightarrow[\text{H}_2\text{O, 150 °C}]{\text{cat.}} \gamma\text{-valerolactone} + CO_2 + H_2O \quad (3\text{-}92)$$

Liu 等[199]报道了 Ru/C 催化水中 5-羟甲基呋喃的氧化反应，获得产物 5-甲酰基-2-呋喃羧酸和 2,5-呋喃二甲酸。研究发现，与 MgO、Ca(OH)$_2$ 和 NaOH 等碱相比，采用水滑石为碱有利于提高选择性，这归因于其碱性适中，通过 XPS 和同位素分析发现，Ru/C 催化剂在反应过程起主导作用的是金属态 Ru，而溶剂水提供了氧源。Zhu 等[200]利用 γ-Al$_2$O$_3$ 载体表面含有丰富的羟基进行 3-氨丙基三乙氧基硅烷修饰，形成了稳定的 Al—O—Si 的结构，而 Ru 活性中心则通过与氨基配位键合在 γ-Al$_2$O$_3$ 的表面，使 Ru 活性位高度分散，且电子密度增加，在 25℃水介质中乙酰丙酸的加氢反应中，获得高得率产物 γ-戊内酯产物，催化 TOF 值达 306 h^{-1}。Bengoechea 等[201]采用 Ru/Al$_2$O$_3$、Rh/Al$_2$O$_3$ 和 Pd/Al$_2$O$_3$ 催化甲酸/水体系中 3 种不同来源的木质素水解，反应温度和时间分别为 340~380℃ 及 2~6 h，其中，Ru/Al$_2$O$_3$ 催化下产物得率高于 92%。Bond 等[202]将四种负载型 Ru 催化剂(Ru/C，Ru/SiO$_2$，Ru/γ-Al$_2$O$_3$，Ru/TiO$_2$)用于水中乙酰丙酸和 2-戊酮的加氢反应，发现载体种类和 Ru 粒子大小对加氢反应活性没有明显影响。近年来，Zahmakiran 等[203]通过气相渗透将 Ru 源负载在 MIL-101 上，然后在 50℃下用 3 bar H$_2$ 还原，获得 Ru/MIL-101 催化剂，Ru 粒径约为 44.2，主要分布在 MIL-101 表面，MOF 结构保存。在水中苯酚加氢生成环己醇反应中[式(3-93)]，Ru/MIL-101 催化剂在较温和条件(50℃和 5 bar H$_2$)下，显示出高活性和选择性，催化剂可重复使用。Wu 等[204]报道在室温下用水合肼还原，获得包裹在有序介孔碳中的 Ru 纳米粒子催化剂，其在硝基苯及其衍生物还原反应中显示出高活性，优于商业化的 Ru/C 催化剂。

$$\text{PhOH} \xrightarrow[\text{H}_2\text{O, 50 °C, 5 bar H}_2]{\text{Ru@MIL-101}} \text{cyclohexanone} + \text{cyclohexanol} \qquad (3\text{-}93)$$

3. 负载型银催化剂

Salam 等[205]将 Ag 纳米粒子负载在石墨烯上，并应用在水中醛、炔与胺的三组分偶联反应(A^3-coupling)及 1,4-二取代-1,2,3-三唑的合成[式(3-94)]。在室温下可获得较高得率的目标产物，催化剂能重复循环使用 5 次以上。

$$\text{R-C}_6\text{H}_4\text{-N}_3 + \text{R}_1\text{-C}_6\text{H}_4\text{-C≡CH} \xrightarrow[\text{H}_2\text{O, rt, 8~10 h}]{0.5 \text{ mol\% cat.}} \text{triazole} \qquad (3\text{-}94)$$

2009 年，Kaneda 等[206]报道羟基磷灰石负载 Ag 纳米粒子催化水中腈水合反应生成酰胺的反应[式(3-95)]，并在其中表现出优越的催化性能，尤其在催化 N、O、S 杂芳环腈反应[式(3-96)]中显示出独特的催化效果，目标产物得率为 94%~99%，催化剂可重复使用。

$$\text{R-C≡N} + \text{H}_2\text{O} \xrightarrow{\text{AgHAP}} \text{R-C(O)NH}_2 \qquad (3\text{-}95)$$

$$\underset{X = \text{N, O, S}}{\text{X-C≡N}} + \text{H}_2\text{O} \xrightarrow{\text{AgHAP}} \text{X-C(O)NH}_2 \qquad (3\text{-}96)$$

Shimizu 等[207]用不同方法制备 Ag/SiO_2 催化剂，Ag 粒径大小为 17~30 nm，考察其在水介质 2-氰基吡啶的水合反应中，发现 700℃下用 H_2 还原制得的催化剂，TOF 值最高，并可以有效地催化其他腈，包括脂肪腈、芳香族腈、杂环芳香族腈和 α,β-不饱和腈的选择性水合反应，均能给出较高得率的酰胺。Xu 等[208]发展了一种功能化载体负载单分散 Ag 催化剂的制备方法。首先将 TiO_2 分散在 Si-H 功能化的聚甲基氢硅氧烷材料中，然后使其还原 $AgNO_3$ 得到单分散 Ag 催化剂，通过优化条件可控制 Ag 大小为 10~30 nm，应用于 2,2,2-三氟-1-苯基乙酮和苯乙炔的反应[式(3-97)]，60℃下反应 2 h，获得目标产物得率为 98%。该催化剂也能成功用于含有不同取代基的 2,2,2-三氟-1-芳基乙酮和芳炔的加成反应，目标产物得率为 90%~98%。在芳醛和芳炔的加成反应[式(3-98)]中也表现出高活性，产物得率中等，若在反应中加入一定量的 2,2,2-三氟-1-苯基乙酮，产物得率可高达 99%。

$$R_1 \underset{}{\overset{O}{\diagdown}} CF_3 + R_2 \underset{}{\diagdown} \xrightarrow[H_2O,\ 60\ ℃,\ 2\ h]{NanoAg@TiO_2@PMHSIPN \atop PPh_3,\ PhCOCF_3} \underset{R_2}{\overset{HO \ \ \diagdown \ R_1}{\diagup CF_3}} \qquad (3\text{-}97)$$

$$R_1 \underset{}{\diagdown} CHO + R_2 \underset{}{\diagdown} \xrightarrow[H_2O,\ 60\ ℃,\ 24\ h]{NanoAg@TiO_2@PMHSIPN \atop PPh_3,\ PhCOCF_3} R_2 \underset{}{\diagdown} \underset{R_1}{\overset{OH}{\diagdown}} \qquad (3\text{-}98)$$

Movahedi 等[209]用离子液体修饰的 ZnO 纳米粒子负载 Ag 纳米粒子(图 3.7)，并将其用于催化水中醛、胺和炔的三组分"一锅"法偶联反应生成炔丙胺。研究发现，该催化剂在传统的有机溶剂(如乙醇、乙腈和 THF 等)中的活性(34%~82%)明显低于水中的活性(92%)。最佳反应条件下，该催化剂可以催化不同底物的偶联反应，除对硝基苯甲醛与吗啉、苯乙炔的反应，其他反应的目标产物得率为 68%~95%，催化剂可重复使用 3 次。

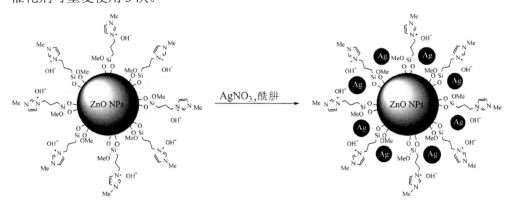

图 3.7 ZnO-IL/Ag 催化剂合成示意图

Mandal 等[210]报道了聚乙二醇和聚对乙烯苯酚存在下，通过液相化学还原法制备 CuNi 合金催化剂，球状合金纳米粒子自组装成链状结构，该催化剂可在室温下有效催化水中或 DMF 中不同炔和叠氮化合物的反应[式(3-99)]，目标产物得率为 56%~83%，且可重复使用 3 次以上。

$$R_1-N_3 + R_2\!\!=\!\!= \xrightarrow[H_2O, rt, 18\ h]{CuNi\text{-}PEG} \underset{R_2}{\overset{N=N}{\diagdown}}\!\!\!N\text{-}R_1 \qquad (3\text{-}99)$$

Ghahremanzadeh 等[211]以大茴香植物提取液作为还原剂和稳定剂，在较温和条件下制备 Ag 纳米粒子催化剂，并将其用于水中螺羟吲哚衍生物的合成[式(3-100)]，产物得率为 77%~96%，如靛红、二胺吡啶与季酮酸在水介质中反应 15 min，给出 87%的目标产物，高于其他有机溶剂如甲醇、THF、甲苯作为反应介质(得率＜65%)，此外，催化剂可重复使用 4 次。

$$\text{(靛红)} + \text{(苯胺)} + \text{(1,3-二酮)} \xrightarrow[H_2O, 回流]{Ag\ NPs} \text{(螺羟吲哚衍生物)} \qquad (3\text{-}100)$$

Zhao 等[212]报道了聚丙烯胺负载的 Ag 纳米粒子催化水中炔-叠氮化合物的环加成反应，如在 80℃水中可高效催化炔-叠氮化合物的环加成产生 1,4-二取代-1,2,3-三唑[式(3-101)]。富电子的芳炔与叠氮苄有利于提高产物得率，而缺电子的芳炔给出的产物得率较低，若延长反应时间，也能给出较高得率的产物，而脂肪炔与叠氮苄反应时需要的时间更长。催化剂具有高稳定性，能够重复使用 10 次以上。

$$R_1\!-\!N_3 + R_2\!\!=\!\!= \xrightarrow[H_2O, 80\ ℃]{Ag\ NPs} \underset{R_2}{\overset{N=N}{\diagdown}}\!\!\!N\text{-}R_1 \qquad (3\text{-}101)$$

2008 年，Sreedhar 等[213]用氨基修饰的 SiO_2 球为载体，通过氨基与 Cu^{2+} 配位固载，再用水合肼还原，得核壳结构 SiO_2@Cu 催化剂(图 3.8)，Cu 粒径为 55~60 nm。可成功催化水中胺[式(3-102)]或二胺[式(3-103)]或硫醇[式(3-104)]与 α,β-不饱和化合物在室温的共轭加成反应，获得 Michael 加成产物，反应 4~30 min，得率达到 70%~98%。

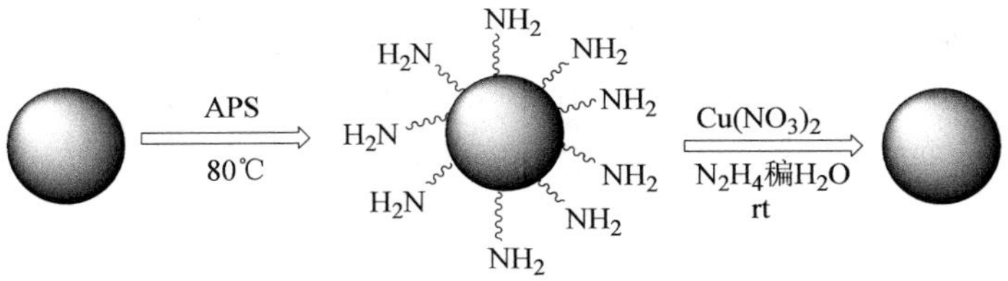

$$\bigcirc \xrightarrow[80\text{℃}]{\text{APS}} \text{H}_2\text{N-}\bigcirc\text{-NH}_2 \xrightarrow[\text{N}_2\text{H}_4 \cdot \text{H}_2\text{O, rt}]{\text{Cu(NO}_3)_2} \bigcirc$$

图 3.8 SiO$_2$@Cu 催化剂合成示意图

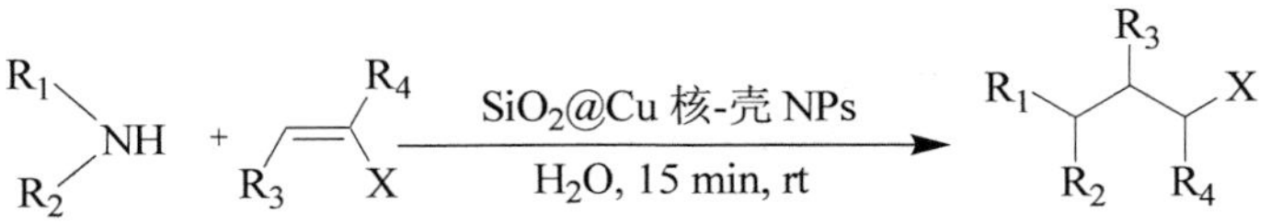

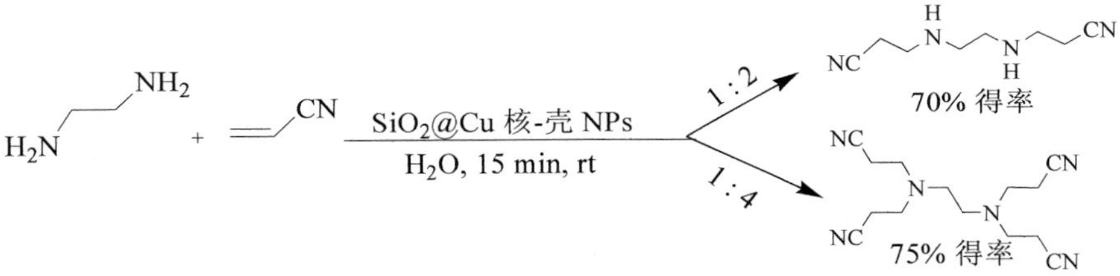

$$\underset{R_2}{\overset{R_1}{\text{N}}}\text{H} + \underset{R_3}{\overset{R_4}{\text{C=C}}}\text{X} \xrightarrow[\text{H}_2\text{O, 15 min, rt}]{\text{SiO}_2\text{@Cu 核-壳 NPs}} R_1\underset{R_2}{\overset{R_3}{\text{CH-C}}}\text{X} \qquad (3\text{-}102)$$

$$\text{H}_2\text{N}\diagup\text{NH}_2 + \diagup\text{CN} \xrightarrow[\text{H}_2\text{O, 15 min, rt}]{\text{SiO}_2\text{@Cu 核-壳 NPs}} \begin{array}{c} 1:2 \rightarrow \text{NC}\diagup\text{N}\diagup\text{N}\diagup\text{CN} \\ \text{70\% 得率} \\ 1:4 \rightarrow \text{(NC}\diagup)_2\text{N}\diagup\text{N(}\diagup\text{CN)}_2 \\ \text{75\% 得率} \end{array} \qquad (3\text{-}103)$$

$$\text{Ph-SH} + \diagup\text{CO}_2\text{R} \xrightarrow[\text{H}_2\text{O, 15 min, rt}]{\text{SiO}_2\text{@Cu 核-壳 NPs}} \text{Ph-S}\diagup\text{CO}_2\text{R} \qquad (3\text{-}104)$$

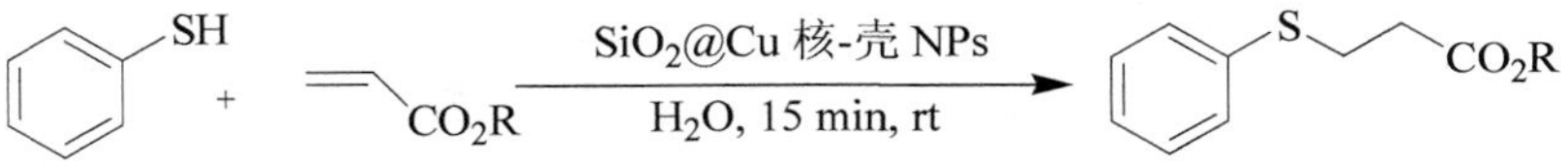

R = CH$_3$, C$_2$H$_5$, i-C$_3$H$_7$, t-C$_4$H$_9$

2011 年，Varma 和 Nasir Baig[214]报道了谷胱甘肽功能化的磁性 Fe$_3$O$_4$ 负载 Cu 纳米粒子催化剂，Cu 粒子大小为 10~25 nm。应用于微波辐射下水中叠氮苄和苯乙炔反应[式(3-105)]，发现催化剂不仅展示出高活性，且产物选择性达 100%，并拓展到炔、溴代烃和叠化钠的环加成反应。

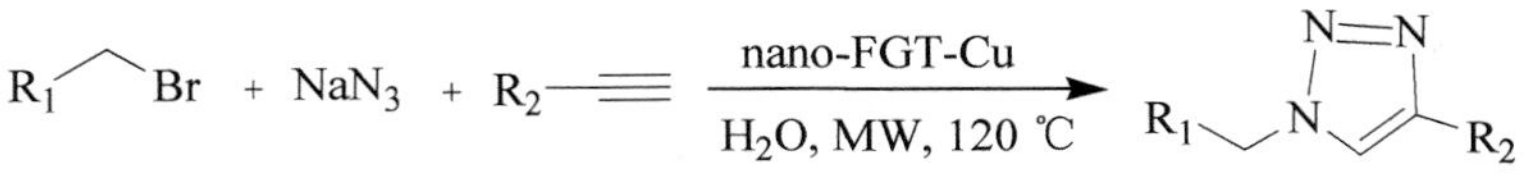

$$R_1\text{-CH}_2\text{Br} + \text{NaN}_3 + R_2\text{-C≡CH} \xrightarrow[\text{H}_2\text{O, MW, 120 ℃}]{\text{nano-FGT-Cu}} R_1\text{-CH}_2\text{-N}\underset{N=N}{\overset{}{\diagdown}}R_2 \qquad (3\text{-}105)$$

Stolle 等[215,216]以多孔玻璃负载金属 Cu，催化微波辐射下水中叠氮与炔的环加成反应[式(3-106)]，在短时间内获得高得率目标产物，与均相 Cu 催化剂的活性相当。

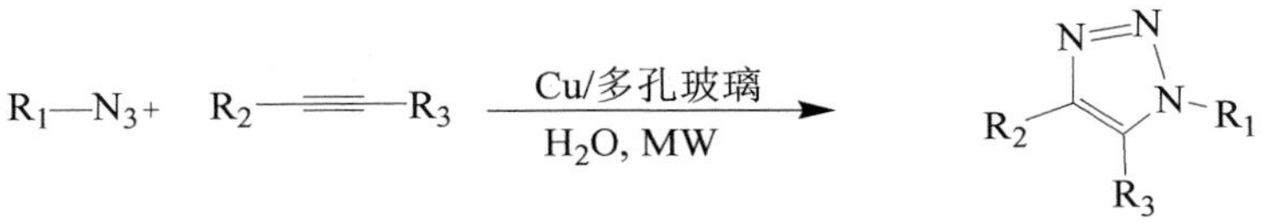

$$R_1\text{-N}_3 + R_2\text{-C≡C-}R_3 \xrightarrow[\text{H}_2\text{O, MW}]{\text{Cu/多孔玻璃}} \underset{R_3}{\overset{R_2}{\diagdown}}\text{N}\underset{N=N}{\overset{}{\diagdown}}R_1 \qquad (3\text{-}106)$$

R$_1$ = Bn, n-decyl; R$_2$ = Ph, 芳基, 脂基, CO$_2$CH$_3$; R$_3$ = H, CO$_2$CH$_3$, TMS

最近，Safa 等[217]报道了 Fe-Cu/ZSM-5 双金属催化剂，在超声作用下，催化水介质中醛与 2-萘酚、1,3-环二酮的三组分缩合反应[式(3-107)]，获得高得率的目标产物。

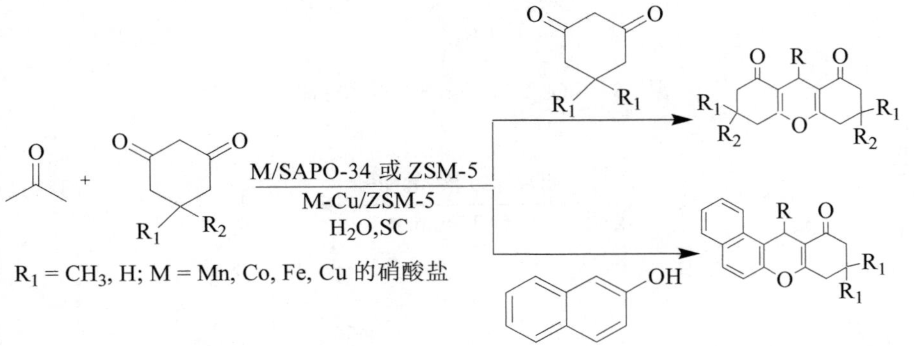

$$R_1 = CH_3, H; M = Mn, Co, Fe, Cu \text{ 的硝酸盐}$$

(3-107)

Paul 等[218]报道了一种新型空气稳定的磁性 Cu 催化剂(Cu(0)-Fe$_3$O$_4$@SiO$_2$/EDAcel)（见彩图 2），载体采用乙二胺功能化的无机/有机杂化体，其中化学官能团氨基可以有效地固定 Cu 纳米粒子，并使其高度分散。该催化剂可高效催化水中叠氮和炔的 1,3-偶极环加成反应以及水中 2-碘苯胺与醛、硫脲的三组分"一锅"法反应合成 2-取代苯并噻唑衍生物，获得较高得率的目标产物。催化剂可用外部磁体快速分离且方便回收，重复使用 6 次活性没有明显降低。

近来，Fu 等[219]报道了将硝酸铜、钨酸铵、硝酸锆溶解在 20%的草酸乙醇溶液中，通过搅拌、洗涤和干燥后，置于 380℃下焙烧 5 h，最后再用 H$_2$ 还原。与其他负载型 Cu 催化剂相比，该催化剂在水中乙酰丙酸加氢生成 γ-戊内酯的反应中催化活性最高，获得目标产物得率为 84%。

4. 负载型铑催化剂

Mukhopadhyay 等[220]报道了用 Rh/C 直接催化卤代芳烃或取代卤代芳烃的自偶联反应。在最佳反应条件下，联苯及其衍生物的得率约为 50%，催化剂可回收并重复使用。Roucoux 等[221]设计了一种新型 SiO$_2$ 负载 Rh 纳米粒子催化剂的方法。该方法基于在 *N,N*-二乙基-*N*-十六烷基-*N*-(2-羟乙基)氯化铵的存在下，通过浸渍方法将 Rh 纳米粒子固载在 SiO$_2$ 上，Rh 粒子粒径大小为 2.4~5.0 nm，以一烷基/二烷基取代的或其他功能化的芳烃衍生物为反应底物，在催化水中加氢反应[式(3-108)]中，TOF 值高达 6430 h^{-1}，且催化剂可重复使用多次。

$$\underset{R_1}{\overset{R}{\text{[arene]}}} \xrightarrow[\text{1~40 bar } H_2, H_2O, \text{rt}]{Rh/SiO_2} \underset{R_1}{\overset{R}{\text{[cyclohexane]}}} \qquad (3\text{-}108)$$

Denicourt-Nowicki 等[222]发展了一种制备 TiO_2 负载 Rh 纳米粒子的简单方法,主要是采用表面活性剂稳定水相胶体,不需焙烧。催化剂应用于常压下水相芳烃的加氢反应,TOF 值达 476 h^{-1},增加氢气压力到 30 bar,TOF 值高达 33333 h^{-1}。在水中邻二甲苯、茴香醚和甲苯的加氢反应[式(3-109)]中,催化剂可重复使用 3 次。该催化剂也可用于芳烃和氯代茴香醚衍生物的加氢反应,TOF 值达 33000 h^{-1},高于传统 Rh@SiO_2 催化剂[223]。

$$\underset{Cl}{\overset{OCH_3}{\text{[arene]}}} \xrightarrow[H_2, \text{rt}, H_2O]{Rh@TiO_2} \overset{OCH_3}{\text{[cyclohexane]}} + \overset{O}{\text{[cyclohexanone]}} \qquad (3\text{-}109)$$

Heeres 等[224]报道了一种用于从 1,2,6-己三醇合成 1,6-己二醇的高效催化剂。1,2,6-己三醇合成 1,6-己二醇有两种方法,一是在分子氧存在下,1,2,6-己三醇直接催化加氢脱氧给出 1,6-己二醇,如采用 Rh-ReO_x/SiO_2,反应物可完全转化,产物得率为 73%;二是先用三氟乙酸催化 1,2,6-己三醇定量地转化成四氢吡喃 2-甲醇,再通过 Rh-ReO_x/SiO_2 催化开环/加氢脱氧给出 1,6-己二醇[式(3-110)],产物选择性达到 96%,该方法适合大批量生产 1,6-己二醇。

$$\underset{OH}{\text{HO}\overset{OH}{\frown}\text{OH}} \xrightarrow{H_2, \text{cat.}} \text{HO}\frown\text{OH} \qquad (3\text{-}110)$$

Kawanami 等[225]以超临界 CO_2-水为反应介质,考察 Rh/C 催化二苯醚的 C—O 键断裂反应[式(3-111)],转化率为 100%,环己醇得率为 96%,若在无水条件下,则主要生成环己基醚,选择性达 98%。作者也将此催化体系应用于其他芳香醚的 C—O 键断裂反应中,转化率为 20%~100%,含有吸电子基团或给电子基团的不对称醚,需要更长的反应时间,且取代基的性质对产物的选择也有着至关重要的影响。在对称醚的反应中,催化活性更高,给出相应的环烷烃和酚产物。

$$\text{Ph-O-Ph} \xrightarrow[H_2, H_2O]{5\%Rh/C} \overset{OH}{\text{[cyclohexanol]}} + \text{[cyclohexane]} \qquad (3\text{-}111)$$

Burri 等[226]用氨基功能化的 SBA-15 负载 Rh 纳米粒子,载体的孔道限制和 NH_2 的稳定作用使 Rh 纳米粒子在载体上高度分散。通过水合肼原位产生的 H_2,在室温下催化水中芳香族硝基化合物的还原,产物胺的得率接近 100%,催化剂的 TOF 值高达 6117 h^{-1},且可重复使用 6 次以上。Arai 等[227]采用 Rh/C 和 Rh/Al$_2$O$_3$ 催化水中苯乙酮的选择性加氢反应生成 1-苯乙醇,结果表明,水是最佳溶剂,这可能是由于 Rh/C 催化剂具有高的氢键给体能力,而 Rh/Al$_2$O$_3$ 催化剂中表现出较低氢键受体性质,从而使转化率提高。

5. 负载型铂催化剂

Xiang 等[228]首先合成了水滑石负载的 Pt 纳米晶催化剂,然后用其催化水中肉桂醛选择性加氢反应。研究表明,Pt 粒子尺寸对加氢产物肉桂醇的选择性具有重要影响,使用较大尺寸的 Pt 催化剂,选择性可达到 85%,添加 NaOH 为碱时,选择性可提高到 90%,催化剂可重复使用 4 次以上。Kaneda 等[229]采用羟基磷灰石负载 Pt 和 Mo 双金属催化剂应用于水中乙酰丙酸选择性加氢生成 1,4-戊二醇[式(3-112)]。Pt 和 MoO$_x$ 相互作用有利于 4-羟基戊酸[式(3-113)]和 γ-戊内酯[式(3-114)]的加氢生成 1,5-戊二醇中间体。催化剂可重复使用多次。最近,他们发展了 H-β-分子筛负载的 Pt 和 Mo 双金属催化剂,在催化乙酰丙酸加氢脱水反应生成 2-甲基四氢呋喃反应中,无需添加剂,目标产物得率达到 86%,这可归因于 Pt 和 MoO$_x$ 相互作用有利于乙酰丙酸形成 1,4-戊二醇中间体,而载体 H-β-分子筛的酸性有利于形成中间体脱水-环化[230]。

$$\text{(3-112)}$$

$$\text{(3-113)}$$

$$\text{(3-114)}$$

第三节 非晶态合金催化剂

非晶态合金通常由金属和类金属(如 B、P、C、Si)组成,是一类具有长程无序、短程有序结构特点,介于晶体和无定形之间的一类新型介观材料,因此具有许多独特的性质,如优良的力学、磁学、电学及化学性质,良好的抗腐蚀、抗辐射性能等，在磁性材料、防腐材料等方面已获得工业应用。长程无序，是指原子在三维空间呈拓扑无序状排列，即不存在一般晶态合金所存在的位错、偏析和晶界等缺点，原子簇堆积相对比较随机且无序，无明显的规律。短程有序，是指在近邻的配位层呈有序状态，一般在几个晶格常数范围内呈短程有序。作为晶体前驱体的非晶态合金，可在较大的组成范围内调变合金的化学组成，从而调控其电子性质；催化活性位可以单一的形式均匀地分布在化学环境中，其表面具有较高浓度的不饱和催化活性中心，且表面能高，这表明非晶态合金具有高浓度、高度配位的不饱和位，使其催化活性和化学选择性通常高于晶态催化剂；因非晶态合金的结构不是多孔性的，因此，非晶态合金催化剂不仅不会存在传统非均相催化剂存在的扩散阻力问题；与晶态合金相比，非晶态合金还具有较好的机械强度。这些优点使非晶态合金催化剂在多相催化反应中具有较好的应用前景。目前已有一些非晶态合金催化剂进行了工业化应用，其工业化前景比较乐观[231-234]。

目前，非晶态合金催化剂主要应用于加氢、氧化、裂解和异构化等反应。在 2005 年，Li 课题组[235]采用 KBH_4 还原不同溶剂的 $CoCl_2$ 溶液，获得了一系列 Co-B 非晶态合金催化剂，并应用于水中呋喃甲醛选择加氢制备呋喃甲醇的反应[式(3-115)]，结果表明，采用乙醇配制的 $CoCl_2$ 溶液制得的 Co-B 非晶态合金催化剂在此加氢反应中的活性最高，而选择性相差不大，仍接近 100%。在不同的反应介质中，糠醛加氢反应初始吸氢速率有很大的区别，水介质中的反应速率最大，正丙醇中反应速率最小(表 3.1)，这可能是由于反应初始吸氢速率很大程度上受溶剂的介电常数所影响，即加氢反应速率随着溶剂介电常数的增大而增大，这是因为从正丙醇到水的极性变化过程中氢气易于吸附在金属催化剂表面[236]。在水介质加氢反应中，糠醛的转化率达到 100%，选择性为 78%。

$$\underset{}{\text{O}}\text{—CHO} \xrightarrow[\text{H}_2\text{O}]{\text{Co-B}} \underset{}{\text{O}}\text{—CH}_2\text{OH} \quad (3\text{-}115)$$

表 3.1 反应过程中溶剂效应的影响

序号	溶剂	介电常数	RHm/[mmol/(h·g)]	转化率/%	选择性/%
1	n-C_3H_7OH	20.3	18.8	47.7	100
2	C_2H_5OH	24.3	41.4	80.7	100
3	CH_3OH	33.6	43.1	90.2	100
4	$HOCH_2CH_2OH$	37.7	48.1	92.5	100
5	H_2O	72.9	72.9	100	78

注：反应条件为 12 mmol 呋喃，$p(H_2)$ = 1.0 MPa，1.0 g Co-B,0.54 g，30 mL 溶剂，80℃，4 h。

Li 等[237]通过低温 KBH$_4$ 还原 InCl$_3$ 制备出热稳定性好和比表面积高的 In-B 非晶态合金催化剂，其在水相 Barbier 反应显示出高活性。从图 3.9a 可见，新制备样品 In-B 在 $2\theta = 32°$ 左右出现一个宽化的弥散峰[238]，表明样品具有非晶态结构，a 中选区电子衍射(SAED)图进一步显示样品 In-B 具有非晶态结构的特征弥散环[232]。图 3.9b 显示，样品经 Ar 保护在 200℃下处理 2.0 h 后出现了金属 In 的特征衍射峰，b 中 SAED 图出现晶态结构电子衍射特有的点阵，表明非晶态结构经过热处理后逐渐晶化。

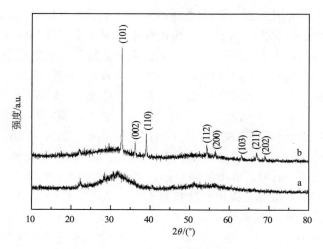

图 3.9　不同样品的 XRD 谱图
a-新鲜样品 In-B SAED 图；b-经 200℃处理的样品 In-B SAED 图

从图 3.10 的 TEM 形貌可以观察到 In-B 细小纳米颗粒，图 3.10(a)中 SAED 图显示 In-B 样品具有非晶态结构的特征弥散环，右图的高分辨率的透射电镜(HRTEM)显现出短程有序长程无序的结构特点。从表 3.2 可见，In-B 非晶态合金在水相苯甲醛烯丙基化反应[式(3-116)]中具有较高的催化活性，一方面归因于其具有独特的短程有序长程无序的结构，造成高度配位不饱和活性位和活性位间强协同作用，有利于反应物的吸附和反应；另一方面主要是由于 In-B 非晶态合金中 In 和 B 间存在强电子相互作用，造成 In 表面富电子而 B 表面缺电子，较高电子密度的 In 活性位有利于单电子转移过程中自由基阴离子中间体的形成，促进了烯丙基化反应。产物 1-苯基-3-丁烯基-1-醇的最高得率可达到 94%，与均相 PdCl$_2$(PPh$_3$)$_2$ 催化剂上的得率相当。

(a) SAED 图

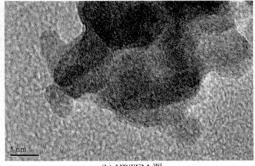
(b) HRTEM 图

图 3.10 In-B 样品的 TEM 图

$$\text{PhCHO} + \text{CH}_2=\text{CHCH}_2\text{Br} \xrightarrow{\text{In-B}} \text{PhCH(OH)CH}_2\text{CH}=\text{CH}_2 \quad (3\text{-}116)$$

表 3.2 不同催化剂的催化性能和结构参数

序号	催化剂	组分/atom%	S_{BET}/(m²/g)	转化率/%	选择性/%	R^S/[mmol/(min·m²)]
1	In-B	In$_{83}$B$_{17}$	136	96	98	0.0023
2	晶态 In-B	In$_{83}$B$_{17}$	88	47	97	0.001
3	粉末	In	109	85	98	0.0019
4	PdCl$_2$(PPh$_3$)$_2$	—	—	96	96	—

注: 反应条件为 2.25 mmol 苯甲醛, 4.5 mmol 3-溴丙烯, 1.12 mmol In-B, 0.54 g Al 粉, 10.0 mL H$_2$O, 50℃, 2 h。

In-B 非晶态合金在催化含不同取代基的苯甲醛与烯丙基卤的 Barbier 反应[式(3-117)]中均表现出高活性(表 3.3)。即使在较稳定的苯乙酮与烯丙基溴的反应中,产物得率也能达到 84%。当采用烯丙基硼试剂进行反应时,给出较高收率的产物,与 Kobayashi 课题组[239]报道的结果基本一致。而用 3-氯丙烯与苯乙酮反应时,得率仅为 63%。在苯甲醛与烯丙基溴的反应中,In-B 催化剂可以重复使用 3 次。

$$\text{R-C}_6\text{H}_4\text{-COR}' + \text{CH}_2=\text{CHCH}_2\text{X} \xrightarrow[\text{H}_2\text{O}]{\text{In-B}} \text{R-C}_6\text{H}_4\text{-C(OH)(R')CH}_2\text{CH}=\text{CH}_2 \quad (3\text{-}117)$$

表 3.3 In-B 催化剂在不同底物的 Barbier 反应中的催化性能

序号	R	X	R′	转化率/%	选择性/%
1	H	Br	H	96	98
2	CH$_3$	Br	H	94	98
3	Cl	Br	H	97	97
4	H	Cl	H	95	98
5	H	Br	CH$_3$	90	94
6	H	Cl	CH$_3$	68	93

注：反应条件为 2.25 mmol 羰基化合物，4.50 mmol 烯丙型试剂，1.12 mmol In-B，0.54 g Al 粉，10 mL H$_2$O，50℃，2 h。

Bai 等[240]报道了 Zr 和聚乙二醇-800 修饰的 Ni-B 非晶态合金催化剂，其性能如表 3.4 所示，在水介质苯甲酸选择性加氢生成环己烷甲酸的反应[式(3-118)]中，修饰后的 Ni-B 催化活性明显高于未修饰的 Ni-B，特别是 Zr 修饰的 Ni-B 催化剂(Zr-Ni-B)，给出最高得率(58.8%)的产物，高于传统的 Raney Ni 催化剂。

$$\text{C}_6\text{H}_5\text{CO}_2\text{H} \xrightarrow[\text{H}_2\text{O, 4 MPa H}_2\text{, 150 ℃, 4 h}]{\text{Ni-Zr-B-PEG-800}} \text{C}_6\text{H}_{11}\text{CO}_2\text{H} + \text{C}_6\text{H}_{11}\text{OH} \quad (3\text{-}118)$$

表 3.4 不同催化剂在水中苯甲酸选择性加氢反应中的催化性能

序号	催化剂	转化率/%	选择性/%	得率/%
1	Ni-B	7.8	82.6	6.4
2	Ni-Ce-B	44.5	99.2	44.8
3	Ni-Ba-B	24.1	91.0	21.9
4	Ni-Fe-B	24.9	98.0	24.4
5	Ni-La-B	33.1	95.3	31.5
6	Ni-Zr-B	64.9	90.6	58.8
7	Raney Ni	58.5	97.5	57.0

注：反应条件为 3.0 g 苯甲酸，0.5 g 催化剂，60 mL H$_2$O，150℃，$p(\text{H}_2)$ = 4 MPa，4 h。

表 3.5 显示，用 PEG-800 修饰的 Zr-Ni-B 非晶态合金催化剂，产物得率明显提高，这主要归因于 Zr 与 PEG-800 的协同作用，降低了活性 Ni 物种的团聚。表 3.6 为不同溶剂对产物选择性的影响，可见在极性大的水介质中，产物中的羰基更容易进入水中，使苯环更易于进行选择性加氢，从而得到高选择性的环己烷甲酸。

表 3.5 不同高分子修饰的 Ni-Zr-B 催化剂的催化性能

序号	催化剂	转化率/%	选择性/%	得率/%
1	Ni-Zr-B	64.9	90.6	58.8
2	Ni-Zr-B-PEG-600	74.5	89.2	66.5
3	Ni-Zr-B-PEG-800	93.5	93.6	87.5

续表

序号	催化剂	转化率/%	选择性/%	得率/%
4	Ni-Zr-B-PEG-1000	70.4	88.1	62.0
5	Ni-Zr-B-PEG-1200	68.5	96.1	65.8
6	Ni-Zr-B-PVP	46.2	78.6	36.3
7	Ni-Zr-B-PAM	30.2	62.1	18.8
8	Ni-Zr-B-PVA	29.7	89.4	26.6

注：反应条件为 3.0 g 苯甲酸，0.5 g 催化剂，60 mL H_2O，150℃，$p(H_2)$ = 4 MPa，4 h。

表 3.6 不同溶剂对 Ni-Zr-B-PEG-800 催化剂活性的影响

序号	溶剂	转化率/%	选择性/%	得率/%
1	水	93.5	93.6	87.5
2	二氧六环	99.7	69.2	69.0
3	THF	87.8	56.1	49.3
4	环己烷	99.6	42.7	42.5
5	乙醇	78.7	1.1	8.7
6	2-丁醇	72.7	58.6	42.6

注：反应条件为 3.0 g 苯甲酸，0.5 g 催化剂，60 mL H_2O，150℃，$p(H_2)$ = 4 MPa，4 h。

他们还采用类似方法制备了 PEG-800 修饰的 Ni-Pd-B 非晶态合金催化剂，在较温和条件下，对苯环表现出了高加氢活性[241]。同时，他们考察了苯甲酸乙酯及苯甲酸在不同极性、质子与非质子性溶剂中的加氢反应活性，结果显示，由于水解作用，虽然苯甲酸乙酯在水中的转化率是 100%，但存在一定量水解加氢副产物——环己基甲酸。而在非质子性的溶剂(如二氧六环、正己烷)中，苯环的选择性加氢反应受到了抑制，这说明质子性溶剂有利于苯环加氢[242]。同时，在苯甲酸加氢反应中，当以乙醇为溶剂时，主要发生酯化副反应，检测不到目标产物环己基甲酸，这也证明，溶剂极性对含有多个官能团化合物的选择性加氢具有重要影响，极性溶剂对极性官能团的"定位作用"使苯环更容易吸附在催化剂表面，进而获得苯环选择性加氢的目标产物[240]。在水介质中，催化剂对其他钝化苯环化合物(如苯乙酸、肉桂酸、苯甲酸、苯甲酸甲酯、苯甲酰胺二苯甲酮等)的选择性加氢反应也表现出高活性。除苯甲酸甲酯和苯甲酸乙酯发生水解反应外，采用其他反应底物时都能获得高产率的目标产物。同时也发现，苯甲酸甲酯在水中比在苯甲酸乙酯中更容易发生水解，因此，以苯甲酸甲酯为底物时，主产物是水解羟加氢产物——环己基甲酸。由于肉桂酸中含有非常容易加氢的 C=C 官能团，往往是先得到双键的加氢产物[243]。在苯乙酮加氢反应中，得到了 C=O 和苯环同时加氢的产物；而在二苯甲酮的加氢反应中，主要得到了羰基氢解的副产物及其加氢产物。

也有报道采用浸渍法将 Ni 源和 Co 源负载在不同载体上，然后用 KBH_4 溶液还原，得到负载型 NiCoB 非晶态合金催化剂，并应用于水相肉桂酸加氢反应[式

(3-119)][244]。由表 3.7 可以看出，不同载体负载的 NiCoB 非晶态合金催化剂对氢化肉桂酸的选择性均达 100%，但活性有所不同，γ-Al_2O_3 负载的 NiCoB 非晶态合金催化剂活性最佳，而采用超声法制备 γ-Al_2O_3 负载的 NiCoB 非晶态合金催化剂活性可以进一步提高，这可能是由于超声使得 Ni 活性位更加分散。此外，催化剂可回收使用，但反应过程中有少量 Ni 从载体上脱落。

$$\text{PhCH=CHCO}_2\text{H} \xrightarrow[\text{2.0 MPa H}_2, \ \text{H}_2\text{O}, \ 120\ ^\circ\text{C}]{\text{NiCoB}/\gamma\text{-Al}_2\text{O}_3} \text{C}_6\text{H}_{11}\text{CH=CHCO}_2\text{H} \qquad (3\text{-}119)$$

表 3.7 不同 NiCoB 非晶态合金催化剂在水中肉桂酸加氢反应中的应用

序号	催化剂	转化率/%	选择性/%
1	NiCoB/TiO_2	58.5	100
2	NiCoB/AC	31.7	100
3	NiCoB/ZSM-5	51.6	100
4	NiCoB/Hb	52.4	100
5	NiCoB/γ-Al_2O_3	71.3	100
6	NiCoB/γ-Al_2O_3-u	83.5	100
7	NiCoB[a]	57.8	100

注：反应条件为 3.0 g 肉桂酸，0.2 g 催化剂(10 wt% NiCoB)，60 mL H_2O，120℃，H_2 初始压力= 2.0 MPa，30 min。
a. 0.02 g NiCoB。

参 考 文 献

[1] Pellón R F, Carrasco R, Milián V, Rodés L. Use of pyridine as cocatalyst for the synthesis of 2-carboxy substituted diphenylethers by Ullmann-Goldberg condensation. Synthetic Communications, 1995, 25: 1077-1083.

[2] Pellón R F, Carrasco R, Rodés L. Synthesis of N-phenylanthranilic acids using water as solvent. Synthetic Communications, 1993, 23: 1447-1453.

[3] Pellón R F, Mamposo T, Carrasco R, Rodés L. Use of pyridine as cocatalyst for the synthesis of N-phenylanthranilic acids. Synthetic Communications, 1996, 26: 3877-3883.

[4] Denicourt-Nowicki A, Romagné M L, Roucoux A. N-(2-Hydroxyethyl)ammonium derivatives as protective agents for Pd(0) nanocolloids and catalytic investigation in Suzuki reactions in aqueous media. Catalysis Communications, 2008, 10: 68-70.

[5] Srimani D, Sawoo S, Sarkar A. Convenient synthesis of palladium nanoparticles and catalysis of Hiyama coupling reaction in water. Organic Letters, 2007, 9: 3639-3642.

[6] Sawoo S, Srimani D, Dutta P, Lahiri R, Sarkar A. Size controlled synthesis of Pd nanoparticles in water and their catalytic application in C-C coupling reactions. Tetrahedron, 2009, 65: 4367-4374.

[7] Nasrollahzadeh M, Sajadi S M, Honarmand E, Maham M. Preparation of palladium nanoparticles using euphorbia thymifolia L. leaf extract and evaluation of catalytic activity in the ligand-free Stille and Hiyama cross-coupling reactions in water. New Journal of Chemistry, 2015, 39: 4745-4752.

[8] Wan J, ding L, Wu T, Ma X, Tang Q. Facile one-pot fabrication of magnetic nanoparticles (MNPs)-supported organocatalysts using phosphonate as an anchor point through direct co-precipitation method. RSC Advances, 2014, 4: 38323-38333.

[9] Song H M, Moosa B A, Khashab N M. Water-dispersable hybrid Au-Pd nanoparticles as catalysts in ethanol oxidation, aqueous phase Suzuki-Miyaura and Heck reactions. Journal of Materials Chemistry A, 2012, 22: 15953-15959.

[10] Monopoli A, Calò V, Ciminale F, Cotugno P, Angelici C, Cioffi N, Nacci A. Glucose as a clean and renewable reductant in the Pd-nanoparticle-catalyzed reductive homocoupling of bromo- and chloroarenes in water. The Journal of Organic Chemistry, 2010, 75: 3908-3911.

[11] Dhital R N, Kamonsatikul C, Somsook E, Bobuatong K, Ehara M, Karanjit S, Sakurai H. Low-temperature carbon-chlorine bond activation by bimetallic gold/palladium alloy nanoclusters: An application to Ullmann coupling. Journal of the American Chemical Society, 2012, 134: 20250-20253.

[12] Dhital R N, Kamonsatikul C, Somsook E, Sakurai H. Bimetallic gold-palladium alloy nanoclusters: An effective catalyst for Ullmann coupling of chloropyridines under ambient conditions. Catalysis Science & Technology, 2013, 3: 3030-3035.

[13] Nasrollahzadeh M, Mohammad Sajadi S, Maham M, Ehsani A. Facile and surfactant-free synthesis of Pd nanoparticles by the extract of the fruits of piper longum and their catalytic performance for the Sonogashira coupling reaction in water under ligand- and copper-free conditions. RSC Advances, 2015, 5: 2562-2567.

[14] Fang P P, Jutand A, Tian Z Q, Amatore C. Au-Pd core-shell nanoparticles catalyze Suzuki-Miyaura reactions in water through Pd leaching. Angewandte Chemie International Edition, 2011, 50: 12184-12188.

[15] Nasrollahzadeh M, Banaei A. Hybrid Au/Pd nanoparticles as reusable catalysts for Heck coupling reactions in water under aerobic conditions. Tetrahedron Letters, 2015, 56: 500-503.

[16] Handa S, Wang Y, Gallou F, Lipshutz B H. Sustainable Fe-ppm Pd nanoparticle catalysis of Suzuki-Miyaura cross-couplings in water. Science, 2015, 349: 1087-1091.

[17] Javaid R, Kawanami H, Chatterjee M, Ishizaka T, Suzuki A, Suzuki T M. Sonogashira C-C coupling reaction in water using tubular reactors with catalytic metal inner surface. Chemical Engineering Journal, 2011, 167: 431-435.

[18] Mévellec V, Roucoux A, Ramirez E, Philippot K, Chaudret B. Surfactant-stabilized aqueous iridium(0) colloidal suspension: An efficient reusable catalyst for hydrogenation of arenes in biphasic media. Advanced Synthesis & Catalysis, 2004, 346: 72-76.

[19] Schulz J, Roucoux A, Patin H. Stabilized rhodium(0) nanoparticles: A reusable hydrogenation catalyst for arene derivatives in a biphasic water-liquid system. Chemistry-A European Journal, 2000, 6: 618-624.

[20] Tsukinoki T, Kanda T, Liu G B, Tsuzuki H, Tashiro M. Organic reaction in water. part 3∶1 A facile method for reduction of aromatic rings using a Raney Ni-Al alloy in dilute aqueous alkaline solution under mild conditions. Tetrahedron Letters, 2000, 41: 5865-5868.

[21] Jason Bonilla R, James B R, Jessop P G. Colloid-catalysed arene hydrogenation in aqueous/supercritical fluid biphasic media. Chemical Communications, 2000, 11: 941-942.

[22] Ishimoto K, Mitoma Y, Nagashima S, Tashiro H, Prakash G K S, Olah G A, Tashiro M. Reduction of carbonyl groups to the corresponding methylenes with Ni-Al alloy in water. Chemical Communications, 2003, 4: 514-515.

[23] Davie M G, Reinhard M, Shapley J R. Metal-catalyzed reduction of N-nitrosodimethylamine with hydrogen in water. Environmental Science & Technology, 2006, 40: 7329-7335.

[24] Callis N M, Thiery E, Bras J L, Muzart J. Palladium nanoparticles-catalyzed chemoselective hydrogenations, a recyclable system in water. Tetrahedron Letters, 2007, 48: 8128-8131.

[25] Ohde H, Wai C M, Kim H, Kim J, Ohde M. Hydrogenation of olefins in supercritical CO_2 catalyzed by palladium nanoparticles in a water-in-CO_2 microemulsion. Journal of the American Chemical Society, 2002, 124: 4540-4541.

[26] Guyonnet Bile E, Denicourt-Nowicki A, Sassine R, Beaunier P, Launay F, Roucoux A. N-methylephedrium salts as chiral surfactants for asymmetric hydrogenation in neat water with rhodium(0) nanocatalysts. ChemSusChem, 2010, 3: 1276-1279.

[27] Denicourt-Nowicki A, Roucoux A. From hydroxycetylammonium salts to their chiral counterparts. A library of efficient stabilizers of Rh(0) nanoparticles for catalytic hydrogenation in water. Catalysis Today, 2015, 247: 90-95.

[28] Wei C, Li C J. Grignard type reaction via C-H bond activation in water. Green Chemistry, 2002, 4: 39-41.

[29] Son S U, Lee S I, Chung Y K, Kim S W, Hyeon T. The first intramolecular Pauson-Khand reaction in water using aqueous colloidal cobalt nanoparticles as catalysts. Organic Letters, 2002, 4: 277-279.

[30] Saha M, Pal A K, Nandi S. Pd(0) NPs: A novel and reusable catalyst for the synthesis of bis(heterocyclyl)methanes in water. RSC Advances, 2012, 2: 6397-6400.

[31] Kim A Y, Bae H S, Park S, Park S, Park K H. Silver nanoparticle catalyzed selective hydration of nitriles to amides in water under neutral conditions. Catalysis Letters, 2011, 141: 685-690.

[32] Bykov V V, Korolev D N, Bumagin N A. Palladium-catalyzed reactions of organoboron compounds with acyl chlorides. Russian Chemical Bulletin, 1997, 46: 1631-1632.

[33] Sakurai H, Tsukuda T, Hirao T. Pd/C as a reusable catalyst for the coupling reaction of halophenols and arylboronic acids in aqueous media. The Journal of Organic Chemistry, 2002, 67: 2721-2722.

[34] Lu G, Franzén R, Zhang Q, Xu Y J. Palladium charcoal-catalyzed, ligandless Suzuki reaction by using tetraarylborates in water. Tetrahedron Letters, 2005, 46: 4255-4259.

[35] Arcadi A, Cerichelli G, Chiarini M, Correa M, Zorzan D. A mild and versatile method for palladium-catalyzed cross-coupling of aryl halides in water and surfactants. European Journal of Organic Chemistry, 2010, 2003: 4080-4086.

[36] Lysén M, Kohler K. Palladium on activated carbon-a recyclable catalyst for Suzuki-Miyaura cross-coupling of aryl chlorides in water. Synthesis, 2006, 2006: 692-698.

[37] Lysén M, Köhler K. Suzuki-Miyaura cross-coupling of aryl chlorides in water using ligandless palladium on activated carbon. Synlett, 2005, 2005: 1671-1674.

[38] Freundlich J S, Landis H E. An expeditious aqueous Suzuki-Miyaura method for the arylation of bromophenols. Tetrahedron Letters, 2006, 47: 4275-4279.

[39] Arvela R K, Leadbeater N E. Suzuki coupling of aryl chlorides with phenylboronic acid in water, using microwave heating with simultaneous cooling. Organic Letters, 2005, 7: 2101-2104.

[40] Bai L. Atom-efficient coupling reaction of aryl bromides with sodium tetraphenylborate catalyzed by reusable Pd/C in water under focused microwave irradiation. Chinese Chemical Letters, 2009, 20: 158-160.

[41] Pal M, Subramanian V, Yeleswarapu K R. Pd/C mediated synthesis of 2-substituted benzo[b] furans/nitrobenzo [b] furans in water. Tetrahedron Letters, 2003, 44: 8221-8225.

[42] Layek M, Dhanunjaya Rao A V, Gajare V, Kalita D, Barange D K, Islam A, Mukkanti K, Pal M. C-N bond forming reaction under copper catalysis: A new synthesis of 2-substituted 5,6-dihydro-4H-pyrrolo[3,2,1-ij]quinolines. Tetrahedron Letters, 2009, 50: 4878-4881.

[43] Raju S, Kumar P R, Mukkanti K, Annamalai P, Pal M. Facile synthesis of substituted thiophenes via Pd/C-mediated sonogashira coupling in water. Bioorganic & Medicinal Chemistry Letters, 2006, 16: 6185-6189.

[44] Reddy E A, Barange D K, Islam A, Mukkanti K, Pal M. Synthesis of 2-alkynylquinolines from 2-chloro and 2,4-dichloroquinoline via Pd/C-catalyzed coupling reaction in water. Tetrahedron, 2008, 64: 7143-7150.

[45] Layek M, Gajare V, Kalita D, Islam A, Mukkanti K, Pal M. A highly effective synthesis of 2-alkynyl-7-azaindoles: Pd/C-mediated alkynylation of heteroaryl halides in water. Tetrahedron, 2009, 65: 4814-4819.

[46] Venkatraman S, Li C J. Carbon-carbon bond formation via palladium-catalyzed reductive coupling in air. Organic Letters, 1999, 1: 1133-1135.

[47] Mukhopadhyay S, Yaghmur A, Baidossi M, Kundu B, Sasson Y. Highly chemoselective heterogeneous Pd-catalyzed biaryl synthesis from haloarenes: Reaction in an oil-in-water microemulsion. Organic Process Research & Development, 2003, 7: 641-643.

[48] Venkatraman S, Li C J. The effect of crown-ether on the palladium-catalyzed Ullmann-type coupling mediated by zinc in air and water. Tetrahedron Letters, 2000, 41: 4831-4834.

[49] Venkatraman S, Huang T, Li C J. Carbon-carbon bond formation via palladium-catalyzed reductive coupling of aryl halides in air and water. Advanced Synthesis & Catalysis, 2002, 344: 399-405.

[50] Mukhopadhyay S, Rothenberg G, Gitis D, Sasson Y. On the mechanism of palladium-catalyzed coupling of haloaryls to biaryls in water with zinc. Organic Letters, 2000, 2: 211-214.

[51] Mukhopadhyay S, Rothenberg G, Sasson Y. Tuning the selectivity of heterogeneous catalysts: A trimetallic approach to reductive coupling of chloroarenes in water. Advanced Synthesis & Catalysis, 2001, 343: 274-278.

[52] Li J H, Xie Y X, Yin D L. New role of CO_2 as a selective agent in palladium-catalyzed reductive Ullmann coupling with zinc in water. The Journal of Organic Chemistry, 2003, 68: 9867-9869.

[53] Sullivan J A, Flanagan K A, Hain H. Suzuki coupling activity of an aqueous phase Pd nanoparticle dispersion and a carbon nanotube/Pd nanoparticle composite. Catalysis Today, 2009, 145: 108-113.

[54] Li Y, Fan X B, Qi J J, Ji J, Wang S L, Zhang G L, Zhang F B. Palladium nanoparticle-graphene hybrids as active catalysts for the Suzuki reaction. Nano Research, 2010, 34: 29-437.

[55] Duan L L, Fu R, Xiao Z G, Zhao Q F, Wang J Q, Chen S J, Wan Y. Activation of aryl chlorides in water under phase-transfer agent-free and ligand-free Suzuki coupling by heterogeneous palladium supported on hybrid mesoporous carbon. ACS Catalysis, 2015, 5: 575-586.

[56] Huang T S, Li C J. Palladium-catalyzed coupling of aryl halides with arylhalosilanes in air and water. Tetrahedron Letters, 2002, 43: 403-405.

[57] Sharavath V, Ghosh S. Palladium nanoparticles on noncovalently functionalized graphene-based heterogeneous catalyst for the Suzuki-Miyaura and Heck-Mizoroki reactions in water. RSC Advances, 2014, 4: 48322-48330.

[58] Kamal A, Srinivasulu V, Seshadri B N, Markandeya N, Alarifi A, Shankaraiah N. Water mediated Heck and Ullmann couplings by supported palladium nanoparticles: Importance of surface polarity of the carbon spheres. Green Chemistry, 2012, 14: 2513-2522.

[59] Ohtaka A, Sansano J M, Nájera C, Miguel-García I, Berenguer-Murcia Á, Cazorla-Amorós D. Palladium and bimetallic palladium-nickel nanoparticles supported on multiwalled carbon nanotubes: Application to carbon carbon bond-gorming reactions in water. ChemCatChem, 2015, 7: 1841-1847.

[60] Lakshmi Kantam M, Roy S, Roy M, Sreedhar B, Choudary B M. Nanocrystalline magnesium oxide-stabilized palladium(0): An efficient and reusable catalyst for Suzuki and Stille cross-coupling of aryl halides. Advanced Synthesis & Catalysis, 2005, 347: 2002-2008.

[61] Monopoli A, Nacci A, Calo V, Ciminale F, Cotugno P, Mangone A, Giannossa L C, Azzone P, Cioffi N. Palladium/zirconium oxide nanocomposite as a highly recyclable catalyst for C-C coupling reactions in water. Molecules, 2010, 15: 4511-4525.

[62] Esmaeilpour M, Sardarian A, Javidi J. Dendrimer-encapsulated Pd(0) nanoparticles immobilized on nanosilica as a highly active and recyclable catalyst for the copper- and phosphine-free Sonogashira-Hagihara coupling reactions in water. Catalysis Science & Technology, 2016, 6: 4005-4019.

[63] Khalafi-Nezhad A, Panahi F. Immobilized palladium nanoparticles on a silica-starch substrate (PNP-SSS): As an efficient heterogeneous catalyst for Heck and copper-free Sonogashira reactions in water. Green Chemistry, 2011, 13: 2408-2415.

[64] Khalafi-Nezhad A, Panahi F. Size-controlled synthesis of palladium nanoparticles on a silica-cyclodextrin substrate: A novel palladium catalyst system for the Heck reaction in water. ACS Sustainable Chemistry & Engineering, 2014, 2: 1177-1186.

[65] Jamwal N, Gupta M, Paul S. Hydroxyapatite-supported palladium (0) as a highly efficient catalyst for the Suzuki Coupling and aerobic oxidation of benzyl alcohols in water. Green Chemistry, 2008, 10: 999-1003.
[66] Massaro M, Riela S, Cavallaro G, Colletti C G, Milioto S, Noto R, Parisi F, Lazzara G. Palladium supported on halloysite-triazolium salts as catalyst for ligand free Suzuki cross-coupling in water under microwave irradiation. Journal of Molecular Catalysis A: Chemical, 2015, 408: 12-19.
[67] Zhang D, Zhou C S, Wang R H. Palladium nanoparticles immobilized by click ionic copolymers: Efficient and recyclable catalysts for Suzuki-Miyaura cross-coupling reaction in water. Catalysis Communications, 2012, 22: 83-88.
[68] Zheng P W, Zhang W Q. Synthesis of efficient and reusable palladium catalyst supported on pH-responsive colloid and its application to Suzuki and Heck reactions in water. Journal of Catalysis, 2007, 250: 324-330.
[69] Ohtaka A, Tamaki Y, Igawa Y, Egami K, Shimomura O, Nomura R. Polyion complex stabilized palladium nanoparticles for Suzuki and Heck reaction in water. Tetrahedron, 2010, 66: 5642-5646.
[70] Nabid M R, Bide Y. H_{40}-PCL-PEG unimolecular micelles both as anchoring sites for palladium nanoparticles and micellar catalyst for Heck reaction in water. Applied Catalysis A: General, 2014, 469: 183-190.
[71] Metin Ö, Durap F, Aydemir M, Özkar S. Palladium(0) nanoclusters stabilized by poly(4-styrenesulfonic acid-co-maleic acid) as an effective catalyst for Suzuki-Miyaura cross-coupling reactions in water. Journal of Molecular Catalysis A: Chemical, 2011, 337: 39-44.
[72] Dell'Anna M M, Mali M, Mastrorilli P, Rizzuti A, Ponzoni C, Leonelli C. Suzuki-Miyaura coupling under air in water promoted by polymer supported palladium nanoparticles. Journal of Molecular Catalysis A: Chemical, 2013, 366: 186-194.
[73] Qiao K, Sugimura R, Bao Q X, Tomida D, Yokoyama C. An efficient Heck reaction in water catalyzed by palladium nanoparticles immobilized on imidazolium-styrene copolymers. Catalysis Communications, 2008, 9: 2470-2474.
[74] Wu L, Zhang X, Tao Z. A mild and recyclable nano-sized nickel catalyst for the Stille reaction in water. Catalysis Science & Technology, 2012, 2: 707-710.
[75] Duan L L, Fu R, Zhang B S, Shi W, Chen S J, Wan Y. An efficient reusable mesoporous solid-based Pd catalyst for selective C_2 arylation of indoles in water. ACS Catalysis, 2016, 6: 1062-1074.
[76] 朱凤霞, 万颖, 李和兴. Pd/MCM-41 催化碘苯 Ullmann 偶合制备联苯. 化学研究与应用, 2005, 17: 403-405.
[77] 陈佳, 万颖, 李和兴. Pd/$NH_2C_3H_6$-MCM-41 催化水介质中的碘苯 Ullmann 反应. 催化学报, 2006, 27: 339-343.
[78] Wan Y, Chen J, Zhang D Q, Li H X. Ullmann coupling reaction in aqueous conditions over the Ph-MCM-41 supported Pd catalyst. Journal of Molecular Catalysis A: Chemical, 2006, 258: 89-94.
[79] Li H X, Chen J, Wan Y, Chai W, Zhang F, Lu Y F. Aqueous medium Ullmann reaction over a novel Pd/Ph-Al-MCM-41 as a new route of clean organic synthesis. Green Chemistry, 2007, 9: 273-280.
[80] Li H X, Chai W, Zhang F, Chen J. Water-medium Ullmann reaction over a highly active and selective Pd/Ph-SBA-15 catalyst. Green Chemistry, 2007, 9: 1223-1228.
[81] Wan Y, Zhang D Q, Zhai Y P, Feng C M, Chen J, Li H X. Periodic mesoporousorganosilicas: A type of hybrid support for water-mediated reactions. Chemistry-An Asian Journal, 2007, 2: 875-881.
[82] Huang J L, Yin J W, Chai W, Liang C, Shen J, Zhang F. Multifunctional mesoporous silica supported palladium nanoparticles as efficient and reusable catalyst for water-medium Ullmann reaction. New Journal of Chemistry, 2012, 36: 1378-1384.
[83] Coccia F, Tonucci L, D'Alessandro N, D'Ambrosio P, Bressan M. Palladium nanoparticles, stabilized by lignin, as catalyst for cross-coupling reactions in water. Inorganica Chimica Acta, 2013, 399: 12-18.
[84] Bhardwaj M, Sahi S, Mahajan H, Paul S, Clark J H. Novel heterogeneous catalyst systems based on Pd(0) nanoparticles onto amine functionalized silica-cellulose substrates [Pd(0)-EDA/SCs]: Synthesis, characterization and catalytic activity toward C-C and C-S coupling reactions in water under limiting basic conditions. Journal of Molecular Catalysis A: Chemical, 2015, 408: 48-59.

[85] Sin E, Yi S S, Lee Y S. Chitosan-*g*-mPEG-supported palladium(0) catalyst for Suzuki cross-coupling reaction in water. Journal of Molecular Catalysis A: Chemical, 2010, 315: 99-104.
[86] Zhou P P, Wang H H, Yang J Z, Tang J, Sun D P, Tang W H. Bio-supported palladium nanoparticles as a phosphine-free catalyst for the Suzuki reaction in water. RSC Advances, 2012, 2: 1759-1761.
[87] Alkordi M H, Liu Y Larsen R W, Eubank J F, Eddaoudi M. Zeolite-like metal-organic frameworks as platforms for applications: On metalloporphyrin-based catalysts. Journal of the American Chemical Society, 2008, 130: 12639-12641.
[88] Dhakshinamoorthy A, Opanasenko M, Cejka J, Garcia H. Metal organic frameworks as heterogeneous catalysts for the production of fine chemicals. Catalysis Science & Technology, 2013, 3: 2509-2540.
[89] Wee L H, Bonino F, Lamberti C, Bordiga S, Martens J A. Cr-MIL-101 encapsulated Keggin phosphotungstic acid as active nanomaterial for catalysing the alcoholysis of styrene oxide. Green Chemistry, 2014, 16: 1351-1357.
[90] Garcia-Garcia P, Muller M, Corma A. MOF catalysis in relation to their homogeneous counterparts and conventional solid catalysts. Chemical Science, 2014, 5: 2979-3007.
[91] Yuan B Z, Pan Y Y, Li Y W, Yin B L, Jiang H F. A highly active heterogeneous palladium catalyst for the Suzuki-Miyaura and Ullmann coupling reactions of aryl chlorides in aqueous media. Angewandte Chemie International Edition, 2010, 49: 4054-4058.
[92] Pascanu V, Yao Q X, Bermejo Gómez A, Gustafsson M, Yun Y F, Wan W, Samain L, Zou X D, Martín-Matute B. Sustainable catalysis: Rational Pd loading on MIL-101Cr-NH$_2$ for more efficient and recyclable Suzuki-Miyaura reactions. Chemistry-A European Journal, 2013, 19: 17483-17493.
[93] Roy A S, Mondal J, Banerjee B, Mondal P, Bhaumik A, Islam S M. Pd-grafted porous metal-organic framework material as an efficient and reusable heterogeneous catalyst for C—C coupling reactions in water. Applied Catalysis A: General, 2014, 469: 320-327.
[94] Li H, Zhu Z H, Zhang F, Xie S H, Li H X, Li P, Zhou X G. Palladium nanoparticles confined in the cages of MIL-101: An Efficient catalyst for the one-pot indole synthesis in water. ACS Catalysis, 2011, 1: 1604-1612.
[95] Xia C H, Xu J, Wu W Z, Liang X M. Pd/C-catalyzed hydrodehalogenation of aromatic halides in aqueous solutions at room temperature under normal pressure. Catalysis Communications, 2004, 5: 383-386.
[96] L X F, Zhang W J, Zhang L M, Yang H Q. Pd nanoparticles confined in fluoro-functionalized yolk-shell-structured silica for olefin hydrogenation in water. Chinese Journal of Catalysis, 2013, 34: 1192-1200.
[97] Dell'Anna M M, Intini S, Romanazzi G, Rizzuti A, Leonelli C, Piccinni F, Mastrorilli P. Polymer supported palladium nanocrystals as efficient and recyclable catalyst for the reduction of nitroarenes to anilines under mild conditions in water. Journal of Molecular Catalysis A: Chemical, 2014, 395: 307-314.
[98] Nakao R, Rhee H, Uozumi Y. Hydrogenation and dehalogenation under aqueous conditions with an amphiphilic-polymer-supported nanopalladium catalyst. Organic Letters, 2005, 7: 163-165.
[99] Levi N, Neumann R. Diastereoselective and enantiospecific direct reductive amination in water catalyzed by palladium nanoparticles stabilized by polyethyleneimine derivatives. ACS Catalysis, 2013, 3: 1915-1918.
[100] Huang Y B, Liu S J, Lin Z J, Li W J, Li X F, Cao R. Facile synthesis of palladium nanoparticles encapsulated in amine-functionalized mesoporous metal-organic frameworks and catalytic for dehalogenation of aryl chlorides. Journal of Catalysis, 2012, 292: 111-117.
[101] Jiang Y J, Li G Z, Li X D, Lu S X, Wang L, Zhang X W. Water-dispersible Fe$_3$O$_4$ nanowires as efficient supports for noble-metal catalyzed aqueous reactions. Journal of Materials Chemistry A, 2014, 2: 4779-4787.
[102] Zhou B W, Song J L, Zhou H C, Wu L Q,Wu, T B, Liu Z M, Han B X. Light-driven integration of the reduction of nitrobenzene to aniline and the transformation of glycerol into valuable chemicals in water. RSC Advances, 2015, 5: 36347-36352.
[103] Oberhauser W, Bartoli M, Petrucci G, Bandelli D, Frediani M, Capozzoli L, Cepek C, Bhardwaj S, Rosi L. Nitrile hydration to amide in water: Palladium-based nanoparticles vs molecular catalyst. Journal of Molecular Catalysis A: Chemical, 2015, 410: 26-33.

[104] Yang H Q, Jiao X, Li S R. Hydrophobic core-hydrophilic shell-structured catalysts: A general strategy for improving the reaction rate in water. Chemical Communications, 2012, 48: 11217-11219.

[105] Chu W, Ooka Y, Kamiya Y, Okuhara T. Reaction path for oxidation of ethylene to acetic acid over Pd/WO$_3$-ZrO$_2$ in the presence of water. Catalysis Letters, 2005, 101: 225-228.

[106] Feng B, Hou Z S, Yang H M, Wang X R, Hu Y, Li H, Qiao Y X, Zhao X G, Huang Q F. Functionalized poly(ethylene glycol)-stabilized water-soluble palladium nanoparticles: Property/activity relationship for the aerobic alcohol oxidation in water. Langmuir, 2010, 26: 2505-2513.

[107] Zhang M M, Sun Q, Yan Z X, Jing J J, Wei W, Jiang D L, Xie J M, Chen M. Ni/MWCNT-supported palladium nanoparticles as magnetic catalysts for selective oxidation of benzyl acohol. Australian Journal of Chemistry, 2013, 66: 564-571.

[108] Dell'Anna M M, Mali M, Mastrorilli P, Cotugno P, Monopoli A. Oxidation of benzyl alcohols to aldehydes and ketones under air in water using a polymer supported palladium catalyst. Journal of Molecular Catalysis A: Chemical, 2014, 386: 114-119.

[109] Ma Z C, Yang H Q, Qin Y, Hao Y J, Li G. Palladium nanoparticles confined in the nanocages of SBA-16: Enhanced recyclability for the aerobic oxidation of alcohols in water. Journal of Molecular Catalysis A: Chemical, 2010, 331: 78-85.

[110] Giachi G, Oberhauser W, Frediani M, Passaglia E, Capozzoli L, Rosi L. Pd-nanoparticles stabilized by pyridine-functionalized poly(ethylene glycol) as catalyst for the aerobic oxidation of α,β-unsaturated alcohols in water. Journal of Polymer Science Part A: Polymer Chemistry, 2013, 51: 2518-2526.

[111] Barthos R, Hegyessy A, May Z, Valyon J. Wacker-oxidation of ethylene over pillared layered material catalysts. Catalysis Letters, 2014, 144: 702-710.

[112] Wang Q, Cai X C, Liu Y Q, Xie J Y, Zhou Y, Wang J. Pd nanoparticles encapsulated into mesoporous ionic copolymer: efficient and recyclable catalyst for the oxidation of benzyl alcohol with O$_2$ balloon in water. Applied Catalysis B: Environmental, 2016, 189: 242-251.

[113] Zou H B, Wang R W, Dai J Y, Wang Y, Wang X, Zhang Z, Qiu S L. Amphiphilic hollow porous shell encapsulated Au@Pd bimetal nanoparticles for aerobic oxidation of alcohols in water. Chemical Communications, 2015, 51: 14601-14604.

[114] Lee I, Joo J B, Yin Y, Zaera F. A yolk@shell nanoarchitecture for Au/TiO$_2$ catalysts. Angewandte Chemie International Edition, 2011, 50: 10208-10211.

[115] Shimizu K I, Kubo T, Satsuma A, Kamachi T, Yoshizawa K. Surface oxygen atom as a cooperative ligand in Pd nanoparticle catalysis for selective hydration of nitriles to amides in water: Experimental and theoretical studies. ACS Catalysis, 2012, 2: 2467-2474.

[116] Wang Y, Liu J H, Xia C G. Oxidative carbonylation of 2-amino-1-alkanols catalyzed by cross-linked polymer dupported palladium in water. Chinese Journal of Catalysis, 2011, 32: 1782-1786.

[117] Tsunoyama H, Sakurai H, Ichikuni N, Negishi Y, Tsukuda T. Colloidal gold nanoparticles as catalyst for carbon-carbon bond formation: Application to aerobic homocoupling of phenylboronic acid in water. Langmuir, 2004, 20: 11293-11296.

[118] Liu C H, Lin C Y, Chen J L, Lai N C, Yang C M, Chen J M, Lu K T. Metal oxide-containing SBA-15-supported gold catalysts for base-free aerobic homocoupling of phenylboronic acid in water. Journal of Catalysis, 2016, 336: 49-57.

[119] Han J, Liu Y, Guo R. Facile synthesis of highly stable gold nanoparticles and their unexpected excellent catalytic activity for Suzuki-Miyaura cross-coupling reaction in water. Journal of the American Chemical Society, 2009, 131: 2060-2061.

[120] Chng L L, Yang J, Wei Y, Ying J Y. Semiconductor-gold nanocomposite catalysts for the efficient three-component coupling of aldehyde, amine and alkyne in water. Advanced Synthesis & Catalysis, 2009, 351: 2887-2896.

[121] Monopoli A, Cotugno P, Palazzo G, Ditaranto N, Mariano B, Cioffi N, Ciminale F, Nacci A. Ullmann homocoupling catalysed by gold nanoparticles in water and ionic liquid. Advanced Synthesis & Catalysis, 2012, 354: 2777-2788.
[122] Nasrollahzadeh M, Banaei A. Hybrid Au/Pd nanoparticles as reusable catalysts for Heck coupling reactions in water under aerobic conditions. Tetrahedron Letters, 2015, 56: 500-503.
[123] Prasad K S, Noh H B, Reddy S S, Reddy A E, Shim Y B. Catalytic properties of Au and Pd nanoparticles decorated on Cu_2O microcubes for aerobic benzyl alcohol oxidation and Suzuki-Miyaura coupling reactions in water. Applied Catalysis A: General, 2014, 476: 72-77.
[124] Boominathan M, Pugazhenthiran N, Nagaraj M, Muthusubramanian S, Murugesan S, Bhuvanesh N. Nanoporous titania-supported gold nanoparticle-catalyzed green synthesis of 1,2,3-triazoles in aqueous medium. ACS Sustainable Chemistry & Engineering, 2013, 1: 1405-1411.
[125] Haruta M, Yamada N, Kobayashi T, Iijima S. Gold catalysts prepared by coprecipitation for low-temperature oxidation of hydrogen and of carbon monoxide. Journal of Catalysis, 1989, 115: 301-309.
[126] Enache D I, Knight D W, Hutchings G J. Solvent-free oxidation of primary alcohols to aldehydes using supported gold catalysts. Catalysis Letters, 2005, 103: 43-52.
[127] Abad A, Concepción P, Corma A, García H. A collaborative effect between gold and a support induces the selective oxidation of alcohols. Angewandte Chemie International Edition, 2005, 44: 4066-4069.
[128] Wolf A, Schüth F. A systematic study of the synthesis conditions for the preparation of highly active gold catalysts. Applied Catalysis A: General, 2002, 226: 1-13.
[129] Wan X Y, Zhou C M, Chen J S, Deng W P, Zhang Q H, Yang Y H, Wang Y. Base-free aerobic oxidation of 5-hydroxymethyl-furfural to 2,5-furandicarboxylic acid in water catalyzed by functionalized carbon nanotube-supported Au-Pd alloy nanoparticles. ACS Catalysis, 2014, 4: 2175-2185.
[130] Mahyari M, Shaabani A, Behbahani M, Bagheri A. Thiol-functionalized fructose-derived nanoporous carbon as a support for gold nanoparticles and its application for aerobic oxidation of alcohols in water. Applied Organometallic Chemistry, 2014, 28: 576-583.
[131] Wang S, Zhao Q F, Wei H, Wang J Q, Cho M, Cho H S, Terasaki O W. Aggregation-free gold nanoparticles in ordered mesoporous carbons: Toward highly active and stable heterogeneous catalysts. Journal of the American Chemical Society, 2013, 135: 11849-11860.
[132] Mahyari M, Shaabani A, Bide Y. Gold nanoparticles supported on supramolecular ionic liquid grafted graphene: A bifunctional catalyst for the selective aerobic oxidation of alcohols. RSC Advances, 2013, 3: 22509-22517.
[133] Wu X, Guo S, Zhang J. Selective oxidation of veratryl alcohol with composites of Au nanoparticles and graphene quantum dots as catalysts. Chemical Communications, 2015, 51: 6318-6321.
[134] Kumar R, Gravel E, Hagege A, Li H, Jawale D V, Verma D, Namboothiri I N N, Doris E. Carbon nanotube-gold nanohybrids for selective catalytic oxidation of alcohols. Nanoscale, 2013, 5: 6491-6497.
[135] Jawale D V, Gravel E, Geertsen V, Li H, Shah N, Kumar R, John J, Namboothiri I N N, Doris E. Size effect of gold nanoparticles supported on carbon nanotube as catalysts in selected organic reactions. Tetrahedron, 2014, 70: 6140-6145.
[136] Liu T, Yang F, Li Y F, Ren L, Zhang L Q, Xu K, Wang X, Xu C M, Gao J S. Plasma synthesis of carbon nanotube-gold nanohybrids: Efficient catalysts for green oxidation of silanes in water. Journal of Materials Chemistry A, 2014, 2: 245-250.
[137] Tsunoyama H, Sakurai H, Negishi Y, Tsukuda T. Size-specific catalytic activity of polymer-stabilized gold nanoclusters for aerobic alcohol oxidation in water. Journal of the American Chemical Society, 2005, 127: 9374-9375.
[138] Biffis A, Minati L. Efficient aerobic oxidation of alcohols in water catalysed by microgel-stabilized metal nanoclusters. Journal of Catalysis, 2005, 236: 405-409.

[139] Mertens P G N, Bulut M, Gevers L E M, Vankelecom I F J, Jacobs P A, De Vos D E. Catalytic oxidation of 1,2-diols to a-hydroxy-carboxylates with stabilized gold nanocolloids combined with a membrane-based catalyst separation. Catalysis Letter, 2005, 102: 57-62.

[140] Biffis A, Cunial S, Spontoni P, Prati L. Microgel-stabilized gold nanoclusters: Powerful "quasi-homogeneous" catalysts for the aerobic oxidation of alcohols in water. Journal of Catalysis, 2007, 251: 1-6.

[141] Saio D, Amaya T, Hirao T. Redox-active catalyst based on poly(anilinesulfonic acid)- supported gold nanoparticles for aerobic alcohol oxidation in water. Advanced Synthesis & Catalysis, 2010, 352: 2177-2182.

[142] Xie T, Lu M, Zhang W W, Li J. A green and efficient oxidation of alcohols by amphiphilic resin-supported gold nanoparticles in aqueous H_2O_2. Journal of Chemical Research, 2011, 35: 397-409.

[143] Kanaoka S, Yagi N, Fukuyama Y, Aoshima S, Tsunoyama H, Tsukuda T, Sakurai H. Thermosensitive gold nanoclusters stabilized by well-defined vinyl ether star polymers: Reusable and furable catalysts for aerobic alcohol oxidation. Journal of the American Chemical Society, 2007, 129: 12060-12061.

[144] Chaki N K, Tsunoyama H, Negishi Y, Sakurai H, Tsukuda T. Effect of Ag-doping on the catalytic activity of polymer-stabilized Au clusters in aerobic oxidation of alcohol. The Journal of Physical Chemistry C, 2007, 111: 4885-4888.

[145] Yuan Y, Yan N, Dyson P J. pH-sensitive gold nanoparticle catalysts for the aerobic oxidation of alcohols. Inorganic Chemistry, 2011, 50: 11069-11074.

[146] Tsunoyama H, Ichikuni N, Sakurai H, Tsukuda T. Effect of electronic structures of Au clusters stabilized by poly(N-vinyl-2-pyrrolidone) on aerobic oxidation catalysis. Journal of the American Chemical Society, 2009, 131: 7086-7093.

[147] Li Y X, Gao Y, Yang C, Sha S S, Hao J F, Wu Y. Facile synthesis of polystyrene/gold composite particles as a highly active and reusable catalyst for aerobic oxidation of benzyl alcohol in water. RSC Advances, 2014, 4: 24769-24772.

[148] Wang S, Wang J, Zhao Q F, Li D, Wang J Q, Cho M, Cho H, Terasaki O, Chen S, Wan Y. Highly active heterogeneous 3 nm gold nanoparticles on mesoporous carbon as catalysts for low-temperature selective oxidation and reduction in water. ACS Catalysis, 2015, 5: 797-802.

[149] Li Y X, Gao Y, Yang C. A facile and efficient synthesis of polystyrene/gold-platinum composite particles and their application for aerobic oxidation of alcohols in water. Chemical Communications, 2015, 51: 7721-7724.

[150] Tuteja J, Nishimura S, Ebitani K. Change in reactivity of differently capped AuPd bimetallic nanoparticle catalysts for selective oxidation of aliphatic diols to hydroxycarboxylic acids in basic aqueous solution. Catalysis Today, 2016, 265: 231-239.

[151] Gorbanev Y Y, Klitgaard S K, Woodley J M, Christensen C H, Riisager A. Gold-catalyzed aerobic oxidation of 5-hydroxymethylfurfural in water at ambient temperature. ChemSusChem, 2009, 2: 672-675.

[152] Gupta N K, Nishimura S, Takagaki A, Ebitani K. Hydrotalcite-supported gold-nanoparticle-catalyzed highly efficient base-free aqueous oxidation of 5-hydroxymethylfurfural into 2,5-furandicarboxylic acid under atmospheric oxygen pressure. Green Chemistry, 2011, 13: 824-827.

[153] Mitsudome T, Noujima A, Mizugaki T, Jitsukawa K, Kaneda K. Supported gold nanoparticlecatalyst for the selective oxidation of silanes to silanols in water. Chemical Communications, 2009: 5302-5304.

[154] An D, Ye A, Deng W, Zhang Q, Wang Y. Selective conversion of cellobiose and cellulose into gluconic acid in water in the presence of oxygen, catalyzed by polyoxometalate-supported gold nanoparticles. Chemistry-A European Journal, 2012, 18: 2938-2947.

[155] Bhattacharya T, Sarma T K, Samanta S. Self-assembled monolayer coated gold-nanoparticle catalyzed aerobic oxidation of α-hydroxy ketones in water: An efficient one-pot synthesis of quinoxaline derivatives. Catalysis Science & Technology, 2012, 2: 2216-2220.

[156] Zhou L P, Yu W J, Wu L, Liu Z, Chen H J, Yang X M, Su Y L, Xu J. Nanocrystalline gold supported on NaY as catalyst for the direct oxidation of primary alcohol to carboxylic acid with molecular oxygen in water. Applied Catalysis A: General, 2013, 451: 137-143.

[157] Zhou L P, Chen M Z, Wang Y Q, Su Y L, Yang X M, Chen C, Xu J. Au/mesoporous-TiO$_2$ as catalyst for the oxidation of alcohols to carboxylic acids with molecular oxygen in water. Applied Catalysis A: General, 2014, 475: 347-354.

[158] Villa A, Janjic N, Spontoni P, Wang D, Su D S, Prati L. Au-Pd/AC as catalysts for alcohol oxidation: Effect of reaction parameters on catalytic activity and selectivity. Applied Catalysis A: General, 2009, 364: 221-228.

[159] Stephenson R, Stuart J, Tabak M. Mutual solubility of water and aliphatic alcohols. Journal of Chemical & Engineering Data, 1984, 29: 287-290.

[160] Ma C Y, Cheng J, Wang H L, Hu Q, Tian H, He C, Hao Z P. Characteristics of Au/HMS catalysts for selective oxidation of benzyl alcohol to benzaldehyde. Catalysis Today, 2010, 158: 246-251.

[161] Liu S S, Sun K Q, Xu B Q. Specific selectivity of Au-catalyzed oxidation of glycerol and other C$_3$-polyols in water without the presence of a base. ACS Catalysis, 2014, 4: 2226-2230.

[162] Demirel-Gülen S, Lucas M, Claus P. Liquid phase oxidation of glycerol over carbon supported gold catalysts. Catalysis Today, 2005, 102-103: 166-172.

[163] Demirel S, Kern P, Lucas M, Claus P. Oxidation of mono- and polyalcohols with gold: Comparison of carbon and ceria supported catalysts. Catalysis Today, 2007, 122: 292-300.

[164] Besson M, Gallezot P, Pinel C. Conversion of biomass into chemicals over metal catalysts. Chemical Reviews, 2014, 114: 1827-1870.

[165] Ohmi Y, Nishimura S, Ebitani K. Synthesis of α-amino acids from glucosamine-HCl and its derivatives by aerobic oxidation in water catalyzed by Au nanoparticles on basic supports. ChemSusChem, 2013, 6: 2259-2262.

[166] Mohammad M, Nishimura S, Ebitani K. Selective aerobic oxidation of 1,3-propanediol to 3-hydroxypropanoic acid using hydrotalcite supported bimetallic gold nanoparticle catalyst in water. AIP Conference Proceedings, 2015, 1649: 58-66.

[167] Yao Y, Xue M, Zhang Z, Zhang M, Wang Y, Huang F. Gold nanoparticles stabilized by an amphiphilic pillar[5]arene: Preparation, self-assembly into composite microtubes in water and application in green catalysis. Chemical Science, 2013, 4: 3667-3672.

[168] Zou L Y, Cui Y Y, Dai W L. Highly efficient Au/TiO$_2$ catalyst for one-pot conversion of nitrobenzene to p-aminophenol in water media. Chinese Journal of Chemistry, 2014, 32: 257-262.

[169] Son P A, Nishimura S, Ebitani K. Production of γ-valerolactone from biomass-derived compounds using formic acid as a hydrogen source over supported metal catalysts in water solvent. RSC Advances, 2014, 4:10525-10530.

[170] Mondal P, Ghosal K, Bhattacharyya S K, Das M, Bera A, Ganguly D, Kumar P, Dwivedi J, Gupta R K, Marti A A, Gupta B K, Maiti S. Formation of a gold-carbon dot nanocomposite with superior catalytic ability for the reduction of aromatic nitro groups in water. RSC Advances, 2014, 4: 25863-25866.

[171] Li G, Zeng C J, Jin R C. Chemoselective hydrogenation of nitrobenzaldehyde to nitrobenzyl alcohol with unsupported Au nanorod catalysts in water. The Journal of Physical Chemistry C, 2015, 119: 11143-11147.

[172] Li G, Zeng C J, Jin R C. Thermally robust Au$_{99}$(SPh)$_{42}$ nanoclusters for chemoselective hydrogenation of nitrobenzaldehyde derivatives in water. Journal of the American Chemical Society, 2014, 136: 3673-3679.

[173] Zhang Q, Wang T J, Xu Y, Zhang Q, Ma L L. Production of liquid alkanes by controlling reactivity of sorbitol hydrogenation with a Ni/HZSM-5 catalyst in water. Energy Conversion and Management, 2014, 77: 262-268.

[174] Lin W W, Cheng H Y, Ming J, Yu Y C, Zhao F Y. Deactivation of Ni/TiO$_2$ catalyst in the hydrogenation of nitrobenzene in water and improvement in its stability by coating a layer of hydrophobic carbon. Journal of Catalysis, 2012, 291: 149-154.

[175] Xu Y, Qiu S B, Long J X, Wang C G, Chang J M, Tan J, Liu Q Y, Ma L L, Wang T J, Zhang Q. In situ hydrogenation of furfural with additives over a RANEY@Ni catalyst. RSC Advances, 2015, 5: 91190-91195.

[176] Mazaheri O, Kalbasi R J. Preparation and characterization of Ni/mZSM-5 zeolite with a hierarchical pore structure by using KIT-6 as silica template: An efficient bi-functional catalyst for the reduction of nitro aromatic compounds. RSC Advances, 2015, 5: 34398-34414.

[177] Rayhan U, Do J H, Arimura T, Yamato T. Reduction of carbonyl compounds by Raney Ni-Al alloy and Al powder in the presence of noble metal catalysts in water. Comptes Rendus Chimie, 2015, 18: 685-692.

[178] Li W, Cheng H Y, Lin W, Liang G F, Zhang C, Zhao F Y. Fabrication of hybrid mesoporous TiO_2-SiO_2(et) supported Ni nanoparticles: An efficient and air/water stable catalyst. Journal of Molecular Catalysis A: Chemical, 2016, 411: 214-221.

[179] Yan N, Zhao C, Luo C, Dyson P J, Liu H C, Kou Y. One-step conversion of cellobiose to C_6-alcohols using a ruthenium nanocluster catalyst. Journal of the American Chemical Society, 2006, 128: 8714-8715.

[180] Luo C, Wang S, Liu H C. Cellulose conversion into polyols catalyzed by reversibly formed acids and supported ruthenium clusters in hot water. Angewandte Chemie International Edition, 2007, 46: 7636-7639.

[181] Deng W P, Liu M, Tan X S, Zhang Q H, Wang Y. Conversion of cellobiose into sorbitol in neutral water medium over carbon nanotube-supported ruthenium catalysts. Journal of Catalysis, 2010, 271: 22-32.

[182] Van de Vyver S, Geboers J, Dusselier M, Schepers H, Vosch T, Zhang L, Van Tendeloo G, Jacobs P A, Sels B F. Selective bifunctional catalytic conversion of cellulose over reshaped Ni particles at the tip of carbon nanofibers. ChemSusChem, 2010, 3: 698-701.

[183] Ding L N, Wang A Q, Zheng M Y, Zhang T. Selective transformation of cellulose into sorbitol by using a bifunctional nickel phosphide catalyst. ChemSusChem, 2010, 3: 818-821.

[184] Ji N, Zhang T, Zheng M Y, Wang A, Wang H, Wang X D, Chen J G. Direct catalytic conversion of cellulose into ethylene glycol using nickel-promoted tungsten carbide catalysts. Angewandte Chemie International Edition, 2008, 47: 8510-8513.

[185] Deng W P, Liu M, Tan X S, Zhang Q H, Wang Y. Conversion of cellulose into sorbitol over carbon nanotube-supported ruthenium catalyst. Catalysis Letters, 2009, 133: 167-174.

[186] Liu M, Deng W P, Zhang Q H, Wang Y L, Wang Y. Polyoxometalate-supported ruthenium nanoparticles as bifunctional heterogeneous catalysts for the conversions of cellobiose and cellulose into sorbitol under mild conditions. Chemical Communications, 2011, 47: 9717-9719.

[187] Deng W P, Wang Y L, Zhang Q H, Wang Y. Development of bifunctional catalysts for the conversions of cellulose or cellobiose into polyols and organic acids in water. Catalysis Surveys from Asia, 2012, 16: 91-105.

[188] Xue W, Song Y, Wang Y J, Wang D D, Li F. Effect of hydrazine hydrate on the Ru-Zn/SiO_2 catalysts performance for partial hydrogenation of benzene. Catalysis Communications, 2009, 11: 29-33.

[189] Manyar H G, Weber D, Daly H, Thompson J M, Rooney D W, Gladden L F, Hugh Stitt E, Jose Delgado J, Bernal S, Hardacre C. Deactivation and regeneration of ruthenium on silica in the liquid-phase hydrogenation of butan-2-one. Journal of Catalysis, 2009, 265: 80-88.

[190] Song L, Li X, Wang H, Wu H, Wu P. Ru nanoparticles entrapped in mesopolymers for efficient liquid-phase hydrogenation of unsaturated compounds. Catalysis Letters, 2009, 133: 63-69.

[191] Gädda T M, Yu Y, Miyazawa A. Ru/C catalyzed cyclization of linear α,ω-diamines to cyclic amines in water. Tetrahedron, 2010, 66: 1249-1253.

[192] Kobayashi H, Komanoya T, Hara K, Fukuoka A. Water-tolerant mesoporous-carbon-supported ruthenium catalysts for the hydrolysis of cellulose to glucose. ChemSusChem, 2010, 3: 440-443.

[193] Komanoya T, Kobayashi H, Hara K, Chun W J, Fukuoka A. Catalysis and characterization of carbon-supported ruthenium for cellulose hydrolysis. Applied Catalysis A: General, 2011, 407: 188-194.

[194] Yen C H, Lin H W, Tan C S. Hydrogenation of bisphenol A- using a mesoporous silica based nano ruthenium catalyst Ru/MCM-41 and water as the solvent. Catalysis Today, 2011, 174: 121-126.

[195] Jiang H J, Jiang H B, Zhu D M, Zheng X L, Fu H Y, Chen H, Li R X. Cooperation between the surface hydroxyl groups of the support and organic additives in the highly selective hydrogenation of citral. Applied Catalysis A: General, 2012, 445-446: 351-358.

[196] Osaka Y, Ikeda Y, Hashizume D, Iwamoto M. Direct hydrodeoxygenation of cellulose and xylan to lower alkanes on ruthenium catalysts in subcritical water. Biomass and Bioenergy, 2013, 56: 1-7.

[197] Chen J Z, Lu F, Zhang J J, Yu W Q, Wang F, Gao J, Xu J. Immobilized Ru clusters in nanosized mesoporous zirconium silica for the aqueous hydrogenation of furan derivatives at room temperature. ChemCatChem, 2013, 5: 2822-2826.

[198] Patankar S C, Yadav G D. Cascade engineered synthesis of γ-valerolactone, 1,4-pentanediol, and 2-methyltetrahydrofuran from levulinic acid using Pd-Cu/ZrO$_2$ Catalyst in water as solvent. ACS Sustainable Chemistry & Engineering, 2015, 3: 2619-2630.

[199] Xie J H, Nie J F, Liu H C. Aqueous-phase selective aerobic oxidation of 5-hydroxymethylfurfural on Ru/C in the presence of base. Chinese Journal of Catalysis, 2014, 35: 937-944.

[200] Artizzu F, Quochi F, Serpe A, Sessini E, Deplano P. Tailoring functionality through synthetic strategy in heterolanthanide assemblies. Inorganic Chemistry Frontiers, 2015, 2: 213-222.

[201] Bengoechea M O, Hertzberg A, Miletić N, Arias P L, Barth T. Simultaneous catalytic de-polymerization and hydrodeoxygenation of lignin in water/formic acid media with Rh/Al$_2$O$_3$, Ru/Al$_2$O$_3$ and Pd/Al$_2$O$_3$ as bifunctional catalysts. Journal of Analytical and Applied Pyrolysis, 2015, 113: 713-722.

[202] Abdelrahman O A, Luo H Y, Heyden A, Román-Leshkov Y, Bond J Q. Toward rational design of stable, supported metal catalysts for aqueous-phase processing: Insights from the hydrogenation of levulinic acid. Journal of Catalysis, 2015, 329: 10-21.

[203] Ertas I E, Gulcan M, Bulu A, Yurderi M, Zahmakiran M. Metal-organic framework (MIL-101) stabilized ruthenium nanoparticles: Highly efficient catalytic material in the phenol hydrogenation. Microporous and Mesoporous Materials, 2016, 226: 94-103.

[204] Hu J, Ding Y, Zhang H, Wu P, Li X. Highly effective Ru/CMK-3 catalyst for selective reduction of nitrobenzene derivatives with H$_2$O as solvent at near ambient temperature. RSC Advances, 2016, 6: 3235-3242.

[205] Salam N, Sinha A, Roy A S, Mondal P, Jana N R, Islam S M. Synthesis of silver-graphene nanocomposite and its catalytic application for the one-pot three-component coupling reaction and one-pot synthesis of 1,4-disubstituted 1,2,3-triazoles in water. RSC Advances, 2014, 4: 10001-10012.

[206] Mitsudome T, Mikami Y, Mori H, Arita S, Mizugaki T, Jitsukawa K, Kaneda K. Supported silver nanoparticle catalyst for selective hydration of nitriles to amides in water. Chemical Communications, 2009, 22: 3258-3260.

[207] Shimizu K I, Imaiida N, Sawabe K, Satsuma A. Hydration of nitriles to amides in water by SiO$_2$-supported Ag catalysts promoted by adsorbed oxygen atoms. Applied Catalysis A: General, 2012, 421-422: 114-120.

[208] Wang H, Yang K F, Li L, Bai Y, Zheng Z J, Zhang W Q, Gao Z W, Xu L W. Modulation of silver-titania nanoparticles on polymethylhydrosiloxane-based semi-interpenetrating networks for catalytic alkynylation of trifluoromethyl ketones and aromatic aldehydes in water. ChemCatChem, 2014, 6: 580-591.

[209] Movahedi F, Masrouri H, Kassaee M Z. Immobilized silver on surface-modified ZnO nanoparticles: As an efficient catalyst for synthesis of propargylamines in water. Journal of Molecular Catalysis A: Chemical, 2014, 395: 52-57.

[210] Biswas M, Saha A, Dule M, Mandal T K. Polymer-assisted chain-like organization of CuNi alloy nanoparticles: Solvent-adoptable pseudohomogeneous catalysts for alkyne-azide click reactions with magnetic recyclability. The Journal of Physical Chemistry, C, 2014, 118: 22156-22165.

[211] Rashid Z, Moadi T, Ghahremanzadeh R. Green synthesis and characterization of silver nanoparticles using ferula latisecta leaf extract and their application as a catalyst for the safe and simple one-pot preparation of spirooxindoles in water. New Journal of Chemistry, 2016, 40: 3343-3349.

[212] Zhao X L, Yang K F, Zhang Y, Xu L W, Guo X Q. Silver nanoparticles embedded in modified polyallylamine resin as efficient catalysts for alkyne-azide 1,3-dipolar cycloaddition in water. Catalysis Communications, 2016, 74: 110-114.

[213] Sreedhar B, Radhika P, Neelima B, Hebalkar N. Synthesis and characterization of silica@copper core-shell nanoparticles: Application for conjugate addition reactions. Chemistry-An Asian Journal, 2008, 3: 1163-1169.

[214] Nasir Baig R B, Varma R S. A highly active magnetically recoverable nano ferrite-glutathione-copper (Nano-FGT-Cu) catalyst for Huisgen 1,3-dipolar cycloadditions. Green Chemistry, 2012, 14: 625-632.

[215] Jacob K, Stolle A, Ondruschka B, Jandt K D, Keller T F. Cu on porous glass: An easily recyclable catalyst for the microwave-assisted azide-alkyne cycloaddition in water. Applied Catalysis A: General, 2013, 451: 94-100.

[216] Jacob K, Stolle A. Microwave-assisted azide-alkyne cycloaddition in water using a heterogeneous Cu-catalyst. Synthetic Communications, 2014, 44: 1251-1257.

[217] Safa K D, Taheri E, Allahvirdinesbat M, Niaei A. Sonochemical syntheses of xanthene derivatives using zeolite-supported transition metal catalysts in aqueous media. Research on Chemical Intermediates, 2016, 42: 2989-3004.

[218] Bhardwaj M, Jamwal B, Paul S. Novel Cu(0)-Fe_3O_4@SiO_2/NH_2cel as an efficient and sustainable magnetic catalyst for the synthesis of 1,4-disubstituted-1,2,3-triazoles and 2-substituted-benzothiazoles via one-pot strategy in aqueous media. Catalysis Letters, 2016, 146: 629-644.

[219] Xu Q, Li X L, Pan T, Yu C G, Deng J, Guo Q X, Fu Y. Supported copper catalysts for highly efficient hydrogenation of biomass-derived levulinic acid and γ-valerolactone. Green Chemistry, 2016, 18: 1287-1294.

[220] Mukhopadhyay S, Joshi A V, Peleg L, Sasson Y. Heterogeneous Rh/C-catalyzed direct reductive coupling of haloaryls to biaryls in water. Organic Process Research & Development, 2003, 7: 44-46.

[221] Mevellec V, Nowicki A, Roucoux A, Dujardin C, Granger P, Payen E, Philippot K. A simple and reproducible method for the synthesis of silica-supported rhodium nanoparticles and their investigation in the hydrogenation of aromatic compounds. New Journal of Chemistry, 2006, 30: 1214-1219.

[222] Hubert C, Denicourt-Nowicki A, Beaunier P, Roucoux A. TiO_2-supported Rh nanoparticles: From green catalyst preparation to application in arene hydrogenation in neat water. Green Chemistry, 2010, 12: 1167-1170.

[223] Hubert C, Bile E G, Denicourt-Nowicki A, Roucoux A. Rh(0) colloids supported on TiO_2: A highly active and pertinent tandem in neat water for the hydrogenation of aromatics. Green Chemistry, 2011, 13: 1766-1771.

[224] Buntara T, Noel S, Phua P H, Melián-Cabrera I, de Vries J G, Heeres H J. From 5-hydroxymethylfurfural (HMF) to polymer precursors: Catalyst screening studies on the conversion of 1,2,6-hexanetriol to 1,6-hexanediol. Topics in Catalysis, 2012, 55: 612-619.

[225] Chatterjee M, Ishizaka T, Suzuki A, Kawanami H. An efficient cleavage of the aryl Ether C—O bond in supercritical carbon dioxide-water. Chemical Communications, 2013, 49: 4567-4569.

[226] Ganji S, Enumula S S, Marella R K, Rao K S R, Burri D R. RhNPs/SBA-NH_2: A high-performance catalyst for aqueous phase reduction of nitroarenes to aminoarenes at room temperature. Catalysis Science & Technology, 2014, 4: 1813-1819.

[227] Yoshida H, Onodera Y, Fujita S I, Kawamori H, Arai M. Solvent effects in heterogeneous selective hydrogenation of acetophenone: Differences between Rh/C and Rh/Al_2O_3 catalysts and the superiority of water as a functional solvent. Green Chemistry, 2015, 17: 1877-1883.

[228] Xiang X, He W H, Xie L S, Li F. A mild solution chemistry method to synthesize hydrotalcite-supported platinum nanocrystals for selective hydrogenation of cinnamaldehyde in neat water. Catalysis Science & Technology, 2013, 3: 2819-2827.

[229] Mizugaki T, Nagatsu Y, Togo K, Maeno Z, Mitsudome T, Jitsukawa K, Kaneda K. Selective hydrogenation of levulinic acid to 1,4-pentanediol in water using a hydroxyapatite-supported Pt-Mo bimetallic catalyst. Green Chemistry, 2015, 17: 5136-5139.

[230] Mizugaki T, Togo K, Maeno Z, Mitsudome T, Jitsukawa K, Kaneda K. One-pot transformation of levulinic acid to 2-methyltetrahydrofuran catalyzed by Pt—Mo/H-β in water. ACS Sustainable Chemistry & Engineering, 2016, 4: 682-685.

[231] 梁娟, 王志群, 闵恩泽. 催化剂新材料. 北京: 化学工业出版社, 1990: 135-136.

[232] 陈洁. 有机波谱分析. 北京: 北京理工大学出版社, 2008: 35-35.

[233] 闵恩泽. 工业催化剂的研制与开发.北京: 中国石化出版社,1997: 195-198.

[234] Molnáar Á, Smith G V, Barták M. New catalytic materials from amorphous metal alloys. In: Eley D D, Paul H P, Eds. Advances in Catalysis. New York: Academic Press, 1989: 329-383.

[235] Chai W M, Luo H S, Li H X. Solvent effects in preparing Co-B catalyst and its application in the selective hydrogenation of furfllral to furfuryl alcohol. Joumal of Shanghai Normal University (Natural Science), 2005, 34: 87-90.

[236] Augustine R L. The stereochemistry of hydrogenation of α,β-unsaturated ketones. Advances in Catalysis,1976, 25: 56-80.

[237] Li H, Dong F X, Xiong M W, Li H X, Li P, Zhou X G. A novelindium-boron amorphous alloy mediator for Barbier-type carbonyl allylation in aqueous medium. Advanced Synthesis & Catalysis, 2011, 353: 2131-2136.

[238] Burkett S L, Sims S D, Mann S. Synthesis of hybrid Inorganic-organic mesoporous silica by co-condensation of siloxane and organosiloxane precursors. Chemical Communications, 1996: 1367-1368.

[239] Schneider U, Ueno M, Kobayashi S. Catalytic ue of indium(0) for carbon-carbon bond transformations in water: General catalytic allylations of ketones with alylboronates. Journal of the American Chemical Society, 2008, 130: 13824-13825.

[240] Bai G Y, Wen X, Zhao Z, Li F, Dong H X, Qiu M D. Chemoselective hydrogenation of benzoic acid over Ni-Zr-B-PEG(800) nanoscale amorphous alloy in water. Industrial & Engineering Chemistry Research, 2013, 52: 2266-2272.

[241] Bai G Y, Zhao Z, Dong H X, Niu L, Wang Y L, Chen Q Z. A NiPdB-PEG(800) amorphous alloy catalyst for the chemoselective hydrogenation of electron-deficient aromatic substrates. ChemCatChem, 2014, 6: 655-662.

[242] Malyala R V, Rode C V, Arai M, Hegde S G, Chaudhari R V. Activity, selectivity and stability of Ni and bimetallic Ni-Pt supported on zeolite Y catalysts for hydrogenation of acetophenone and its substituted derivatives. Applied Catalysis A: General, 2000, 193: 71-86.

[243] Handjani S, Marceau E, Blanchard J, Krafft J M, Che M, Mäki-Arvela P, Kumar N, Wärnå J, Murzin D Y. Influence of the support composition and acidity on the catalytic properties of mesoporous SBA-15, Al-SBA-15, and Al_2O_3-supported Pt catalysts for cinnamaldehyde hydrogenation. Journal of Catalysis, 2011, 282: 228-236.

[244] Bai G Y, Dong H X, Zhao Z, Chu H L, Wen X, Liu C, Li F. Effect of support and solvent on the activity and stability of NiCoB amorphous alloy in cinnamic acid hydrogenation. RSC Advances, 2014, 4: 19800-19805.

第四章 负载型有机金属催化水介质有机合成

第一节 有序介孔有机金属催化剂

受酶催化生化反应的启发，水介质有机反应主要采用均相催化剂，其中绝大多数是有机金属催化剂，虽然其具有比较高的催化活性和选择性，但由于难分离、回收及重复使用，从而成本高且容易对环境造成重金属污染[1,2]。解决方法是固载均相催化剂实现多相催化。所谓均相催化剂的固载化，就是把均相催化剂用物理或化学方法与固体载体相结合形成非均相催化剂。在这种固载化催化剂中，活性组分往往与均相催化剂具有同样的性质和结构。固载化均相催化剂容易从反应体系中分离并反复使用，但活性和选择性往往不如均相催化剂，主要是由于活性位分散度下降及存在扩散阻力，同时固载化导致活性中心微化学环境的变化。

有序介孔材料具有高比表面积、可控孔径尺寸和孔道结构、多样化的骨架组成，在吸附与分离、催化、传感、能量转化与储能等领域应用前景广阔，受到了来自化学、材料、物理、生物等领域众多研究者的关注。从催化角度看，介孔材料有利于反应物分子以及产物分子的扩散，从而提高催化效率。1998 年，Zhao 等[3]合成了 SBA 系列有序介孔分子筛，不仅保留了 MCM 系列分子筛的结构特点，而且增加了介孔分子筛的孔壁厚度，从而提高了分子筛的水热稳定性。1999 年，科学家首次报道了用带有双功能基团的有机硅源代替正硅酸乙酯，通过表面活性剂自组装来合成周期性介孔有机硅(PMOs)[4]。PMOs 的孔壁是四配位连接的硅和氧原子以及桥键连接的有机基团[5]。不同于有机基团锚定在孔墙上或孔墙内的功能化介孔二氧化硅材料，PMOs 中的有机氧化硅成分使它既具有牢固的结构又具有一定的疏水性质，同时有机基团赋予材料类似于有机分子的神奇特征[6]。

目前，固载介孔有机金属催化剂的方法主要有两种：①后嫁接法，即通过一定的化学作用力，将活性组分后修饰在所用的载体上；②共聚法，即在形成载体的过程中原位加入具有活性中心的组分，从而一步形成非均相催化剂。

一、嫁接法固载有机金属催化剂

嫁接法主要是通过形成共价键、物理吸附、包裹法和静电作用等实现均相有机金属催化剂固载在有序介孔载体上。早在 2006 年，Li 等通过后嫁接法，将 $RuCl_2(PPh_3)_3$ 催化剂固载到氨基或二苯基膦功能化的介孔氧化硅如 SBA-15 和

FDU-12 上(图 4.1)，在水相高烯丙醇异构化反应[式(4-1)]中，其催化性能不仅大大优于用普通 SiO$_2$ 固载的催化剂，产物得率也与均相催化剂相当。重要的是，该催化剂可以重复循环使用 7 次[7]。同样将 RuCl$_2$(PPh$_3$)$_3$ 固载在氨基功能化的 SBA-15 上，也得到相似的催化效果[8]。随后，他们制备了二苯基膦功能化 FDU-12 固载的 RuCl$_2$(PPh$_3$)$_3$[9]，由于具有三维连通的有序介孔结构，不易出现局部孔堵塞的情况，有利于反应物与产物的脱附，进一步提高了催化效率。

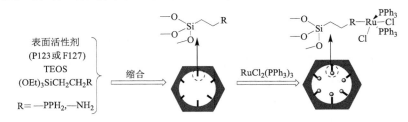

图 4.1 嫁接法制备有序介孔有机金属钌催化剂的示意图

(4-1)

大量的理论和实验数据证实，传统介孔硅材料如 MCM-41 和 SBA-15 因其不规则形貌及较长孔道，不利于反应底物和产物的扩散[10]，因此，固载后的有机金属催化剂活性受到一定限制。Li 等[11]报道了将 PdCl$_2$(PPh$_3$)$_2$ 固载在二苯基膦功能化的介孔二氧化硅纳米小球上，得到一种短孔道(约 100 nm)的有序介孔有机金属 Pd 催化剂(图 4.2)。在水相 Suzuki 反应[式(4-2)]以及 Sonogashira 反应[式(4-3)]中，其催化性能明显改善，达到 PdCl$_2$(PPh$_3$)$_2$ 均相催化剂的水平，催化剂具高稳定性，能够重复使用。

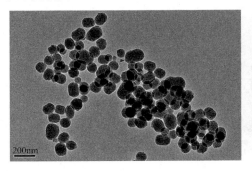
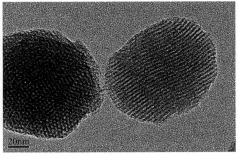

图 4.2 Pd-PPh$_2$-MSN-2 样品的 FETEM 图

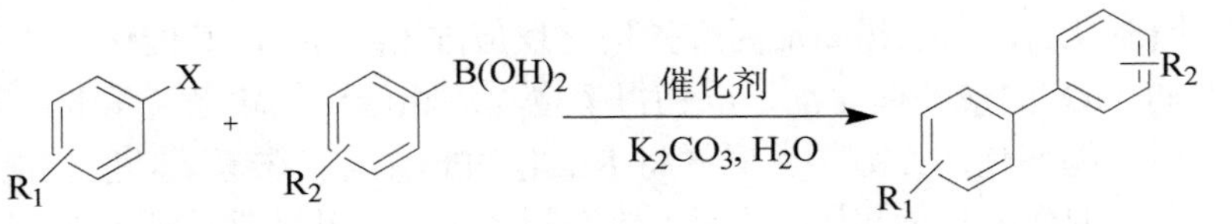

$$\underset{R_1}{\text{Ar}}-X + \underset{R_2}{\text{Ar}}-B(OH)_2 \xrightarrow[K_2CO_3, H_2O]{\text{催化剂}} \underset{R_1}{\text{Ar}}-\underset{R_2}{\text{Ar}} \quad (4\text{-}2)$$

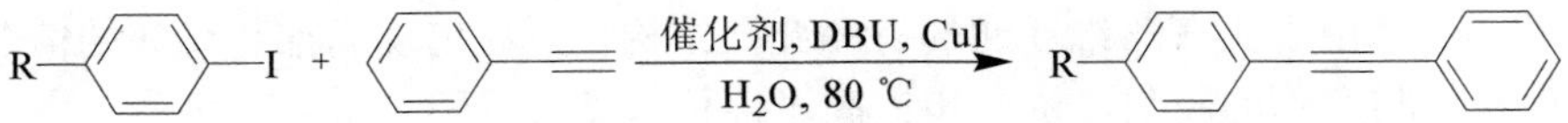

$$R\text{-}C_6H_4\text{-}I + HC\equiv C\text{-}Ph \xrightarrow[H_2O, 80\ ^\circ C]{\text{催化剂, DBU, CuI}} R\text{-}C_6H_4\text{-}C\equiv C\text{-}Ph \quad (4\text{-}3)$$

虽然用功能化的有序介孔氧化硅固载有机金属催化剂可以提高催化活性位的分散度及反应底物的扩散速率，但是氧化硅本身疏水性较差，不利于有机反应底物在催化剂表面的吸附，从而影响催化性能。为了提高催化剂的疏水性，在 SBA-15 的表面修饰聚乙二醇，再固载 Pd(PPh$_3$)$_4$（图 4.3）并应用于水介质 Suzuki 反应。由于提高了催化剂疏水性，能够增加对有机底物的吸附，从而提高催化效率。同时，该催化剂能够多次重复使用[12]。

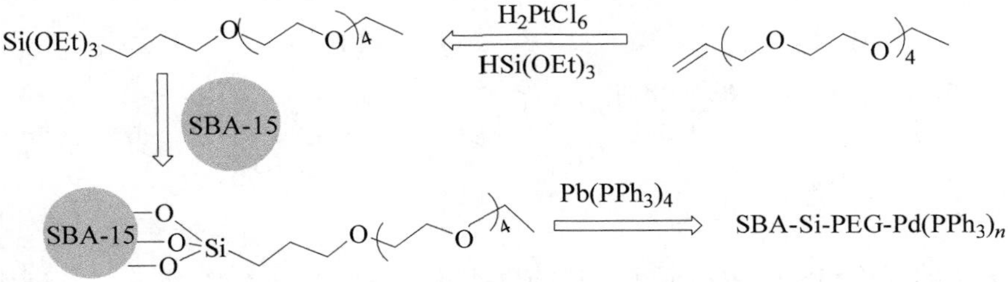

图 4.3 SBA-Si-PEG-Pd(PPh$_3$)$_n$ 催化剂合成示意图

对介孔材料进行功能化的缺点是修饰基团容易堵塞介孔孔道，不利于反应底物和产物的扩散[10]。Li 等[13]发展了一种新方法，即用乙基桥联有机硅烷代替正硅酸乙酯，通过共聚法制得氨基功能化的有序介孔有机硅材料，由于材料的骨架中含有均匀分布的有机基团乙基，其疏水性大大提高，以此为载体固载 Pd(PPh$_3$)$_2$Cl$_2$ 均相催化剂，得到的催化剂在水相 Barbier 反应[式(4-4)]中，催化性能与 Pd(PPh$_3$)$_2$Cl$_2$ 相近，其优点是可重复使用。

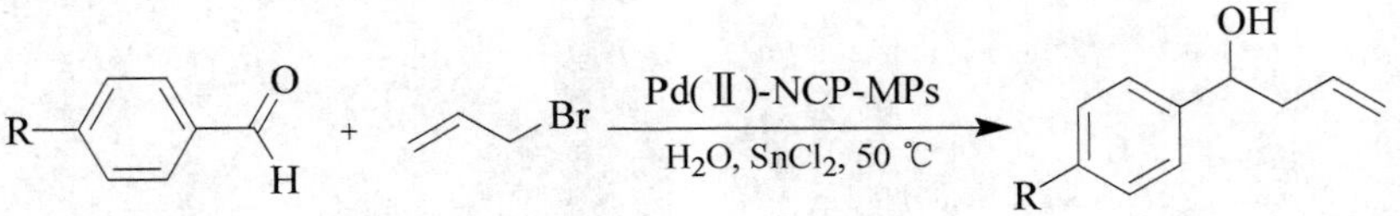

$$R\text{-}C_6H_4\text{-}CHO + CH_2=CHCH_2Br \xrightarrow[H_2O, SnCl_2, 50\ ^\circ C]{\text{Pd(II)-NCP-MPs}} R\text{-}C_6H_4\text{-}CH(OH)CH_2CH=CH_2 \quad (4\text{-}4)$$

随后，他们用共聚法合成了二苯基膦功能化及乙基桥联的有序介孔有机硅材料，再通过配位键固载有机金属 Pd 催化剂，在水相 Barbier 反应中显示出高活性，催化剂可多次循环使用。Zhang 等[14]用溶剂挥发自组装(EISA)技术合成了二苯基膦功能化及乙基桥联的有序介孔有机硅，由此固载的 Pd(PPh$_3$)$_2$Cl$_2$ 有机金属催化剂，在水相 Barbier 和 Sonogashira 反应中均表现出与均相催化剂相似的催化效率,催化剂可重复使用。最近,他们用类似方法合成了二苯基膦功能化及苯基桥联的有机硅,再由此固载 Pd(PPh$_3$)$_2$Cl$_2$ 有机金属催化剂,在水相 Babier 反应中,不仅表现出与均相催化剂 Pd(PPh$_3$)$_2$Cl 相当的催化性能,且可以重复使用 5 次以上。采用同样载体,也可以固载 Ni(PPh$_3$)$_2$Cl$_2$ 有机金属催化剂,在水相对甲氧基碘苯与苯乙炔的偶联反应中,显示出与均相催化剂 Ni(PPh$_3$)$_2$Cl$_2$ 相当的催化活性,且优于其他介孔材料固载的有机金属 Ni 催化剂[15]。

Rostamnia 等[16]采用含有双电荷的功能化 SBA-15 络合 PdCl$_2$,得到有序介孔有机金属催化剂(图 4.4)。在 80℃下催化水介质碘代和溴代芳烃与芳基硼酸的 Suzuki 偶联反应,在较短的时间内给出非常高得率的联苯型产物,TON 值高达 1710,且催化剂表现出优异稳定性,能够重复使用 9 次以上。

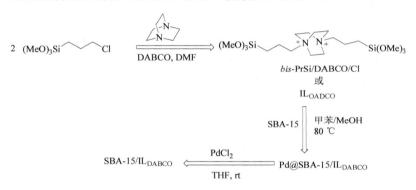

图 4.4 催化剂 Pd@SBA-15/IL$_{DABCO}$ 合成示意图

近年来,介孔酚醛树脂材料引起人们的广泛关注,一方面具有介孔材料的优点(如大比表面积、有序的孔结构),另一方面,疏水性骨架有利于有机物的吸附,提高有机反应催化活性[17-19]。Wu 等[20]用一氯二苯基膦功能化的介孔酚醛树脂 FDU 固载 Pd(OAC)$_2$ 有机金属催化剂,发现其在水相 Suzuki 偶联反应中显示出优良的催化性能,采用不同芳基卤代物作为底物均得到了较高得率的目标产物,且它也是一种可回收循环使用的催化剂。Li 等[21]利用溶剂 EISA 的方法合成出了二苯基膦功能化的有序介孔高分子材料,随后利用配位作用固载各种有机金属催化剂(见彩图 3)。在水介质中的 C—C 偶联反应中,如 Barbier、Sonogashira、Heck [式(4-5)] 和 A^3-偶联[式(4-6)]等,这些催化剂均表现出优良的性能,催化效率与相应的均相催化剂相当,催化剂均能够重复使用 6 次。

$$\text{PhB(OH)}_2 + C_4H_9O\text{-CH=CH}_2 \xrightarrow[\text{H}_2\text{O, 100 ℃}]{\text{Rh(I)-NCP-MPs}} \text{Ph-CH=CH-COOC}_4\text{H}_9 \quad (4\text{-}5)$$

$$\text{piperidine} + \text{PhCHO} + \text{PhC≡CH} \xrightarrow[\text{H}_2\text{O, 70 °C}]{\text{Au(I)-NCP-MPs}} \text{Ph-CH(N-piperidinyl)-C≡C-Ph} \qquad (4-6)$$

二、共聚法固载有机金属催化剂

虽然采用嫁接法可以固载均相有机金属催化剂,但随着负载量增大,可能会造成孔道的堵塞,使反应物分子不能很好地在孔道内扩散,不利于反应的进行,且该方法制得的催化剂活性中心分布不均一,因此,在水相有机反应中催化活性、选择性及使用寿命等受到一定限制。共聚法可制备活性位高度分散于介孔材料的固载化有机金属催化剂。先合成含有催化活性中心的有机金属硅烷,然后以表面活性剂为模板剂,将其与另一种硅烷进行共缩聚,再除去模板剂,得到有序介孔结构的有机金属催化剂,活性位均匀地嵌入介孔材料的孔墙中,这不仅不会引起孔道的堵塞,而且还可以使催化活性位高度分散,有利于催化剂活性的提高[22, 23],同时可以抵御活性位的流失[24]。

Li 等利用上述方法合成一系列新型介孔有机金属催化剂,如图 4.5 所示,将表面活性剂(如 P123[25]、F127[26]、Brij76[27])、骨架硅烷分子[如正硅烷乙酯[28]、1,2-双(三乙氧基硅基)乙烷[29]、1,4-双(三乙氧基硅基)苯[30]]和 Pd 或 Rh 等有机金属硅源分别作为介孔孔道、骨架结构以及催化活性位的构筑单元。硅烷分子水解缩聚后的硅胶体团簇与表面活性剂分子组装,疏水性作用使得包含的有机金属活性分子趋于存在于表面活性剂疏水区域,硅胶体团簇/表面活性剂聚集体继续缩聚形成有序液晶介观相的纳米复合物,除去表面活性剂即可制得介孔有机金属催化剂,其中活性位富集于孔壁表面,该方法可有效保持活性位分子的分子构型,且由于表面活性剂和骨架构筑硅烷的种类容易调变,因此可以精确控制孔结构和骨架单元化学组成,从而设计出有利于催化反应的微环境。

图 4.5 共聚法有序介孔有机金属催化剂合成示意图

如表 4.1 所示，采用共聚组装合成的有序介孔 Ru 有机金属催化剂，在水介质高烯丙醇异构化反应中，产物得率为 71%，远高于采用嫁接法制得的催化剂(产物得率只有 48%)，可能是由于两者活性位的分散均匀性和化学环境不同[26]。ICP 分析显示，共聚组装催化剂中的每个 Ru 与三个 PPh$_2$-配体相络合，而嫁接催化剂中 Ru 和一个 PPh$_2$-或两个 PPh$_3$-配体相络合。此外，XPS 谱显示，共聚组装的有序介孔 Ru 有机金属催化剂中，P 物种的电子结合能比嫁接法合成的催化剂中的 P 低约 0.5 eV，这归因于 PPh$_3$ 配体比 PPh$_2$ 有更强的给电子能力。另外，共聚法组装的催化剂具有更好的重复使用能力，这归因于镶嵌入孔壁的催化活性位具有强流失能力。为了测试该制备方法的普适性，他们还分别合成了 Pd(Ⅱ)[31]、Au(Ⅰ)[27]或 Rh(Ⅰ)[25]有机金属催化剂，在水介质中，以有机金属 Au 催化的炔烃水合反应[式(4-7)]、Rh 催化的 Heck 反应[式(4-5)]以及 Ru 催化的异构化反应[式(4-1)]为探针，考察各催化剂性能，发现共聚组装催化剂的效率和使用寿命均优于嫁接催化剂，说明共聚组装是制备高效介孔有机金属催化剂的普适性方法。

$$R_1 \!-\!\!\!\equiv\!\!\!- R_2 \xrightarrow[H_2O]{cat.} \underset{R_1}{\overset{O}{\|}}\!\!-\!\! R_2 \tag{4-7}$$

表 4.1 不同类型组装和后嫁接催化剂催化性能比较

催化剂	负载量/wt%	有机反应	得率/%
Pd(PPh$_3$)$_2$Cl$_2$	—		90
Pd-PMO(Ph)	1.7	Barbier 反应	89
Pd-PPh$_2$-PMO(Ph)	1.7		71
Au(PPh$_3$)Cl	—		26
Au-PMO(Ph)	0.22	水合反应	91
Au-PPh$_2$-PMO(Ph)	0.26		24
Rh(PPh$_3$)$_3$Cl	—		86
Rh-PMO(Ph)	0.42	Heck 反应	84
Rh-PPh$_2$-PMO(Ph)	0.47		71
Ru(PPh$_3$)$_3$Cl	—		75
Ru-PMO(Ph)	0.40	异构化反应	77
Ru-PPh$_2$-PMO(Ph)	0.43		58

注：反应条件：Barbier 反应为含有 0.040 mmol Pd(Ⅱ)催化剂，0.050 mL 苯甲醛，0.15 mL 烯丙基溴，5.0 mL H$_2$O，2.0 mmol SnCl$_2$，50℃，12 h；水合反应为含有 5.0 mmol Au(Ⅰ)催化剂，0.25 mmol 苯乙炔，25 μmol H$_2$SO$_4$，100℃，7 h；Heck 反应为含有 0.0078 mmol Rh(Ⅰ)催化剂，0.013 g 苯硼酸，75 μL 丙烯酸丁酯，6.0 mL H$_2$O，100℃，5 h；异构化反应为含有 0.0034 mmol Ru(Ⅱ)催化剂，0.025 mL 1-(4-甲基苯基)-3-丁烯-1-醇，5.0 mL H$_2$O，100℃，8 h。

Zhao 等[32]报道了共聚法合成三维周期性有序介孔有机金属 Pd 催化剂，并应用于水介质 Sonogashira 反应[式(4-3)]。实验证明，三维周期性有序介孔催化剂优于二维有序介孔结构催化剂，这可归因于三维笼状结构有利于有机底物和产物的扩散。该催化剂的效率达到均相催化剂的水平，且可重复使用 5 次。随后，他们又用共聚组装制备了具有三维介孔结构的有机金属 Rh 催化剂[33]，介孔孔壁中镶嵌苯基基团，可提高催化剂疏水性，进一步改善在水介质有机合成反应中的催化效率和使用寿命。

气溶胶辅助自组装法是一种采用表面活性剂或嵌段共聚物作为模板剂合成有序介孔结构的纳米微球和薄膜的比较有效快捷的方法。用这种方法可以将多种有机金属硅烷原位组装进二氧化硅的骨架中，不仅其介孔的孔结构和孔大小可以通过表面活性剂与硅酸盐的组装进行调变，而且可用于连续高效、快捷地制备纳米球状负载型有机金属催化剂[34]。Zhang 等[35]利用该方法，在模板剂 CTAB 和 NaCl 的作用下，采用"一步法"快速合成具有有序二维六方孔道和腔内空穴结构的介孔微球状有机金属 Pd、Ru 和 Rh 催化剂(图 4.6)[36]，在相对应的水介质有机反应中，它们的催化活性和选择性均高于嫁接法和不加 NaCl 制备的催化剂，这归因于它的大比表面积以及可作为微反应器的晶体空腔，提高了有机反应物的吸附和扩散速率，从而提高催化活性；同时也促进产物释放，在一定程度上减少副产物生成，提高了选择性。这些催化剂的催化效率与均相有机金属催化剂相当，且能够重复使用，显示出良好的应用前景。

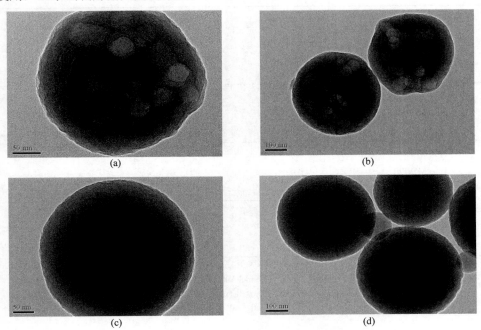

图 4.6 有 NaCl 参与制备的功能化材料负载有机金属前(a)后(b)和无 NaCl 参与制备的功能化材料负载有机金属前(c)后(d)的 TEM 图

第二节 负载型磁性有机金属催化剂

纳米介孔材料负载有机金属催化剂展现出优异的催化活性和选择性，然而，将这些纳米异相催化剂从反应体系中分离和回收过程烦琐。磁性有机金属催化剂可以通过外加磁场，快速分离催化剂[37]。近年来，人们将均相有机金属催化剂以物理或化学的方法与磁性纳米材料载体相结合，获得的多相催化剂[38, 39]不仅保留了均相催化剂的微环境，显示出高活性和选择性，且可快速分离和回收及循环使用；另外，磁性材料易于调节形貌和粒径等[40, 41]，因此，在催化领域备受青睐。

Reiser 等[42]将芘修饰的 Pd-N-杂环卡宾固载在磁性 Co/石墨烯纳米粒子上，并应用于催化水介质卤代芳烃的羟基羰基化反应[式(4-7)]，在 100℃和 1 atm CO 下反应，就能获得高得率的产物。有趣的是，Pd 活性物种可溶解到反应介质中均相催化羟基羰基化反应，当反应结束后冷却到室温时，再次被 Co/石墨烯纳米粒子吸附。磁性载体能使 Pd 催化剂快速分离，催化剂可重复循环使用 16 次以上。

Liu 等[43]首先通过正硅酸乙酯在碱性条件下水解将二氧化硅包裹在磁性 Fe_3O_4 纳米粒子上，然后采用嫁接法将有机配体($1S,2S$)-(+)-N-对甲苯磺酰基-1,2-二苯基乙二胺[(S,S)-TsDPEN]固载到磁性纳米上，最后通过络合作用将二氯双(五甲基环戊二烯)二氯化铑([Cp*RhCl$_2$]$_2$)固载(图 4.7)。在水介质取代苯乙酮的不对称转移氢化反应[式(4-8)]中，显示出与均相催化剂相当的活性和对映选择性，磁性催化剂易于回收，可反复使用 10 次以上。

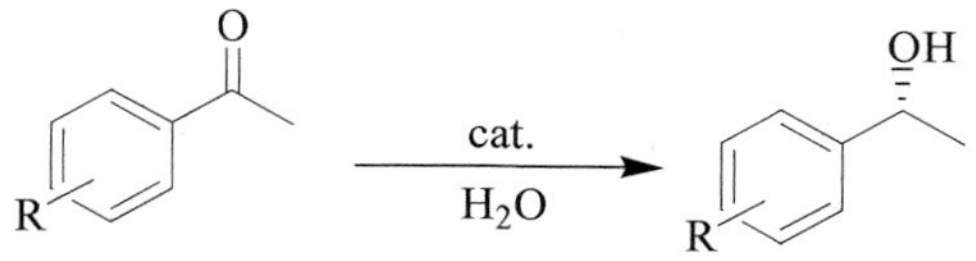

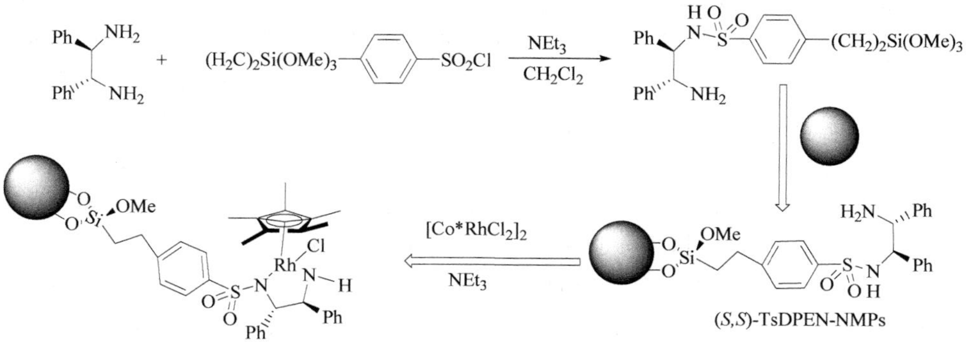

图 4.7 磁性纳米材料固载催化剂的制备示意图

最近，Li 等[44]用 PPh_2 功能化的 SiO_2 包裹 Fe_3O_4 磁性粒子，再嫁接有机金属

Pd 催化剂(图 4.8)。其在水介质 Suzuki 和 Sonogashira 反应中展现出高效率,与均相催化剂相当,这可归因于高度分散的 Pd 活性位,能像均相催化剂一样全面接触底物参与反应。如在水介质 4-甲基苯硼酸和碘苯之间的 Suzuki 反应中,其催化活性和选择性均与均相 $PdCl_2(PPh_3)_2$ 催化剂相似。使用碘苯和溴苯作底物时都能取得高催化效率,但使用氯苯时,活性较低。苯环上带有供电子的—NH_2 和—OCH_3 时,对活性没有明显的影响,但相对来说,苯环上带有吸电子基团有利于提高反应活性。同样地,在水介质 Sonogashira 反应中,该催化剂也展现出与均相 $PdCl_2(PPh_3)_2$ 相当的催化效率。催化剂可重复使用 8 次以上。

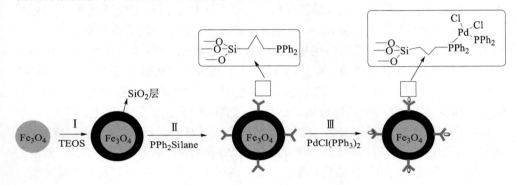

图 4.8 催化剂 Pd-PPh_2-SiO_2@Fe_3O_4 合成示意图

他们又通过 PPh_2 功能化的有序介孔氧化硅包裹 Fe_3O_4 核,再嫁接 Pd 和 Rh 有机金属催化剂[45]。在水相 Suzuki、Sonogashira、Miyaura-Michael 和 Heck 等反应中,有序介孔 Pd 和 Rh 有机金属催化剂的性能优于无介孔催化剂,这可归因于高比表面积,有利于吸附反应底物,从而提高反应活性。催化剂能够循环使用 8 次以上,优于无介孔催化剂,这说明介孔结构的存在能抑制活性中心流失。

Liu 等[46]先采用苯基有机基团功能化的氧化硅包裹磁性 Fe_3O_4 纳米粒子,再嫁接手性有机金属 Rh 催化剂(图 4.9)。该催化剂在水相不对称转移氢化反应中显示高活性和立体选择性,超疏水性苯基修饰在保持高立体选择性的同时,能够有效提高活性,TOF 值高达 480 h^{-1},反应速率的提高归因于催化剂的超疏水性骨架,催化效率甚至优于均相催化剂。例如,在水相 4-甲基苯乙酮不对称转移氢化反应中,该催化剂可在 1.5 h 内完成反应,而对应的均相催化剂则需要 4 h 完成。重要的是该催化剂可通过外加磁铁方便回收,可循环使用 10 次以上。

Tajbakhsh 等[47,48]以联吡唑功能化的氧化硅包裹 Fe_3O_4 负载有机金属 Cu 催化剂 MNP@Biim Cu(Ⅰ) (见彩图 4),在水中炔丙胺[式(4-9)]、1,4-二取代 1,2,3-三唑[式(4-10)]和咪唑并[1,2-a]吡啶[式(4-11)]合成反应中,该催化剂均显示优良的催化性能,且可重复使用 10 次以上。

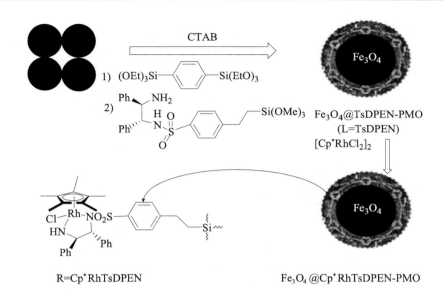

图 4.9 Fe₃O₄@Cp*RhTsDPEN-PMO 催化剂的合成示意图

$$R_1CHO + R_2NHR_3 + Ph-C\equiv CH \xrightarrow[H_2O, \text{回流}]{MNP@Biim\ Cu(\text{I})} \begin{array}{c} R_3\ \ R_2 \\ \diagdown\ \ N\diagup \\ R_1-CH \\ | \\ C\equiv C-Ph \end{array} \quad (4\text{-}9)$$

$$\underset{X}{H_2C-R_2} + NaN_3 + R_1-C\equiv CH \xrightarrow[H_2O,\ rt]{MNP@Biim\ Cu(\text{I})} \text{triazole} \quad (4\text{-}10)$$

$$R_1CHO + Ph-\!\!\equiv\!\! + R_2\text{-}\underset{N}{\text{Py}}\text{-}NH_2 \xrightarrow[CTAB,\ H_2O,\ \text{回流}]{MNP@Biim\ Cu(\text{I})} \text{imidazopyridine} \quad (4\text{-}11)$$

Ghorbani-Choghamarani 等[49]合成了一种对空气和水稳定的磁性催化剂，如图4.10 所示，在水介质四苯基硼化钠与卤代烃的 Suzuki 反应[式(4-12)]以及 Heck 反应[式(4-13)]中，均表现出优异的催化性能，即使氯代烃的 C—C 偶联也能给出良好的产物得率，而且催化剂可重复使用 9 次以上。

$$R\text{-}C_6H_4\text{-}X + NaBPh_4 \xrightarrow[Na_2CO_3,\ H_2O,\ 80\ ^\circ C]{Pd\text{-}ATBA\text{-}MNPs} R\text{-}C_6H_4\text{-}Ph \quad (4\text{-}12)$$

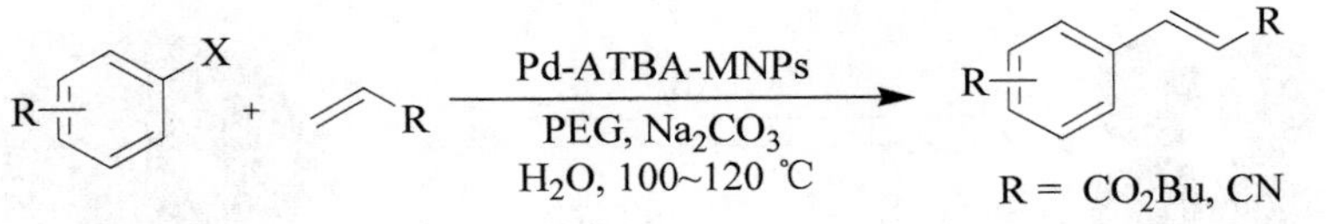

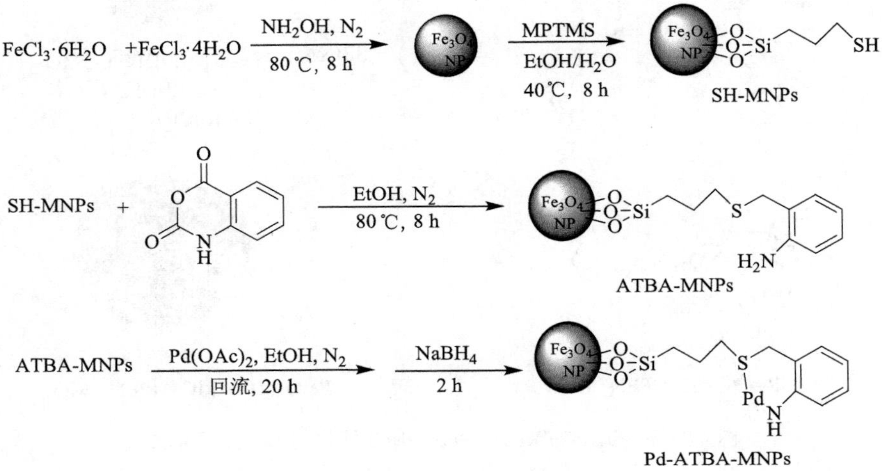

图 4.10 Pd-ATBA-MNPs 催化剂合成示意图

第三节 负载型手性催化剂

手性是自然界的基本属性之一，很多物种在自然界漫长的演化过程中出现了手性特征，如海螺的螺纹、缠绕植物的缠绕方向等。而对于生物界来讲，手性更是一切生命的基础。目前世界上研究的千余种新药中，有 70%以上具有手性，另外，农药、除草剂、植物生长调节剂、甜味剂和香料也都表现出手性特征。用于手性化合物制备的反应称为不对称合成化学。所谓不对称合成，就是指在手性环境中把非对称原料转化为手性产物的方法，其中手性催化剂的使用起着至关重要的作用。传统不对称合成中主要使用均相手性催化剂，虽然其具有高活性和对映选择性，但催化剂难以分离和重复使用，解决这个问题的有效途径是均相催化剂固载化。本节主要介绍不同载体固载的手性有机金属催化剂及其在水相不对称合成中的催化性能。

一、无机载体固载手性催化剂

无机载体由于优异的化学、机械及热稳定性而常被用来固载均相催化剂，其中，氧化硅载体由于易得、廉价、无毒、稳定以及方便表面化学修饰而备受青睐。2004 年，Tu 等[50]用有序介孔氧化硅如 SBA-15 和 MCM-41 固载 Ru-TsDPEN 手性催化剂，在水相不同芳醛的不对称加氢反应中表现出高活性、选择性和 ee 值，催化剂易回收并可重复使用 11 次以上。作者研究发现,加入一定量的表面活性剂可增加催化剂的循环使用次数。2006 年,Li 等[51]采用包裹法将 TsDPEN-Ru 负载在 SBA-16 孔道中(见彩图 5),在水/甲酸钠体系中催化取代苯乙酮的手性还原,转化率超过 94%,ee 值为 87%~93%,催化剂可循环使用 6 次以上。采用类似方法制备了 SBA-16 孔道包裹的 Co(Salen)手性催化剂,在水相取代环氧乙烷的开环反应[式(4-14)]中,与均相催化剂的活性和立体选择性相当,催化剂可重复使用 12 次。

$$\underset{R}{\triangle}\!\!\!\!\!\!\!\!\overset{O}{} + H_2O \xrightarrow{Co(Salen)/SBA-16-C8} \underset{R}{\triangle}\!\!\!\!\!\!\!\!\overset{O}{} + \underset{R}{\overset{OH}{}}\!\!\!\!\!\!\!\!OH \qquad (4\text{-}14)$$

Jacobsen 和 Belser[52]通过与正十硫醇的交换反应在粒径为 3.4 nm 的 Au 纳米粒子表面引入 Salen 的末端巯基,然后,Co(Ac)$_2$ 通过与 Salen 末端巯基的络合作用而固载在金纳米粒子表面。这种催化剂在水介质外消旋己烷-1-过氧化物的水解动力学拆分反应中展现出高活性,反应 5 h 后,目标产物的 ee 值接近 100%,催化剂可重复使用 7 次。2008 年,Li 等固载化有机金属 Ru 手性催化剂(图 4.11) [53]及有机金属 Ir 手性催化剂[54],在芳香酮不对称加氢反应中显示出高活性及选择性。以苯乙酮为例,经不对称加氢反应得到的 R 构型苯乙醇 ee 值大于 99%,转化率也大于 99.5%,均高于对应的均相催化剂,在不同取代基的苯乙酮及 1-萘乙酮加氢反应中,产物得率均大于 98%,且催化剂可重复使用 8 次以上。

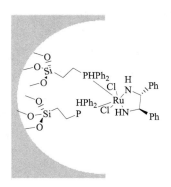

图 4.11　Ru-SBA-15-(*R*,*R*)-DPEN 手性催化剂结构示意图

Liu 等[55]通过灌装法将有机金属 Rh 手性催化剂包裹在功能化介孔氧化硅中,得到离子对型催化剂(图 4.12),在超声作用下水相芳香酮不对称转移氢化反应中,表现出与均相催化剂相当的活性和对映选择性,表明包裹不影响原有的手性化学环境,此外,离子对作用有利于手性催化,催化剂可重复使用 5 次以上,归因于载体的超大笼状结构以及优异的热稳定性。

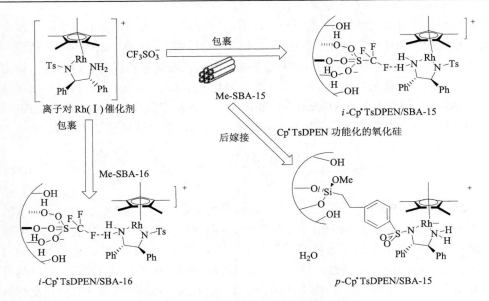

图 4.12 负载型手性铑催化剂结构示意图

后来他们以 CTAC 为模板剂，通过共聚法首先将功能化手性硅源固载于核壳介孔氧化硅球的内核，再加入正硅酸乙酯生长外壳，最后与有机金属 Rh 进行配位，制得基于无机硅的核壳型手性催化剂（图 4.13）。在水相中苯乙酮及其他芳香酮的不对称加氢反应中，显示与均相催化剂相当的活性和立体选择性，当采用苯乙酮为底物时，转化率超过 99%，同时，ee 值达到 97%。当芳香酮的对位为—NO_2 和—CN 等吸电子基时，对映选择性明显降低；当间位有取代基时，可获得较高的 ee 值。此外，催化剂可循环利用 12 次以上，显示出良好的应用前景[56]。

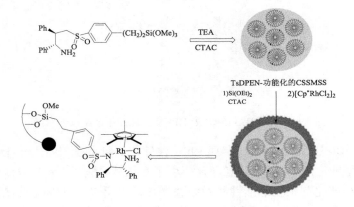

图 4.13 Cp*RhTsDPEN-CSSMSS 催化剂合成示意图

2013 年，Liu 等[57]分别采用共聚和嫁接法制备了固载于介孔氧化硅的有机金属 Rh 手性催化剂(图 4.14)，其在水介质芳香酮不对称加氢反应中都表现出了高活性及选择性，且可多次循环使用。研究表明，共聚法制备的有机金属 Rh 手性催化剂，由于保留了手性微环境，且催化活性位高度分散，其催化活性和选择性优于其他方法制得的催化剂。

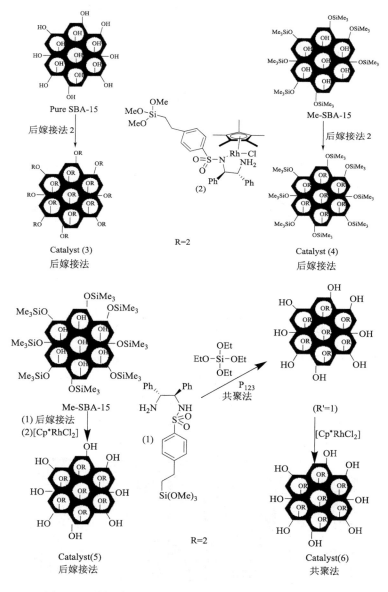

图 4.14　嫁接法和共聚法合成负载型手性铑催化剂示意图

随后，他们将有机金属 Ir 手性催化剂和功能化的花状氧化硅进行自组装(图 4.15)，所获固载化催化剂在水相 α-氰基和 α-硝基取代芳香酮[式(4-15)]反应中展示出高活性和对映选择性，甚至高于均相催化剂，催化剂可重复利用至少 10 次[58]。他们采用类似的方法制备了负载型有机金属 Rh 手性催化剂，该催化剂具有三维有序球状花型介孔结构，使其在水相取代苯乙酮的不对称加氢反应中显示出高活性和对映选择性。如在苯乙酮的不对称加氢反应中，反应物的转化率达 99%，明显高于均相催化剂，且产物(S)-1-苯基-1-乙醇的 ee 值也能达 97%，更重要的是，催化剂可重复使用 10 次以上[59]。

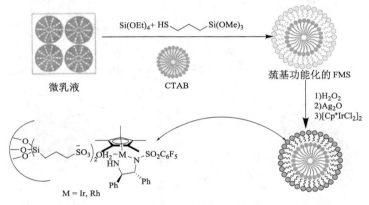

图 4.15　花状 Cp*IrPFPSDPEN-FMS 催化剂合成示意图

$$\text{Ar}\overset{O}{\underset{}{\|}}X \xrightarrow[\text{HCO}_2\text{H, H}_2\text{O, rt}]{\text{Cp*IrPFPSDPEN-FMS}} \text{Ar}\overset{OH}{\underset{}{|}}X \qquad (4\text{-}15)$$

将均相手性有机金属催化剂固载在纳米粒子上，既能保证活性位高度分散，又能避免孔道中的扩散阻力[60,61]。例如，Liu 等[62]合成了一种三维类菊花型介孔氧化硅(图 4.16)，然后固载有机金属 Ru 手性催化剂，如图 4.17 所示。在水介质苯乙酮系列化合物、大分子芳香酮系列化合物和喹啉系列化合物的加氢反应中，转化率均达到 99% 以上，ee 值也均高于 95%，有的甚至高达 99%，且催化剂可重复利用 10 次以上。

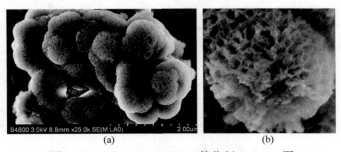

图 4.16　ArRuTsDPEN-CMS 催化剂 HRTEM 图

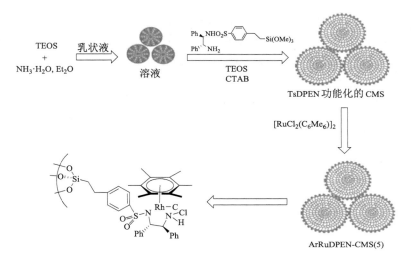

图 4.17 ArRuTsDPEN-CMS 催化剂合成示意图

最近，Cheng 等[63]设计了一种固载在氧化玻璃片上的有机金属 Rh 手性催化剂，在水相芳酮的不对称加氢反应中，催化活性和对映选择性达到均相催化剂的水平，含有给电子基团的苯乙酮给出产物的对映选择性明显高于含有吸电子基团的底物。

二、有机高分子载体固载型手性催化剂

高分子负载在手性催化剂中的应用越来越受到人们的关注，Xiao[64]设计了一种聚苯乙二醇负载的有机金属 Ru 手性催化剂，在甲酸钠水溶液中的芳香酮不对称转移氢化反应[式(4-8)]中，该催化剂显示出高催化效率，目标产物的对映选择性达到 85%~94%，催化剂可回收并能重复使用 14 次以上。Uozumi 等[65]采用含有手性膦配体的功能化聚苯乙烯–聚乙二醇(图 4.18)固载有机金属 Pd 手性催化剂，在水中环烯烃基酸酯与仲胺的反应[式(4-16)]中表现出高活性以及对映选择性(92%~95%)，且催化剂可多次重复使用。

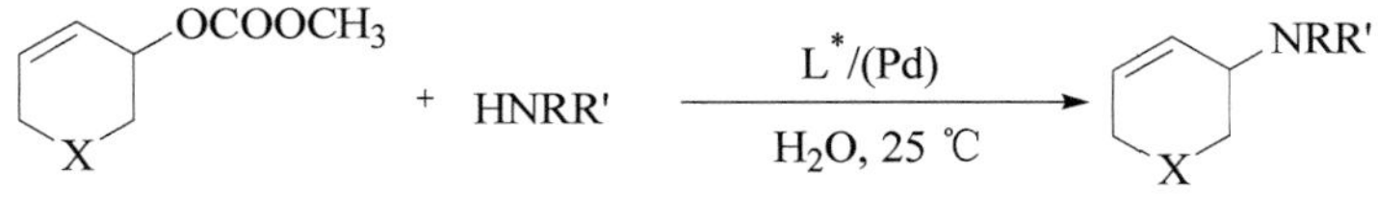

$$\underset{X}{\text{OCOOCH}_3} + \text{HNRR}' \xrightarrow[\text{H}_2\text{O, 25 }^\circ\text{C}]{L^*/(\text{Pd})} \underset{X}{\text{NRR}'} \quad (4\text{-}16)$$

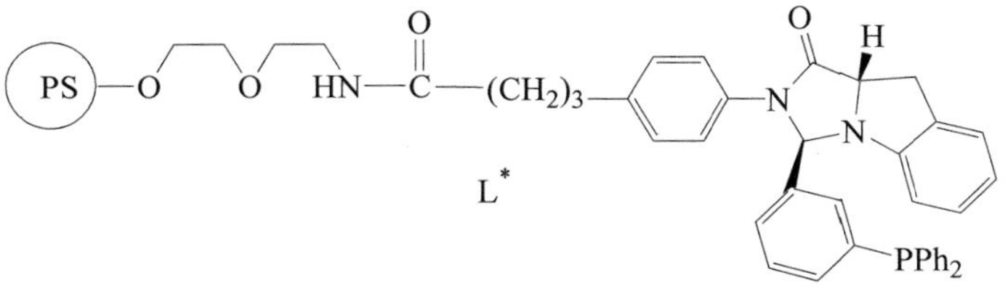

图 4.18 聚苯乙烯-聚乙二醇(PS)负载手性膦配体结构示意图

Uozumi 和 Suzuka[66]采用双亲载体聚苯乙烯-聚乙烯固载咪唑并吲哚手性 Pd 催化剂(图 4.19)，并用于水中环烯烃基酸酯的硝基甲基化反应[式(4-17)]，以 Li_2CO_3 为碱，在 60℃下反应就能给出 91%的目标环烯烃基硝基甲烷，且 ee 值高达 98%。

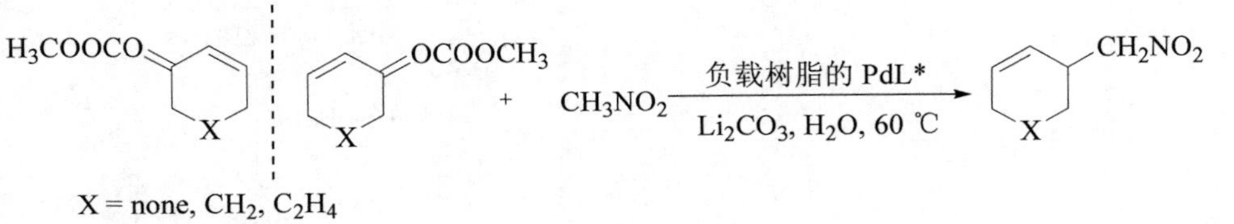
H3COOCO
X
OCOOCH3
X
+
CH3NO2
负载树脂的 PdL*
Li2CO3, H2O, 60 ℃
CH2NO2
X
X = none, CH2, C2H4

(4-17)

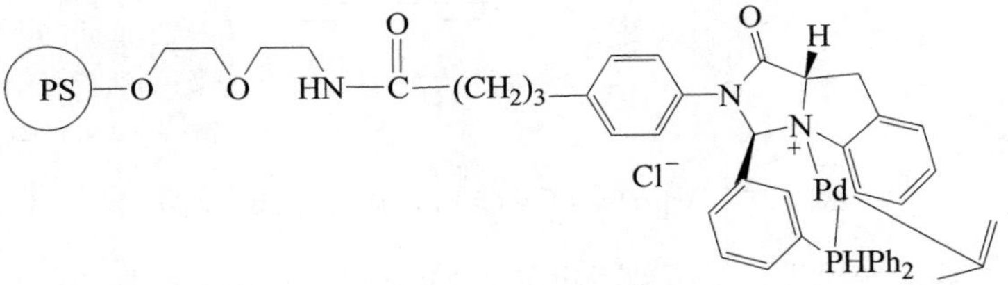
PS
O
O
HN
C
(CH2)3
N
N
H
Cl−
Pd
PHPh2

图 4.19　负载型手性 Pd 催化剂结构示意图

Itsuno 等[67~69]设计了一类新型的对磺化聚苯乙烯支撑的 TsDPEN 类配体(图 4.20)，并通过配位键固载有机金属 Ru 手性催化剂，在水相苯乙酮的不对称加氢反应中显示高活性和对映选择性。通过改变聚合物各种单体的比例可调节配体的亲水亲脂系数，发现恰当的亲水亲脂系数对于水相反应的催化活性和对映选择性具有促进作用。另外，使用季铵盐作为抗衡离子比钠离子更有效，不仅大大加快了反应速

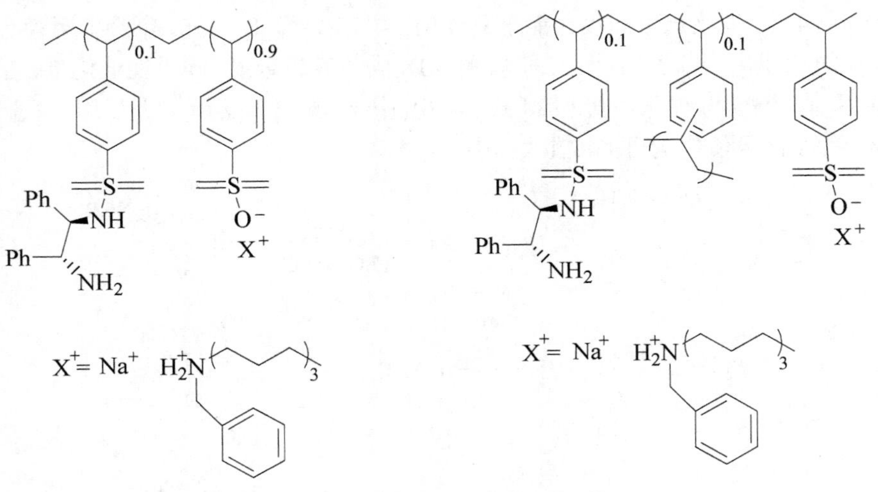
0.1
0.9
S
Ph
NH
Ph
NH2
S
O−
X+
X+= Na+
H2N+
3
0.1
0.1
S
Ph
NH
Ph
NH2
S
O−
X+
X+= Na+
H2N+
3

图 4.20　聚合物负载手型二胺配体结构示意图

率，而且能有效地提高对映选择性，充分说明两亲性催化剂所形成的微观反应环境对反应效率至关重要。此类配体也成功地应用于亚胺的水相不对称转移氢化反应中[67]。由于配体负载于高分子载体上，因此催化剂可方便回收并循环使用。

树状大分子是近年来蓬勃发展的一类具有特殊结构与性能的新型有机大分子化合物。Fréchet 聚芳醚型树状大分子具有精确可控和可剪裁的三维结构及特殊的物理化学性质。2006 年，Deng 等[70]首次将手性环己二胺配体负载于疏水性的 Fréchet 聚芳醚型树状大分子上(图 4.21)，并通过配位反应固载各种有机金属手性催化剂，发现在水相苯乙酮不对称加氢反应[式(4-18)]中，Rh 基手性催化剂活性最高。由于手性催化剂连接在强疏水性的树状大分子上，在水相反应中，催化剂易溶于呈液态的底物相中，在剧烈搅拌下，反应在乳浊态下进行，从而表现出很高的活性，ee 值也达到 95%。催化剂经沉淀后能循环使用 6 次以上。

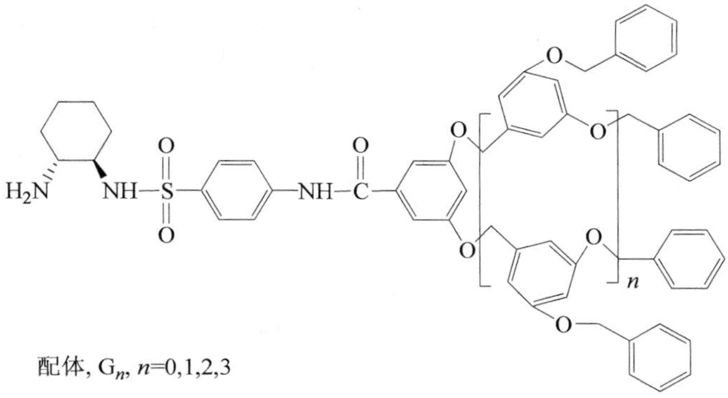

配体, G_n, $n=0,1,2,3$

图 4.21　树状大分子负载手型配体结构示意图

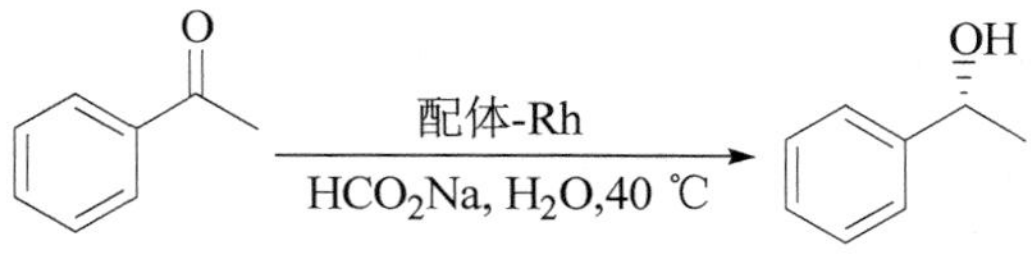

$$(4-18)$$

Schomacker 等[71,72]将有机金属 Rh 手性催化剂固载在表面功能化的聚乙烯载体上(图 4.22)，在水相芳香酮不对称加氢反应[式(4-18)]中，不仅具有优良催化效果和立体选择性，且在合适的 pH 下可以循环使用。

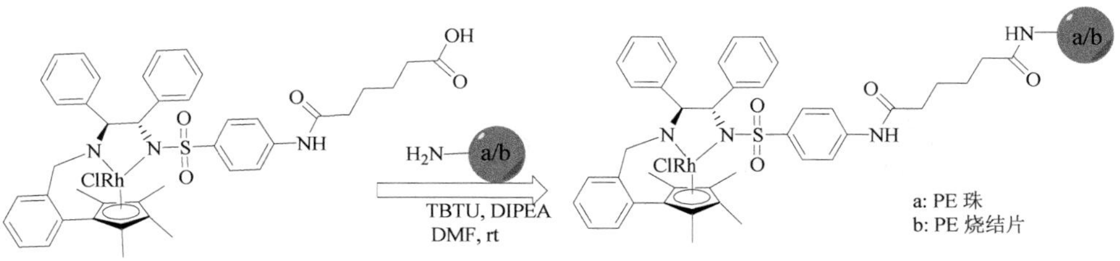

a: PE 珠
b: PE 烧结片

图 4.22　固载有机金属 Rh (Ⅲ) (S,S)催化剂合成示意图

Zhou 和 Ma 采用相似的方法，以含有磺酰胺的聚乙二醇为载体固载有机金属 Ru 手性催化剂，并应用于水相苯乙酮不对称加氢反应中，6 h 后转化率达 100%，产物 ee 值达 85%。在催化苯丙酮的反应中，虽然转化率达 100%，但选择性仅为 72%。此外，随着空间位阻增加，ee 值随之降低。如间氧基苯乙酮和对甲氧基苯乙酮加氢反应，虽然转化率达 100%，但 ee 值只有 85%和 79%。同样地，对于邻位取代的苯乙酮，虽然转化率较高，但产物 ee 值较低，例如，邻甲基苯乙酮加氢产物的 ee 值仅为 53.5%，催化剂可重复使用 3 次[73]。Ma 等[74]通过自由基共聚法合成了含有手性基团(1R,2R)-(+)-N_1-甲苯磺酰基-1,2-二苯乙烯-1,2-二胺的聚合物膦酸酯-聚苯乙烯，然后以此为载体固载有机金属 Ru 手性催化剂，在水介质取代苯乙酮的不对称加氢反应中，显示出高催化活性，产物得率为 94%~98%，对映选择性为 93.9%~97.8%，化学选择性达到 100%。

碳材料广泛用于固载有机金属催化剂，其结构与催化性能密切相关。例如，氧化石墨具有规整的片状表面结构，表面丰富的羧基、羟基和环氧基等活性基团，有助于进行材料功能化，疏水表面有利于提高有机反应的速率。Liu 等[75]采用 Hummers 方法合成了氧化石墨片并修饰手性二胺基团，然后通过配位固载有机金属 Rh 手性催化剂，在水介质芳香酮的不对称加氢反应中显示出高活性和立体选择性。如在甲酸钠水溶液中进行苯乙酮不对称加氢反应，40℃下反应就能给出 96%的转化率和 92%的 ee 值，均呈现出高活性和对映选择性，优于原位配位的催化剂。有趣的是，苯环取代基对反应活性没有明显影响，即无论是供电子和吸电子的取代基都呈现出高的催化活性和对映选择性。

Uozumi[76]报道了含有 P 和 N 配体的聚乙二醇–聚苯乙烯树脂固载有机金属 Pd 手性催化剂(图 4.23)，在水介质环烯烃基碳酸甲酯与氧或碳亲核试剂的偶联反应[式(4-19)]中显示出较好的催化效率和对映选择性，如环庚烯基碳酸甲酯与二苄基胺的反应中，产物得率高达 91%，ee 值也达到了 98%，但该催化剂几乎用于有机溶剂介质(如 THF 和 CH_2Cl_2 等)中的偶联反应。

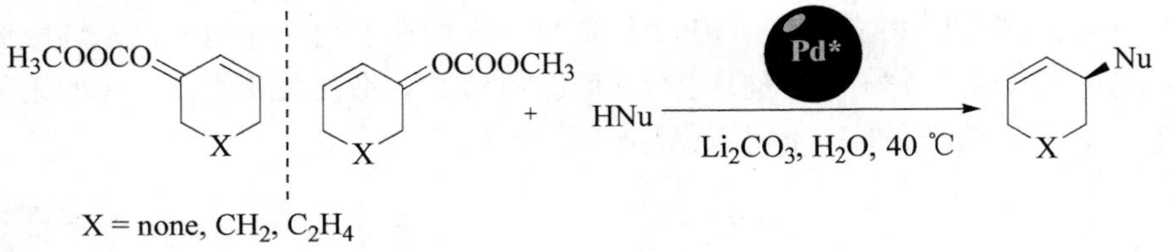

$$\begin{array}{c}\text{H}_3\text{COOCO}\overset{}{\underset{X}{\bigcirc}} \mid \overset{\text{OCOOCH}_3}{\underset{X}{\bigcirc}} + \text{HNu} \xrightarrow[\text{Li}_2\text{CO}_3, \text{H}_2\text{O}, 40\,°\text{C}]{\text{Pd*}} \overset{\text{Nu}}{\underset{X}{\bigcirc}}\end{array} \quad (4\text{-}19)$$

X = none, CH_2, C_2H_4

Pd* ≡ PS−O−(─O─)$_n$NH─CO─(CH$_2$)$_3$─ [结构式]

M=PdCl(η^3-C_3H_5)

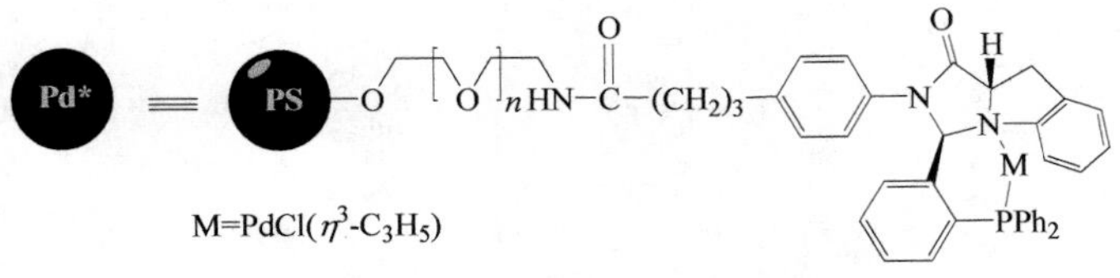

图 4.23　固载[PdCl(η^3-C_3H_5)]$_2$ 催化剂结构示意图

温敏性聚合物在水相中的溶解度具有独特现象，取决于温度、pH 和盐溶液浓度等外界环境[77]。如聚 N-异丙基丙烯酰胺，在低温时通过氢键作用溶于水，但在高温时，与水分子间的氢键断裂，表现为疏水特性而不溶于水，可用于催化剂的分离。而且这种温敏特性具有可逆性，聚 N-异丙基丙烯酰胺的临界温度大约在 32℃，且发生相变的温度范围窄，引入亲水性单体的比例越大，临界温度越低。基于温敏特性，Yin 等[78]通过共价键法将有机金属 Ti 手性催化剂嫁接到温敏材料聚(N-异丙基丙烯酰胺-co-N,N-二甲基丙烯酰胺)上，制备了非均相温敏性手性 Salen Ti (Ⅳ)温敏催化剂，并利用红外、紫外、热重等分析方法对其进行表征。该催化剂在 25℃的反应体系中可以自组装形成微乳，催化活性、化学选择性和对映选择性明显提高。反应完成后，通过简单的温控分离可以达到催化剂的回收且可重复使用 8 次以上。此外，该催化剂的优异性能还表现为其在水介质中的催化效率大于有机溶剂中的效率，且优于均相催化剂，应用前景广阔。

三、有机无机杂化材料固载手性催化剂

有机无机杂化材料是近年来的研究热点。2012 年，Liu 等[79]用季铵盐基团和 TsDPEN 修饰八乙烯基低聚倍半硅氧烷，制备出一种双功能固载型催化剂(图 4.24)。在取代苯乙酮的不对称加氢反应中，催化效果优于同类固载型催化剂，这归因于修饰的季铵盐基团起到了相转移催化作用。

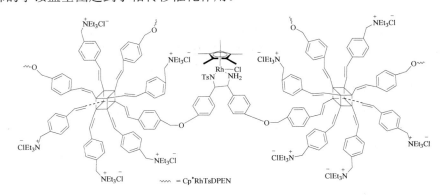

图 4.24 八面体低聚倍半硅氧烷负载 TsDPEN-Rh 催化剂结构示意图

后来，他们用三种不同的骨架硅源(乙基、苯基、乙烯基)作为催化剂的骨架结构，通过溶胶-凝胶法合成了负载型有机金属 Rh 手性催化剂(图 4.25)，并考察了在水介质苯乙酮系列化合物加氢反应中的催化性能[80]。实验结果表明，以乙烯基为骨架的催化剂活性和产物 ee 值最高。产生这种差异的原因可能是催化剂中的电子堆积效应。另外，在未加入相转移催化剂的情况下，催化剂的活性明显好于其相应的均相催化剂，这也表明载体的亲油性和亲水性基团的重要性。另外，催化剂

能循环使用 12 次以上。

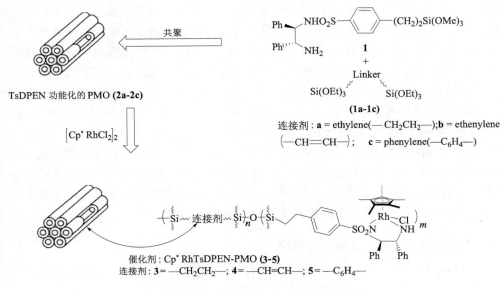

图 4.25　Cp*RhTsDPEN-PMO 合成示意图

然后，他们以 1,2-双三乙氧基硅基乙烷作为骨架硅源，手性磺酰氯硅源作为配体硅源，通过共聚法合成了一种新型载体，再通过配位金属离子，获得一种介孔空心小球结构的有机金属 Rh 手性催化剂(图 4.26)，其直径约为 20 nm，催化剂固载于介孔内。在水相芳香酮酯系列化合物的还原加成反应[式(4-20)]中，显示出优异的催化性能，能够在 8~12 h 之内将反应底物完全转化，产物的 ee 值达 99%，均高于相应的均相催化剂。这一方面是由于催化剂能快速与反应物接触，另一方面催化剂的手性化学环境可以加快有机反应的进行。通过 Pickering 实验可以看出，在有机/无机的催化剂体系中催化剂可以形成一种水包油的体系，极大增加催化剂与反应物的接触面积，从而使反应在较短时间内完成。此外，催化剂能够循环使用 10 次以上[81]。

2014 年，Liu 等[82]合成了一种新型的有机-无机氧化硅负载的有机金属 Ir 手性催化剂(图 4.27)，在水相对 α-硝基芳香酮及 α-氰基芳香酮系列化合物的不对称加氢反应[式(4-15)]中，显示出高活性和立体选择性。这可归因于催化剂特有的有机骨架结构，这种疏水性骨架可增加底物与催化剂的接触概率，从而提高反应速率、反应转化率及立体选择性，也可抑制活性位流失，催化剂可循环利用 10 次以上。

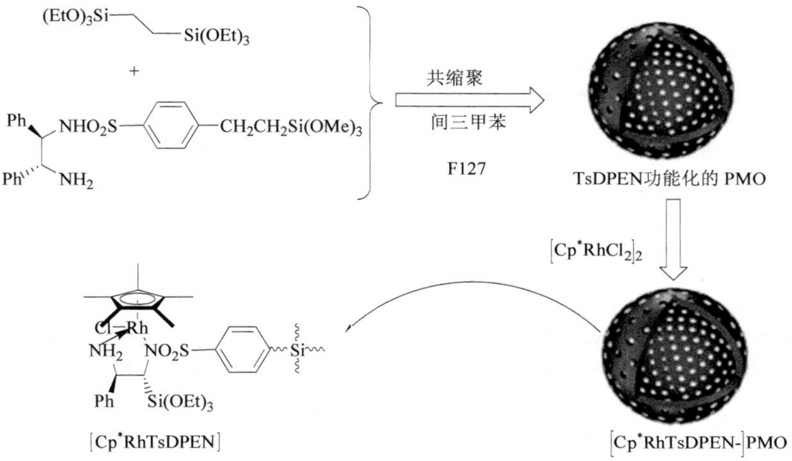

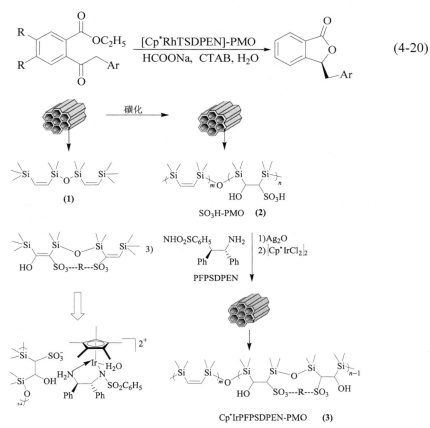

图 4.26 [Cp*RhTSDPEN]-PMO 催化剂合成示意图

$$\text{(4-20)}$$

图 4.27 Cp*IrPFPSDPEN-PMO 催化剂合成示意图

参 考 文 献

[1] Sheldon R A, Bekkum H V. Fine chemicals through heterogeneous catalysis. Germany: Wiley-VCH Verlag Gmb H, 2007: 1-11.
[2] Sheldon R A. The E factor: fifteen years on. Green Chemistry, 2007, 9: 1273-1283.
[3] Zhao D Y, Huo Q S, Feng J L, Chmelka B F, Stucky G D. Nonionic triblock and star diblock copolymer and oligomeric surfactant syntheses of highly ordered, hydrothermally stable, mesoporous silica structures. Journal of the American Chemical Society, 1998, 120: 6024-6036.
[4] Asefa T, MacLachlan M J, Coombs N, Ozin G A. Periodic mesoporous organosilicas with organic groups inside the channel walls. Nature, 1999, 402: 867-871.
[5] Inagaki S, Guan S, Fukushima Y, Ohsuna T, Terasaki O. Novel mesoporous materials with a uniform distribution of organic groups and inorganic oxide in their frameworks. Journal of the American Chemical Society, 1999, 121: 9611-9614.
[6] Park S S, Santha Moorthy M, Ha C S. Periodic mesoporous organosilicas for advanced applications. NPG Asia Materials, 2014, 6: 96-99.
[7] Li H X, Zhang F, Wan Y, Lu Y F. Homoallylic alcohol isomerization in water over an immobilized Ru(II) organometallic catalyst with mesoporous structure. The Journal of Physical Chemistry, B, 2006, 110: 22942-22946.
[8] Wan Y, Zhang F, Lu Y F, Li H X. Immobilization of Ru(II) complex on functionalized SBA-15 and its catalytic performance in aqueous homoallylic alcohol isomerization. Journal of Molecular Catalysis A: Chemical, 2007, 267: 165-172.
[9] Li H X, Zhang F, Yin H, Wan Y, Lu Y F. Water-medium isomerization of homoallylic alcohol over a Ru(II) organometallic complex immobilized on FDU-12 support. Green Chemistry, 2007, 9: 500-505.
[10] Mihalcik D J, Lin W. Mesoporous silica nanosphere supported ruthenium catalysts for asymmetric hydrogenation. Angewandte Chemie International Edition, 2008, 47: 6229-6232.
[11] Zhang F, Yang X S, Zhu F X, Huang J L, He W H, Wang W, Li H X. Highly active and reusable organometallic catalysts covalently bonded to an ordered mesoporous polymer. Chemical Science, 2012, 3: 476-484.
[12] Yang Q, Ma S M, Li J, Xiao F S, Xiong H. A water-compatible, highly active and reusable PEG-coated mesoporous silica-supported palladium complex and its application in Suzuki coupling reactions. Chemical Communications, 2006: 2495-2497.
[13] Zhu F X, Yang X S, Li H X. Ethyl-bridged periodic mesoporous organic silica supported palladium(II) organometallic catalyst for water-medium Barbier reaction. Acta Chimica Sinica, 2010, 68: 217-221.
[14] Kang C M, Huang J L, He W H, Zhang F. Periodic mesoporous silica-immobilized palladium(II) complex as an effective and reusable catalyst for water-medium carbon-carbon coupling reactions. Journal of Organometallic Chemistry, 2010, 695: 120-127.
[15] Yin J W, Chai W, Zhang F, Li H X. Periodic mesoporous silica-supported Ni(II) organometallic complex as an active and reusable nanocatalyst for water-medium Sonogashira coupling reaction. Applied Organometallic Chemistry, 2013, 27: 512-518.
[16] Rostamnia S, Doustkhah E, Zeynizadeh B. Cationic modification of SBA-15 pore walls for Pd supporting: Pd@SBA-15/ILDABCO as a catalyst for Suzuki coupling in water medium. Microporous and Mesoporous Materials, 2016, 222: 87-93.
[17] Yan S Y, Gao Y, Xing R, Shen Y, Liu Y M, Wu P, Wu H H. An efficient synthesis of (E)-nitroalkenes catalyzed by recoverable diamino-functionalized mesostructured polymers. Tetrahedron, 2008, 64: 6294-6299.
[18] Li X H, Shen Y L, Xing R, Liu Y M, Wu H H, He M Y, Wu P. Effective and reusable Pt catalysts supported on periodic mesoporous resols for chiral hydrogenation. Catalysis Letters, 2008, 122: 325-329.

[19] Yang Z L, Wang J W, Huang K, Ma J, Yang Z Z, Lu Y F. Functional mesoporous polymers from phenolic building oligomers. Macromolecular Rapid Communications, 2008, 29: 442-446.

[20] Yao C F, Li H G, Wu H H, Liu Y M, Wu P. Mesostructured polymer-supported diphenylphosphine-palladium complex: An efficient and recyclable catalyst for Heck reactions. Catalysis Communications, 2009, 10: 1099-1102.

[21] Zhang F, Yang X S, Zhu F X, Huang J L, He W H, Wang W, Li H X. Highly active and reusable organometallic catalysts covalently bonded to an ordered mesoporous polymer. Chemical Science, 2012, 3: 476-484.

[22] Taguchi A, Schüth F. Ordered mesoporous materials in catalysis. Microporous and Mesoporous Materials, 2005, 77: 1-45.

[23] Lu Y F. Surfactant-templated mesoporous materials: From inorganic to hybrid to organic. Angewandte Chemie International Edition, 2006, 45: 7664-7667.

[24] Whitesides G M, Grzybowsk B. Self-assembly at all scales. Science, 2002, 295: 2418-2421.

[25] Yang X S, Zhu F X, Huang J L, Zhang F, Li H X. Phenyl@Rh(Ⅰ)-bridged periodic mesoporous organometalsilica with high catalytic efficiency in water-medium organic reactions. Chemistry of Materials, 2009, 21: 4925-4933.

[26] Li H X, Yin H, Zhang F, Li H, Huo Y N, Lu Y F. Water-medium clean organic reactions over an active mesoporous Ru(Ⅱ) organometallic catalyst. Environmental Science & Technology, 2009, 43: 188-194.

[27] Zhu F X, Zhang F, Yang X S, Huang J L, Li H X. Periodic mesoporous organogold(Ⅰ)silica as an active and reusable catalyst for alkyne hydration. Journal of Molecular Catalysis A: Chemical, 2011, 336: 1-7.

[28] Zhu F X, Yang X S, Li H X. Ethyl-bridged periodic mesoporous organic silica supported palladium(Ⅱ) organometallic catalyst for water-medium Barbier reaction. Journal of Molecular Catalysis(China), 2010, 24: 217-221.

[29] Zhu F X, Yang X S, Yang D D, Li H X. Preparation of ordered mesoporous organometallic Pd(Ⅱ) and its catalytic performance for Suzuki reactions in water medium. Chinese Journal of Catalysis, 2010, 31: 1388-1392.

[30] Zhu F X, Zhou J F, Zhu H Q, Li H X. Periodic mesoporous organogold(Ⅰ)silica catalysts for hydration of alkyne to methyl ketone in water medium. Chinese Journal of Catalysis, 2012, 33: 1061-1066.

[31] Huang J L, Zhu F X, He W H, Zhang F, Wang W, Li H X. Periodic mesoporous organometallic silicas with unary or binary organometals inside the channel walls as active and reusable catalysts in aqueous organic reactions. Journal of the American Chemical Society, 2010, 132: 1492-1493.

[32] Zhu F X, Zhu L L, Sun X J, An L T, Zhao P S, Li H X. Synthesis of Pd(Ⅱ) organometal incorporated ordered *Im3m* mesostructural silica and its catalytic activity in a water-medium Sonogashira reaction. New Journal of Chemistry, 2014, 38: 4594-4599.

[33] Zhu F X, Sun X F, Zhou J F, Zhao P S. An active and recyclable bicontinuous cubic *Ia3d* mesostructural Rh(Ⅰ) organometal catalyst for 1,4-conjugate addition reaction in aqueous medium. Green Chemistry Letters and Reviews, 2014, 7: 250-256.

[34] Carné-Sánchez A, Imaz I, Cano-Sarabia M, Maspoch D. A spray-drying strategy for synthesis of nanoscale metal-organic frameworks and their assembly into hollow superstructures. Nature Chemistry, 2013, 5: 203-211.

[35] Zhang F, Kang C M, Wei Y Y, Li H X. Aerosol-spraying synthesis of periodic mesoporous organometalsilica spheres with chamber cavities as active and reusable catalysts in aqueous organic reactions. Advanced Functional Materials, 2011, 21: 3189-3197.

[36] Jiang X, Brinker C J. Aerosol-assisted self-assembly of single-crystal core/nanoporous shell particles as model controlled release capsules. Journal of the American Chemical Society, 2006, 128: 4512-4513.

[37] Yang Q H, Liu J, Zhang L, Li C. Functionalized periodic mesoporous organosilicas for catalysis. Journal of Materials Chemistry, 2009, 19: 1945-1955.

[38] Gibson V, SKimberley B J P, White A J, William D C, Gibson V, Howard P. High activity ethylene polymerisation catalysts based on chelating diamide ligands. Chemical Communications, 1998: 313-314.

[39] Zeidan R K, Davis M E. The effect of acid-base pairing on catalysis: An efficient acid-base functionalized catalyst for aldol condensation. Journal of Catalysis, 2007, 247: 379-382.

[40] Polshettiwar V, Varma R S. Green chemistry by nano-catalysis. Green Chemistry, 2010, 12: 743-754.
[41] Lu A H, Salabas E L, Schüth F. Magnetic nanoparticles: Synthesis, protection, functionalization, and application. Angewandte Chemie International Edition, 2007, 46: 1222-1244.
[42] Wittmann S, Schätz A, Grass R N, Stark W J, Reiser O. A recyclable nanoparticle-supported palladium catalyst for the hydroxycarbonylation of aryl halides in water. Angewandte Chemie International Edition, 2010, 49: 1867-1870.
[43] Sun Y Q, Liu G H, Gu H Y, Huang T Z, Zhang Y, Li H X. Magnetically recoverable SiO_2-coated Fe_3O_4 nanoparticles: A new platform for asymmetric transfer hydrogenation of aromatic ketones in aqueous medium. Chemical Communications, 2011, 47: 2583-2585.
[44] Chen M Z, Zhang F, Li H X. Water-medium organic reactions on an active and easily recycled Pd(II) organometallic catalyst bonded to magnetic $SiO_2@Fe_3O_4$ support. Current Organic Chemistry, 2013, 17: 1051-1057.
[45] Zhang F, Chen M Z, Wu X T, Wang W, Li H X. Highly active, durable and recyclable ordered mesoporous magnetic organometallic catalysts for promoting organic reactions in water. Journal of Material Chemistry, A, 2014, 2: 484-491.
[46] Gao X S, Liu R, Zhang D C, Wu M, Cheng T Y, Liu G H. Phenylene-coated magnetic nanoparticles that boost aqueous asymmetric transfer hydrogenation reactions. Chemistry-A European Journal, 2014, 20: 1515-1519.
[47] Tajbakhsh M, Farhang M, Hosseinzadeh R, Sarrafi Y. Nano Fe_3O_4 supported biimidazole Cu(I) complex as a retrievable catalyst for the synthesis of imidazo[1,2-a]pyridines in aqueous medium. RSC Advances, 2014, 4: 23116-23124.
[48] Tajbakhsh M, Farhang M, Baghbanian S M, Hosseinzadeh R, Tajbakhsh M. Nano magnetite supported metal ions as robust, efficient and recyclable catalysts for green synthesis of propargylamines and 1,4-disubstituted 1,2,3-triazoles in water. New Journal of Chemistry, 2015, 39: 1827-1839.
[49] Ghorbani-Choghamarani A, Tahmasbi B, Moradi P. Palladium-S-propyl-2-aminobenzothioate immobilized on Fe_3O_4 magnetic nanoparticles as catalyst for Suzuki and Heck reactions in water or poly(ethylene glycol). Applied Organometallic Chemistry, 2016, 30: 422-430.
[50] Liu P N, Deng J G, Tu Y Q, Wang S H. Highly efficient and recyclable heterogeneous asymmetric transfer hydrogenation of ketones in water. Chemical Communications, 2004: 2070-2071.
[51] Yang H Q, Li J, Yang J, Liu Z M, Yang Q H, Li C. Asymmetric reactions on chiral catalysts entrapped within a mesoporous cage. Chemical Communications, 2007: 1086-1088.
[52] Belser T, Jacobsen E N. Cooperative catalysis in the hydrolytic kinetic resolution of epoxides by chiral [(salen)Co(III)] complexes immobilized on gold colloids. Advanced Synthesis & Catalysis, 2008, 350: 967-971.
[53] Liu G H, Yao M, Zhang F, Gao Y, Li H X. Facile synthesis of a mesoporous silica-supported catalyst for Ru-catalyzed transfer hydrogenation of ketones. Chemical Communications, 2008: 347-349.
[54] Liu G H, Yao M, Wang J Y, Lu X Q, Liu M M, Zhang F, Li H X. Enantioselective hydrogenation of aromatic ketones catalyzed by a mesoporous silica-supported iridium catalyst. Advanced Synthesis & Catalysis, 2008, 350: 1464-1468.
[55] Xu Y L, Cheng T Y, Long J, Liu K, Qian Q Q, Gao F, Liu G H, Li H X. An ion-pair immobilization strategy in rhodium-catalyzed asymmetric transfer hydrogenation of aromatic ketones. Advanced Synthesis and Catalysis, 2012, 354: 3250-3258.
[56] Zhang H S, Jin R H, Yao H, Tang S, Zhuang J L, Liu G H, Li H X. Core-shell structured mesoporous silica: A new immobilized strategy for rhodium catalyzed asymmetric transfer hydrogenation. Chemical Communications, 2012, 48: 7874-7876.
[57] Long J, Liu G H, Cheng T Y, Yao H, Qian Q Q, Zhuang J L, Gao F, Li H X. Immobilization of rhodium-based transfer hydrogenation catalysts on mesoporous silica materials. Journal of Catalysis, 2013, 298: 41-50.
[58] Deng B X, Cheng T Y, Wu M, Wang J Y, Liu G H. Enantioselective reduction of α-cyano and α-nitro substituted acetophenones promoted by a bifunctional mesoporous silica. ChemCatChem, 2013, 5: 2856-2860.

[59] Gao F, Jin R H, Zhang D C, Liang Q X, Ye Q Q, Liu G H. Flower-like mesoporous silica: A bifunctionalized catalyst for rhodium-catalyzed asymmetric transfer hydrogenation of aromatic ketones in aqueous medium. Green Chemistry, 2013, 15:2208-2214.

[60] Schätz A, Reiser O, Stark W J. Nanoparticles as semi-heterogeneous catalyst supports. Chemistry-A European Journal, 2010, 16: 8950-8967.

[61] Astruc D, Lu F, Aranzaes J R. Nanoparticles as recyclable catalysts: The frontier between homogeneous and heterogeneous catalysis. Angewandte Chemie International Edition, 2005, 44: 7852-7872.

[62] Liu R, Cheng T Y, Kong L Y, Chen C, Liu G H, Li H X. Highly recoverable organoruthenium-functionalized mesoporous silica boosts aqueous asymmetric transfer hydrogenation reaction. Journal of Catalysis, 2013, 307: 55-61.

[63] Cheng T Y, Zhuang J L, Yao H, Zhang H S, Liu G H. Immobilization of chiral Rh catalyst on glass and application to asymmetric transfer hydrogenation of aryl ketones in aqueous media. Chinese Chemical Letters, 2014, 25: 613-616.

[64] Li X G, Wu X F, Chen W P, Hancock F E, King F, Xiao J L. Asymmetric transfer hydrogenation in water with a supported Noyori-Ikariya catalyst. Organic Letters, 2004, 6: 3321-3324.

[65] Uozumi Y, Tanaka H, Shibatomi K. Asymmetric allylic amination in water catalyzed by an amphiphilic resin-supported chiral palladium complex. Organic Letters, 2004, 6: 281-283.

[66] Uozumi Y, Suzuka T. π-Allylic C_1-substitution in water with nitromethane using amphiphilic resin-supported palladium complexes. Journal of Organic Chemistry, 2006, 71: 8644-8646.

[67] Haraguchi N, Tsuru K, Arakawa Y, Itsuno S. Asymmetric transfer hydrogenation of imines catalyzed by a polymer-immobilized chiral catalyst. Organic & Biomolecular Chemistry, 2009, 7: 69-75.

[68] Arakawa Y, Haraguchi N, Itsuno S. Design of novel polymer-supported chiral catalyst for asymmetric transfer hydrogenation in water. Tetrahedron Letters, 2006, 47: 3239-3243.

[69] Arakawa Y, Chiba A, Haraguchi N, Itsuno S. Asymmetric transfer hydrogenation of aromatic ketones in water using a polymer-supported chiral catalyst containing a hydrophilic pendant group. Advanced Synthesis & Catalysis, 2008, 350: 2295-2304.

[70] Jiang L, Wu T F, Chen Y C, Zhu J, Deng J G. Asymmetric transfer hydrogenation catalysed by hydrophobic dendritic DACH-rhodium complex in water. Organic & Biomolecular Chemistry, 2006, 4: 3319-3324.

[71] Dimroth J, Keilitz J, Schedler U, Schomacker R, Haag R. Immobilization of a modified tethered rhodium (Ⅲ)-p-toluenesulfonyl-1,2-diphenylethylenediamine catalyst on soluble and solid polymeric supports and successful application to asymmetric transfer hydrogenation of ketones. Advanced Synthesis & Catalysis, 2010, 352: 2497-2506.

[72] Dimroth J, Schedler U, KeilitzJ, Haag R, Schomaecker R. New polymer-supported catalysts for the asymmetric transfer hydrogenation of acetophenone in water-kinetic and mechanistic investigations. Advanced Synthesis & Catalysis, 2011, 353: 1335-1344.

[73] Zhou Z Q, Ma Q, Sun Y, Zhang A Q, Li L. Ruthenium(Ⅱ)-catalyzed asymmetric transfer hydrogenation of aromatic ketones in water using novel water-soluble chiral monosulfonamide ligands. Heteroatom Chemistry, 2010, 21: 505-514.

[74] Xu X, Wang R, Wan J W, Ma X B, Peng J D. Phosphonate-containing polystyrene copolymer-supported Ru catalyst for asymmetric transfer hydrogenation in water. RSC Advances, 2013, 3: 6747-6751.

[75] 刘克堂, 徐圆圆, 段梦圆, 魏玢莹, 陈倩芸, 程探宇, 刘国华. 氧化石墨负载的手性二胺铑催化剂催化芳香酮的不对称氢转移反应. 上海师范大学学报(自然科学版), 2013, 42: 106-111.

[76] Uozumi Y. Heterogeneous asymmetric catalysis in water with amphiphilic polymer-supported homochiral palladium complexes. Bulletin of the Chemical Society of Japan, 2008, 81: 1183-1195.

[77] 赵江峰, 银董红. 温控相转移手性 salen Mn(Ⅲ)配合物的设计及其在水相不对称催化反应中的催化性能研究. 湖南: 湖南师范大学, 2013: 17-32.

[78] Zhang Y, Tan R, Zhao G, Luo X, Xing C, Yin D. Thermo-responsive self-assembled metallomicelles accelerate asymmetric sulfoxidation in water. Journal of Catalysis, 2016, 335: 62-71.

[79] Tang S, Jin R H, Zhang H S, Yao H, Zhuang J L, Liu G H, Li H X. Recoverable organorhodium-functionalized polyhedral oligomeric silsesquioxane: A bifunctional heterogeneous catalyst for asymmetric transfer hydrogenation of aromatic ketones in aqueous medium. Chemical Communications, 2012, 48: 6286-6288.

[80] Liu R, Jin R H, Kong L Y, Wang J Y, Chen C, Cheng T Y, Liu G H. Organorhodium-functionalized periodic mesoporous organosilica: High hydrophobicity promotes asymmetric transfer hydrogenation in aqueous medium. Chemistry- An Asian Journal, 2013, 8: 3108-3115.

[81] Liu R, Jin R H, An J Z, Zhao Q K, Cheng T Y, Liu G H. Hollow-shell-structured nanospheres: A recoverable heterogeneous catalyst for rhodium-catalyzed tandem reduction/lactonization of ethyl 2-acylarylcarboxylates to chiral phthalides. Chemistry-An Asian Journal, 2014, 9: 1388-1494.

[82] Chen C, Kong L Y, Cheng T Y, Jin R H, Liu G H. Facile construction of functionalized periodic mesoporous organosilica for Ir-catalyzed enantioselective reduction of α-cyanoacetophenones and α-nitroacetophenones. Chemical Communications, 2014, 50: 10891-10893.

第五章 负载型多功能催化水介质有机合成

精细化学品合成通常需要多个不同的催化过程,目前催化反应大部分是单步骤,每个过程均涉及烦琐的分离和提纯,导致大量资源浪费及废弃物的产生,从而增加了生产成本并造成环境污染[1,2]。"一锅"串联反应是解决上述问题的有效手段[3]。主要优点是:①减少了溶剂和洗脱剂用量,从而降低了有毒有害溶剂的排放;②减少了酸碱反应和氧化还原反应的后处理过程,从而消除了废弃物的排放;③避免了反应中间体的分离与纯化,简化了操作过程;④提高了反应选择性,有效减少了副产物的产生[4]。目前,"一锅"串联反应已被广泛应用于精细合成过程中,如加成反应、取代反应、重排反应、环化反应、碳-碳偶联反应、酸-碱协同反应、氧化还原反应、氧化-加氢反应以及不对称合成反应等[5-8]。其中,非均相催化水介质中"一锅"串联反应已成为近年来的新兴研究方向之一[9]。然而,非均相催化剂活性位并不像均相催化剂一样均匀溶解在反应体系中,因此催化活性和选择性比均相催化低[10]。纳米催化剂合成技术的发展有效地解决了上述问题,通过纳米材料结构、形貌和组成等的调控,从而赋予非均相纳米催化剂独特的物理化学特性[11-15]。例如,通过控制金属颗粒催化剂纳米尺度,提高活性位暴露比表面积,增大催化剂与反应物的接触表面积,从而大大提高反应活性[16,17]。此外,通过控制纳米催化剂载体结构、形貌等来有效提高活性位的分散度[18];调控有序孔道结构来加快反应物和产物与催化剂的接触和扩散速率[19];纳米孔道内限域效应的构筑可以作为微反应器,增加反应物的浓度,从而提高反应活性和选择性[20]。本章主要介绍负载型多功能催化剂的合成及其在水介质串联有机反应中的应用(图 5.1),主要包括有机金属催化剂、固体酸–碱催化剂以及金属-酸等多功能催化剂。

图 5.1 负载型多功能催化剂催化"一锅"串联反应示意图

第一节　负载型双有机金属催化剂

有机金属催化剂广泛应用于有机合成，然而，对于将不同催化功能的两种或两种以上有机金属同时负载在一个载体上，实现多功能催化的报道相对较少。2010 年本课题组成功合成了一系列有序介孔结构双有机金属多功能催化剂等(见彩图 6)，并实现了催化水介质中"一锅"串联有机反应[21]。其合成方法主要是通过将两种或两种以上有机金属硅烷试剂在表面活性剂作用下进行共组装，形成有序孔道结构，然后除去表面活性剂，得到有序介孔结构双功能有机金属催化剂。如通过此方法获得的有机金属 Pd 和 Ru 的催化剂(Pd-Rh@PMO)，其在水介质中 Pd 催化的 Bariber 反应和 Ru 催化的高丙烯醇异构化的串联反应[式(5-1)]中，给出产物烯丙醇的得率和选择性分别为 67%和 71%，高于后嫁接法制备的双功能催化剂，这归因于均匀分布的活性位和保持完整的活性位化学微环境。且催化剂可重复使用多次。此方法也可合成同时含有机金属 Pd 和 Rh 的双功能催化剂[22]，在水介质醛烯化反应和 Heck 串联反应[式(5-2)]中，表现出与均相催化剂相当的催化性能，转化率、选择性和得率分别为 90%、71%和 64%，催化剂可回收和重复利用多次。

$$\text{PhCHO} + \text{CH}_2=\text{CHCH}_2\text{Br} \xrightarrow[\text{Pd(II)}]{\text{Barbier 反应}} [\text{Ph-CH(OH)-CH}_2\text{-CH=CH}_2] \xrightarrow[\text{Ru(II)}]{\text{异构化反应}} \text{Ph-CH=CH-CH(OH)-CH}_3 \quad (5\text{-}1)$$

$$\text{Ph-CH=CH-CHO} \xrightarrow[\text{Rh(I)}]{\text{TMS-CHN}_2 \text{ 醛烯基化反应}} \text{Ph-CH=CH-CH=CH}_2 \xrightarrow[\text{Pd(II)}]{\text{PhI, NEt}_3 \text{ Heck 反应}} \text{Ph-CH=CH-CH=CH-Ph} \quad (5\text{-}2)$$

"一锅"不对称串联反应不仅需要解决两种不同类型有机金属催化剂的内在不兼容性，而且要调整内、外各影响因素的匹配性[23, 24]。最近，Liu 等[25]报道了同时含有机金属 Ru 和 Pd 手性催化剂的双功能催化剂(图 5.2)，这种策略可以有效消除有机金属配合物之间的不兼容性，其在水相芳基硼酸与卤代芳烃化合物"一锅"手性联芳醇的串联反应[式(5-3)]中，无论芳基硼酸含有供电子基团还是吸电子基团，均能得到较高得率的目标产物，且产物 ee 值大于 94%，催化剂可重复利用 8 次以上。

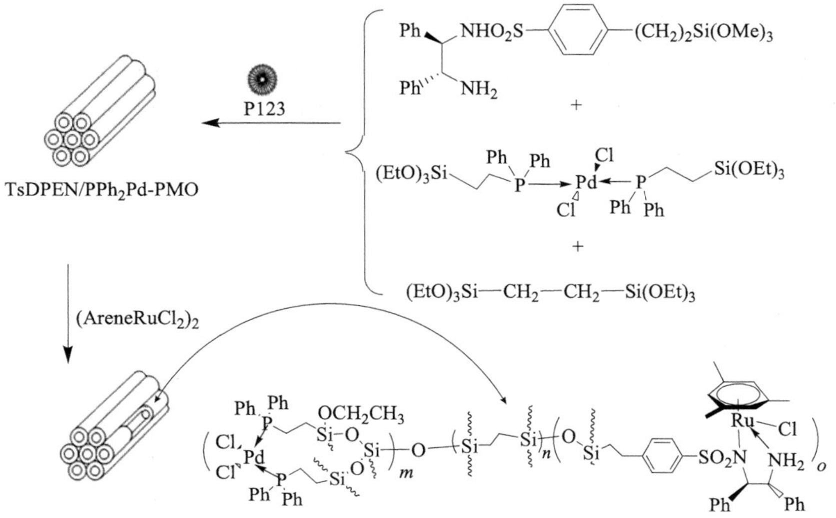

图 5.2 催化剂 AreneRuTsDPEN/PdPPh₂-PMO 合成示意图

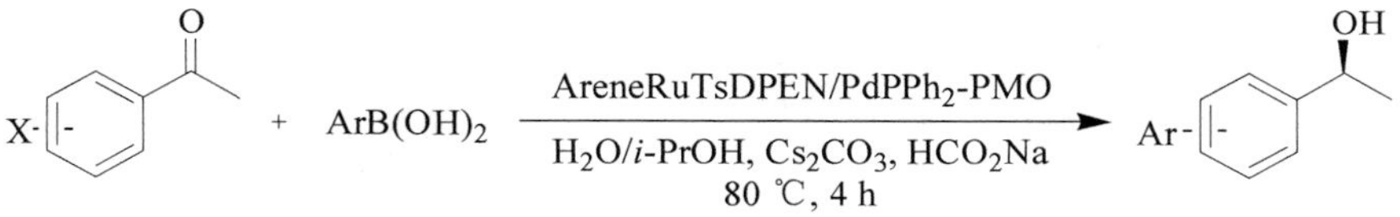

$$\text{X-}\underset{}{\bigcirc}\text{-}\overset{\text{O}}{\underset{}{\text{C}}}\text{-} + \text{ArB(OH)}_2 \xrightarrow[\text{H}_2\text{O}/i\text{-PrOH, Cs}_2\text{CO}_3, \text{HCO}_2\text{Na}]{\text{AreneRuTsDPEN/PdPPh}_2\text{-PMO}} \text{Ar-}\underset{}{\bigcirc}\text{-}\overset{\text{OH}}{\underset{}{\text{CH}}}\text{-} \quad (5\text{-}3)$$

第二节　负载型酸-碱催化剂

酸-碱双功能催化剂是指同时含有酸性及碱性活性中心的催化剂[26]，该类催化剂通过电子对的接收或者质子的传递而起催化作用，已被广泛应用于酯化反应、偶联反应及缩合反应等。在酸-碱协同催化的反应过程中，作为亲电试剂的酸以及作为亲核试剂的碱分别与底物发生催化反应，利用底物与酸碱中心的相互作用对质子或电子对进行转移，生成活泼的正负离子后再进一步结合生成产物，底物与催化剂分离后，催化剂可循环使用[27]。图 5.3 显示了酸-碱双功能催化剂分别活化反应底物的模型。

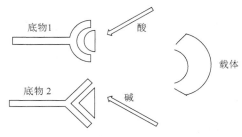

图 5.3　酸-碱双功能催化剂分别活化反应底物的模型

由于酸碱中和反应，酸碱催化很难在均相反应体系中实现，构筑负载型酸-碱纳米双功能催化剂可以有效解决这一矛盾。早在 2006 年，Alauzun 等[28]利用介孔二氧化硅材料高比表面积、孔隙大小分布均匀和可调形貌及丰富的羟基官能团等特点[29]，通过共聚法固载酸碱基团，制得有序介孔结构的双功能酸-碱催化剂，如图 5.4 所示。

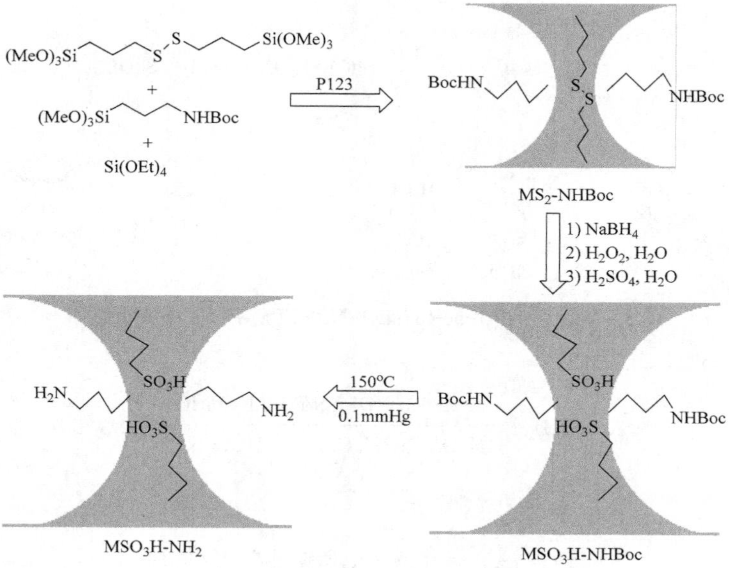

图 5.4 周期有序介孔结构双功能催化剂(NH_2/SO_3H-PMOs)合成示意图

Shylesh 等[30]采用相似的合成技术，制备了一种新颖的有序介孔氧化硅固载双功能酸-碱催化剂，酸性官能团(—SO_3H)主要位于苯基硅烷试剂疏水的一端，而碱性官能团(—NH_2)主要位于苯基硅烷试剂亲水的一端。在催化水相"一锅"缩醛的水解及随后的 Henry 反应[式(5-4)]中表现出优异的催化效率，这主要归因于酸与碱活性位之间的偶合、协同催化作用以及有序介孔结构促进反应物和产物的扩散。

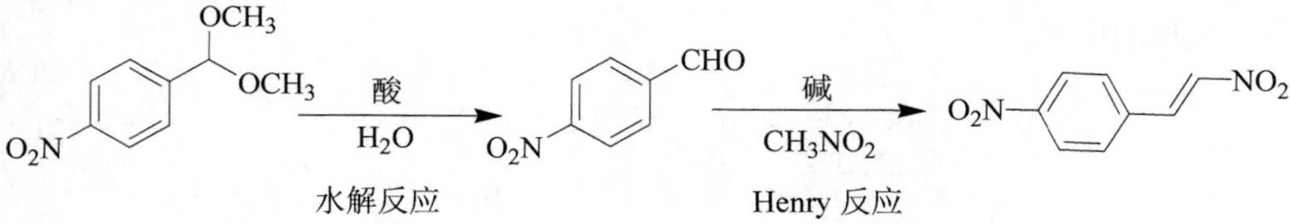

(5-4)

Huang 等[31]报道了一种周期有序介孔氧化硅载体同时固载的"Brønsted"酸和碱催化剂(图 5.5)，该催化剂在催化"一锅"串联有机反应[式(5-4)]中，转化率达到

100%。若分别使用"Brønsted"酸或碱催化剂时,检测不到产物,这可能是由于酸和碱之间发生了中和反应,使其丧失催化活性。这种酸-碱双功能催化剂还可循环使用 5 次以上。

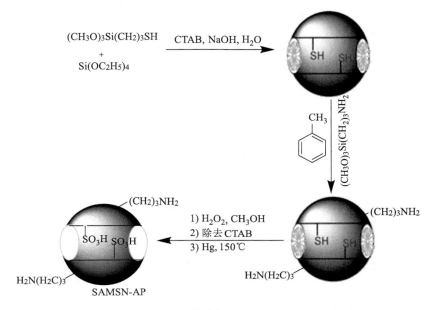

图 5.5 双功能酸-碱催化剂 SAMSN-AP 合成示意图

除了有序介孔硅材料作为载体外,也有报道将酸、碱活性官能团同时嫁接到聚合物[32]或氧化石墨烯[33]或有机金属框架络合物(MOFs)[34]等载体上。例如,Helms 等将酸性基团(—$C_6H_4SO_3H$)和碱性基团(吡啶基)固载在聚合物上构成双功能催化剂(图 5.6),在"一锅"串联有机反应中,虽然酸碱活性位点完全分离,但小分子反应物比较容易穿过聚合物表面层,到达核内部进行催化反应。从反应结果看,酸催化的 4-硝基苯甲醛缩二甲醇脱保护和碱催化的 Baylis-Hillman 反应[式(5-5)]给出了 65%得率的目标产物,对比实验也证明,当没有用聚合物保护层,或者直接采用均相催化剂时,不能获得理想产物,这是因为在反应体系中酸和碱的中和反应使催化剂失活[32]。

聚丙烯纤维是一种合成纤维,具有成本低、张力强以及对化学试剂稳定等优点。Zhang 等[35]以聚丙烯纤维(PPF)为载体,通过连续嫁接法,将丙烯酸和 4-乙烯基吡啶固载,得到双功能催化剂,如图 5.7 所示。在水相对硝基苯甲醛与硝基甲烷缩合反应[式(5-6)]中,由于有机碱和有机酸的协同催化作用,所获得的产物得率为 87%。当酸和碱摩尔比为 2∶1 时,催化性能达到最佳,也可应用于水相 Gewald 反应[式(5-7)](得率:91%~97%)、串联 Michael-Henry 反应[式(5-8)](得率:

87%~99%)、醛、丙二腈与 1-萘酚的三组分反应[式(5-10)] (得率：91%~97%)、醛与氰乙酸乙酯的 Knoevenagel 反应[式(5-10)] (产率 84%)，且催化剂可重复使用 10 次以上。

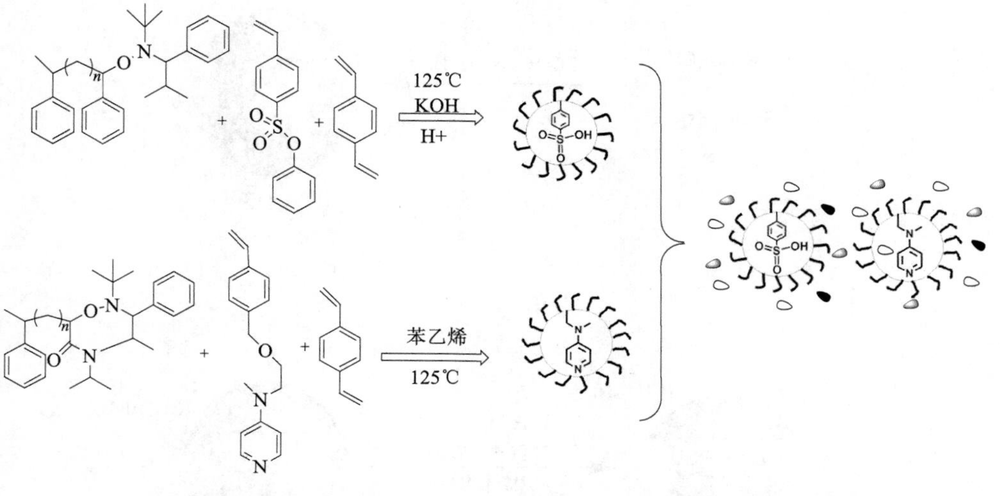

图 5.6 核内包含酸、碱活性位聚合物双功能催化剂合成示意图

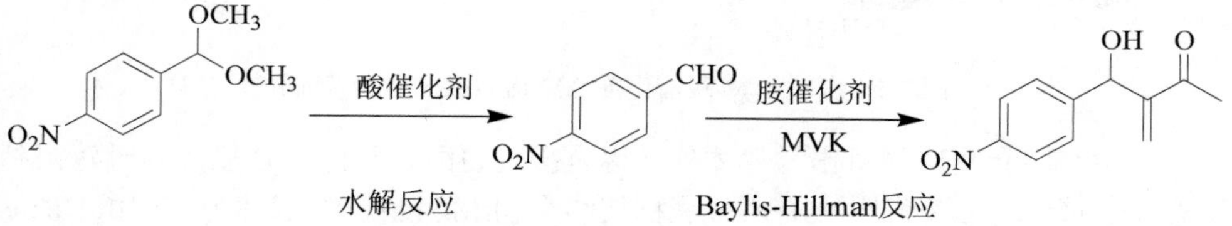

$$\text{(5-5)}$$

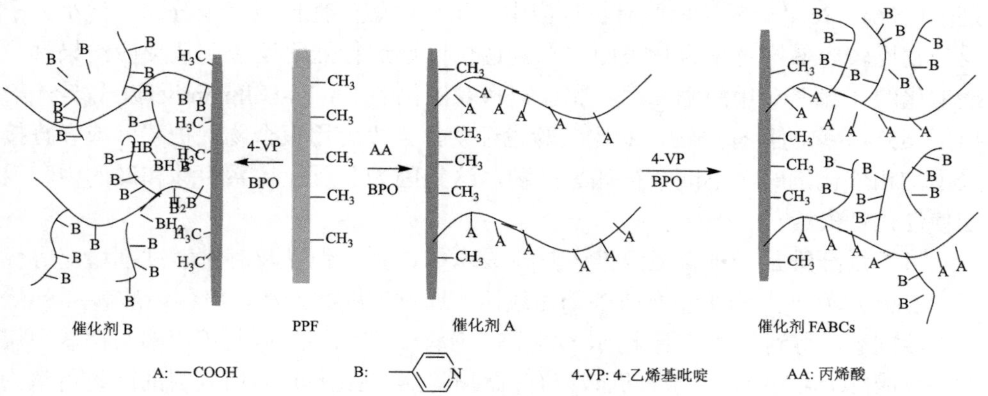

图 5.7 催化剂 FABCs 合成示意图

A: —COOH   B: 4-吡啶基   4-VP: 4-乙烯基吡啶   AA: 丙烯酸

$$\text{PhCHO} + \text{CH}_3\text{NO}_2 \xrightarrow{\text{FABCs}} \text{Ph-CH(OH)-CH}_2\text{NO}_2 \quad (5\text{-}6)$$

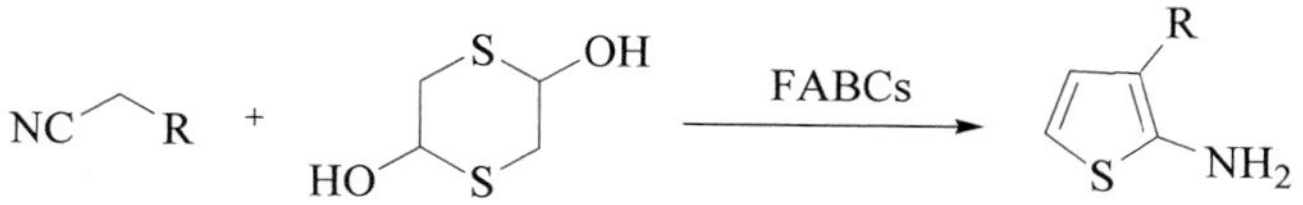

$$\text{NC-CH}_2\text{R} + \text{HO-(S,S cyclic)-OH} \xrightarrow{\text{FABCs}} \text{thiophene-NH}_2\text{-R} \quad (5\text{-}7)$$

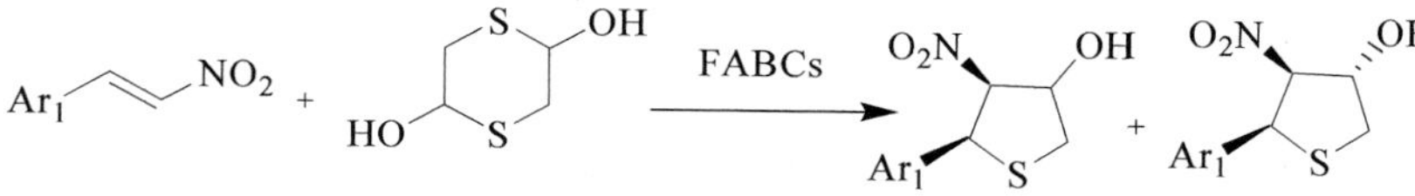

$$\text{Ar}_1\text{-CH=CH-NO}_2 + \text{HO-(S,S cyclic)-OH} \xrightarrow{\text{FABCs}} \text{(tetrahydrothiophene products)} \quad (5\text{-}8)$$

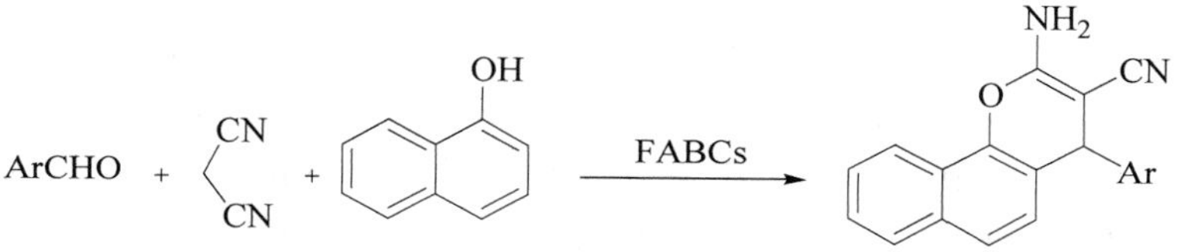

$$\text{ArCHO} + \text{CH}_2(\text{CN})_2 + \text{naphthol} \xrightarrow{\text{FABCs}} \text{(benzochromene product)} \quad (5\text{-}9)$$

$$\text{PhCHO} + \text{NC-CH}_2\text{-CO}_2\text{C}_2\text{H}_5 \xrightarrow{\text{FABCs}} \text{Ph-CH=C(CN)-CO}_2\text{C}_2\text{H}_5 \quad (5\text{-}10)$$

Li 等[33]采用硅烷化反应一步将有机胺嫁接于氧化石墨烯表面(图 5.8)，与氧化石墨烯边缘的羧酸构成新型酸-碱双功能催化剂，在 Hydrolysis-Knoevenagel "一锅"串联反应[式(5-11)]中，显示出高催化效率，且催化剂可以重复使用 5 次以上。以氧化石墨烯作为载体，催化活性远远高于以活性炭、碳纳米管和 SBA-15 作载体，这主要归因于：①石墨烯固载的酸或碱催化剂在反应过程中存在的状态类似于 1~2nm 厚度的石墨烯，大大提高了催化剂在溶液中的分散度，形成一种"类均相"反应体系，而其他载体的催化剂在溶剂中的分散度由于受到其结构的影响而大大降低，且还可能影响传质；②石墨烯表面丰富的官能团，尤其是其中含有的一定量的羧酸基团，在第一步酸催化反应中非常重要，其他载体表面的官能团不如氧化石墨烯表面丰富，使其在第一步反应中的活性比石墨烯负载的催化剂要低。他们还发现氧化石墨烯边缘的羧酸可协助伯胺激活底物，有利于提高其在 Michael-Henry 反应[式(5-12)]中的催化活性，催化剂可套用 5 次以上[36]。

$$\text{GO} \xrightarrow[\text{硅烷偶联}]{\text{(C}_2\text{H}_5\text{O})_3\text{Si-(CH}_2)_3\text{-NR}_2,\ \text{NR}_2 = \text{NH}_2, \text{NHCH}_3, \text{N(CH}_3)_2} \text{有机胺-GO}$$

图 5.8 有机胺/氧化石墨烯双功能催化剂的合成示意图

$$\text{(5-11)}$$

$$\text{(5-12)}$$

最近,Sobhani 和 Zarifi 利用氧化石墨烯表面中的羧基,通过嫁接碱性基团吡啶(图 5.9),获得一种酸-碱双功能催化剂。由于氧化石墨烯独特的二维结构和适量的酸性位,其在水中醛、二烷基亚磷酸酯和丙二腈的 Knoevenagel-Phospha Michael 加成反应[式(5-13)]中展示出高活性,产物得率为 70%~90%。催化剂可重复使用 6 次以上[37]。

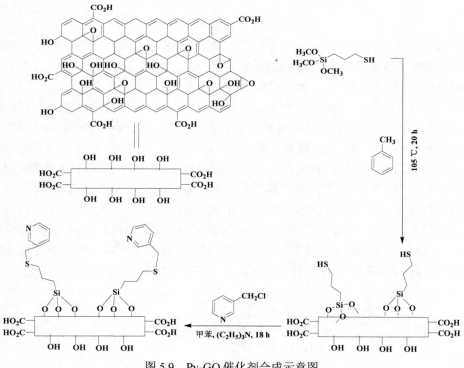

图 5.9　Py-GO 催化剂合成示意图

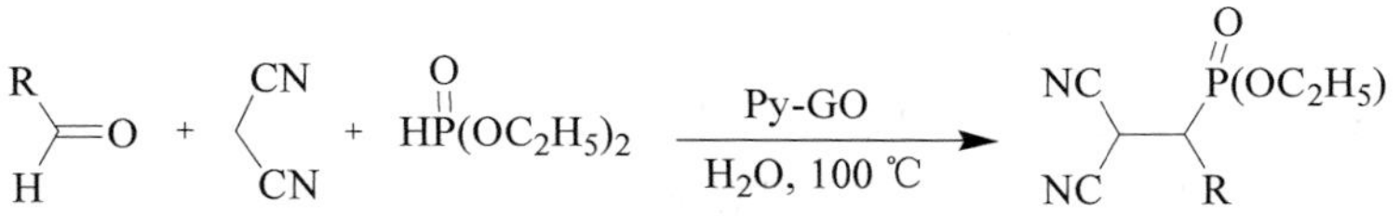

$$\underset{H}{\overset{R}{>}}=O + \underset{CN}{\overset{CN}{>}} + \underset{\parallel}{\overset{O}{HP(OC_2H_5)_2}} \xrightarrow[H_2O,\ 100\ ℃]{Py\text{-}GO} \underset{NC}{\overset{NC}{>}}\underset{R}{\overset{P(OC_2H_5)}{\underset{\parallel}{O}}} \quad (5\text{-}13)$$

Knoevenagel-phospha Michael 反应

有机金属骨架材料(MOFs)是一类由金属离子或金属簇与有机配体通过自组装形成的多孔晶态材料，具有比表面积和孔容积大、孔结构可调、功能可调等多种优点[38]，近年来在催化等领域获得了广泛应用[39]。MOFs 材料的金属位点及有机配体都可作为潜在的催化活性中心[40]。此外，MOFs 材料孔结构比较丰富，还可将纳米金属分散于孔道内而引入新的催化活性中心[41]。MIL-101(Cr)是一类具有良好热稳定性和水稳定性的 MOFs 材料，更重要的是它含有不饱和配位的金属 Cr 位点，其可作为 Lewis 酸催化剂。Shi 等[34]利用 MIL-101 的多孔性等特点，通过嫁接法将酸(—SO_3H)和碱(—NH_2)官能团引入 MIL-101 载体中，制备出多孔结构双功能酸-碱催化剂(图 5.10)，并应用于水中水解-Henry 缩合反应[式(5-4)]，90℃下反应 24 h，目标产物的得率高达 90%，催化剂可重复使用多次。

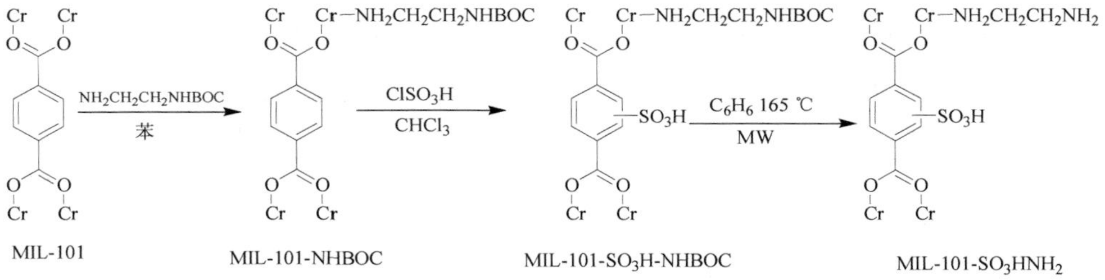

图 5.10　MIL-101-SO_3H-NH_2 催化剂合成示意图

第三节　负载型金属-酸双功能催化剂

负载型金属-酸双功能催化剂在水相有机反应中应用的一个典型例子就是纤维素水解。一方面，纤维素具有结晶度高、物理和化学性能稳定、玻璃化转变温度较高等特性；另一方面，极强的氢键也使纤维素不溶于常用的溶剂，进而难以被直接利用。近年来随着石油、煤炭储量的下降以及石油价格的飞速增长，以及各国对环境污染问题的日益关注和重视，人们越来越重视对纤维素这种可持续发展、可再生资源的应用。传统的酸催化纤维素水解存在腐蚀性强、水解生成的葡萄糖会进一步降解、废酸液难处理等缺点，酶催化降解纤维素虽然效率高，但成本也高，反应时间长，且产物和酶不易分离。近年来，开发的超临界水处理纤维素，可在高温高压下用很短时间将纤维素转化成葡萄糖及其他产物，避免了产物的进一步降解和脱水,但存在产物选择性低、能耗高等缺点,限制了它的应用。

早在 2006 年,Fukuoka 和 Dhepe[42]利用固体酸(γ-Al_2O_3 或 Al_2O_3-SiO_2)等载体负载贵金属 Ru 或 Pt 为催化剂,在 190℃下,首次实现了水相纤维素到多元醇的一步催化转化过程。如以 Pt/Al_2O_3 为催化剂,六元醇的最高产率为 31%,延长反应时间,六元醇的产率并没有提高。他们进一步提出了反应机理,认为氢气在金属表面发生解离吸附,并在催化剂的表面形成 H^+,然后水中的纤维素接近催化剂表面,糖苷键在氢离子质子化作用下发生断裂生成葡萄糖,而后迅速被金属催化氢化生成山梨醇[43]。2010 年,Wang 等[44]首次报道了 Keggin 类型 $Cs_xH_{3-x}PW_{12}O_{40}$ 负载的 Ru 纳米粒子催化剂,在 160℃的条件下催化水中纤维二糖或纤维素水解和加氢串联反应,给出产物果糖醇得率为 43%,作者证实了 Brønsted 酸性位产生的 H_2 对于纤维二糖和纤维素转化成果糖醇发挥了关键作用。

将杂多酸或无机酸与负载型催化剂相结合,同样可以实现纤维素的"一锅"催化转化。例如,Onda 等[45]首先通过浸渍法将 H_2PtCl_6 负载在活性炭上,然后在 300℃下氢气还原 6 h,最后将所得的负载型 Pt 催化剂在浓硫酸中进行酸化,即得到双功能催化剂。该催化剂在空气氛围下可以有效地催化纤维素或淀粉水解氧化反应(图 5.11),在 120℃下反应 24 h 可以获得得率为 46%的葡萄糖酸。

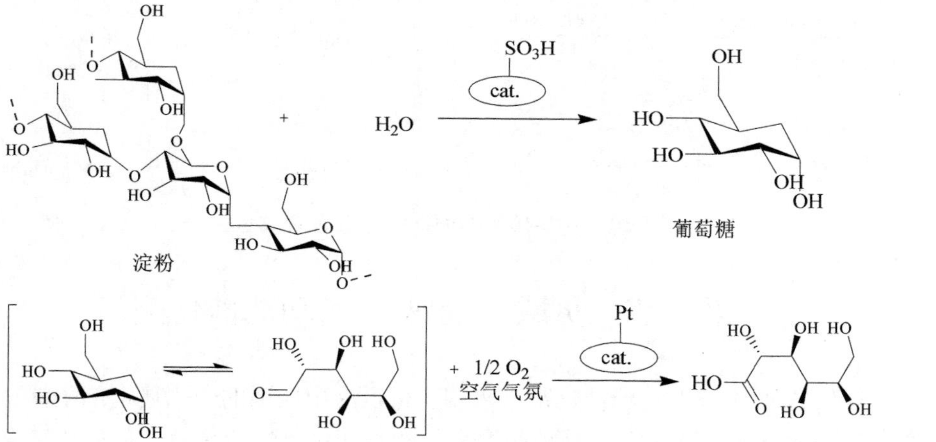

图 5.11 Pt/AC-SO_3H 催化淀粉水解氧化反应过程图

Han 等[46]通过浸渍法将 1~3 nm Au 纳米粒子负载在 Keggin 类型 $Cs_2HPW_{12}O_{40}$ 上(见彩图 7),在 0.5 MPa 的氧气条件下,于 145℃催化纤维二糖水解氧化得到的葡萄糖酸[式(5-14)],产物得率高达 96%。而通过沉积-浸渍法制得的催化剂也显示相近的催化效率,产物得率为 97%。这种催化剂也可催化纤维素水解氧化制备葡萄糖酸,得率为 60%。由于在反应过程中杂多酸中 H^+ 容易流失,催化剂难以循环使用。

$$\left[\begin{array}{c}\text{纤维素结构}\end{array}\right]_{\frac{n}{2}} \xrightarrow[\text{酸催化剂}]{H_2O} \xrightarrow{O_2} \text{氧化剂} \quad HO\text{-葡萄糖酸结构-}OH \qquad (5\text{-}14)$$

Wang 等[47]采用聚多金属氧酸盐($Cs_xH_{3-x}PW_{12}O_{40}$)负载型的 Au 纳米粒子,催化水中纤维二糖或纤维素水解-氧化反应转化成葡萄糖酸,催化性能明显优于典型金属氧化物(SiO_2,Al_2O_3,TiO_2)、碳纳米管以及分子筛(HZSM-5 和 HY)为载体的催化剂,这可能是由于催化剂中载体具有较强的酸性以及较小的 Au 纳米粒子,强酸性不仅有利于纤维二糖的转化,且使葡萄糖酸易解析阻止其降解从而提高其选择性,而较小 Au 纳米粒子可加速葡萄糖中间体氧化,从而提高纤维二糖转化和葡萄糖酸的生成。遗憾的是,由于长时间水热反应导致 H^+ 流失,催化剂不能重复使用。Chen 等[48]报道了一种金属-酸双功能催化剂应用于水介质纤维素转化成己糖醇的"一锅"串联有机反应[式(5-15)]中,得到己糖醇产率为 97%和山梨醇选择性为 95%,纤维素转化率为 100%。

$$\text{纤维二糖/纤维素} \xrightarrow[\text{酸性活性位}]{POM/MOF} \text{D-葡萄糖} \xrightarrow[\text{金属活性位}]{H_2, Ru} \text{山梨醇} + \text{甘露醇} + \text{其他醇} \qquad (5\text{-}15)$$

Li 等[49]采用共聚法制得巯基功能化的有序介孔有机杂化氧化硅,由巯基原位还原 Au(Ⅲ),获得固载 1.0 nm 左右 Au 纳米粒子和—SO_3H 的双功能催化剂(彩图 8),顺利地催化水相中的苯乙炔转化成苯乙酮[式(5-16)],如果采用不含—SO_3H 的 Au 催化剂,则反应不能进行,而当向这个反应体系中再加入一定量硫酸后,反应可以进行,但在同样的反应条件下转化率只有 36%。由此可以看出,炔烃水合反应不仅需要金属催化活性中心,而且需要酸作为共催化剂。在该催化剂作用下,无论是脂肪炔还是芳香炔均能顺利地和水发生反应,生成相应的酮。当苯乙炔中苯环上含有给电子基时,炔烃水合反应进行地较为顺利,而连有吸电子基团则不利于这个反应进行。催化剂可重复使用 10 次以上。

$$R_1-\!\!\!\equiv\!\!\!-R_2 + H_2O \xrightarrow[H_2O,\ 80\ ^\circ\!C]{Au\text{-}SH/SO_3H\text{-}PMO(Et)} R_1\overset{O}{\underset{}{\text{–}}}\!\!-R_2 \qquad (5\text{-}16)$$

参 考 文 献

[1] Albrecht Ł, Jiang H, Jørgensen K A. A Simple recipe for sophisticated cocktails: Organocatalytic one-pot reactions-concept, nomenclature, and future perspectives. Angewandte Chemie International Edition, 2011, 50: 8492-8509.
[2] Ambrosini L M, Lambert T H. Multicatalysis: Advancing synthetic efficiency and inspiring discovery. ChemCatChem, 2010, 2: 1373-1380.
[3] Denmark S E, Thorarensen A. Tandem [4+2]/[3+2] cycloadditions of nitroalkenes. Chemical Reviews, 1996, 96: 137-166.
[4] Bruggink A, Schoevaart R, Kieboom T. Concepts of nature in organic synthesis: Cascade catalysis and multistep conversions in concert. Organic Process Research & Development, 2003, 7: 622-640.
[5] Tietze L F, Brasche G, Gericke K M. Front matter. In domino reactions in organic synthesis. Germany: Wiley-VCH Verlag GmbH & Co. KGaA, 2006: I-XIV.
[6] Fuentes N, Kong W, Fernández-Sánchez L, Merino E, Nevado C. Cyclization cascades via N-amidyl radicals toward highly functionalized heterocyclic scaffolds. Journal of the American Chemical Society, 2015, 137: 964-973.
[7] Kazem Shiroodi R, Gevorgyan V. Metal-catalyzed double migratory cascade reactions of propargylic esters and phosphates. Chemical Society Reviews, 2013, 42: 4991-5001.
[8] Wang Y, Lu H, Xu P F. Asymmetric catalytic cascade reactions for constructing diverse scaffolds and complex molecules. Accounts of Chemical Research, 2015, 48: 1832-1844.
[9] Climent M J, Corma A, Iborra S, Sabater M J. Heterogeneous catalysis for tandem reactions. ACS Catalysis, 2014, 4: 870-891.
[10] Schlögl R. Heterogeneous catalysis. Angewandte Chemie International Edition, 2015, 54: 3465-3520.
[11] Polshettiwar V, Luque R, Fihri A, Zhu H B, Bouhrara M, Basset J M. Magnetically recoverable nanocatalysts. Chemical Reviews, 2011, 111: 3036-3075.
[12] Blaser H U, Pugin B, Studer M. Enantioselective heterogeneous catalysis: Academic and industrial challenges // Chiral Catalyst Immobilization and Recycling. Weinheim: Wiley-VCH Verlag GmbH, 2007: 1-17.
[13] Smith G V, Notheisz F. Chapter 1- introduction to catalysis. In Heterogeneous catalysis in organic chemistry. San Diego: Academic Press, 1999: 1-28.
[14] Shylesh S, Schünemann V, Thiel W R. Magnetically separable nanocatalysts: Bridges between homogeneous and heterogeneous catalysis. Angewandte Chemie International Edition, 2010, 49: 3428-3459.
[15] Corma A. From microporous to mesoporous molecular sieve materials and their use in catalysis. Chemical Reviews, 1997, 97: 2373-2420.
[16] Somorjai G A, Frei H, Park J Y. Advancing the frontiers in nanocatalysis, biointerfaces, and renewable energy conversion by innovations of surface techniques. Journal of the American Chemical Society, 2009, 131: 16589-16605.
[17] Li Y, Somorjai G A. Nanoscale advances in catalysis and energy applications. Nano Letters, 2010, 10: 2289-2295.
[18] Polshettiwar V, Varma R S. Green chemistry by nano-catalysis. Green Chemistry, 2010, 12: 743-754.
[19] Taguchi A, Schüth F. Ordered mesoporous materials in catalysis. Microporous and Mesoporous Materials, 2005, 77: 1-45.
[20] Goettmann F, Sanchez C. How does confinement affect the catalytic activity of mesoporous materials?. Journal of Materials Chemistry, 2007, 17: 24-30.
[21] Huang J L, Zhu F X, He W H, Zhang F, Wang W, Li H X. Periodic mesoporous organometallic silicas with unary or binary organometals inside the channel walls as active and reusable catalysts in aqueous organic reactions. Journal of the American Chemical Society, 2010, 132: 1492-1493.

[22] Huang J, Zhang F, Li H. Organometal-bridged PMOs as efficiency and reusable bifunctional catalysts in one-pot cascade reactions. Applied Catalysis A: General, 2012, 431-432: 95-103.

[23] Grondal C, Jeanty M, Enders D. Organocatalytic cascade reactions as a new tool in total synthesis. Nature Chemistry, 2010, 2: 167-178.

[24] Enders D, Grondal C, Hüttl M R M. Asymmetric organocatalytic domino reactions. Angewandte Chemie International Edition, 2007, 46: 1570-1581.

[25] Zhang D C, Xu J Y, Zhao Q K, Cheng T Y, Liu G H. A site-isolated organoruthenium-/organopalladium-bifunctionalized periodic mesoporous organosilica catalyzes cascade asymmetric transfer hydrogenation and Suzuki cross-coupling. ChemCatChem, 2014, 6: 2998-3003.

[26] 甄开吉, 王国甲, 毕颖丽, 李荣生, 阚秋斌. 催化作用基础. 3 版. 北京: 科学出版社, 2005: 280-360.

[27] Motokura K, Fujita N, Mori K, Mizugaki T, Ebitani K, Kaneda K. An acidic layered clay is combined with a basic layered clay for one-pot sequential reactions. Journal of the American Chemical Society, 2005, 127: 9674-9675.

[28] Alauzun J, Mehdi A, Reyé C, Corriu R J P. Mesoporous materials with an acidic framework and basic pores a successful cohabitation. Journal of the American Chemical Society, 2006, 128: 8718-8719.

[29] Diaz U, Brunel D, Corma A. Catalysis using multifunctional organosiliceous hybrid materials. Chemical Society Reviews, 2013, 42: 4083-4097.

[30] Shylesh S, Wagener A, Seifert A, Ernst S, Thiel W R. Mesoporous organosilicas with acidic frameworks and basic sites in the pores: An approach to cooperative catalytic reactions. Angewandte Chemie International Editio, 2010, 49: 184-187.

[31] Huang Y L, Xu S, Lin V S Y. Bifunctionalized mesoporous materials with site-separated Brønsted acids and bases: Catalyst for a two-step reaction sequence. Angewandte Chemie International Edition, 2011, 50: 661-664.

[32] Helms B, Guillaudeu S J, Xie Y, McMurdo M, Hawker C J, Fréchet J M J. One-pot reaction cascades using star polymers with core-confined catalysts. Angewandte Chemie International Edition, 2005, 44: 6384-6387.

[33] Zhang F, Jiang H Y, Li X Y, Wu X T, Li H X. Amine-functionalized GO as an active and reusable acid-base bifunctional catalyst for one-pot cascade reactions. ACS Catalysis, 2014, 4: 394-401.

[34] Li B Y, Zhang Y M, Ma D X, Li L, Li G H, Li G D, Shi Z, Feng S H. A strategy toward constructing a bifunctionalized MOF catalyst: Post-synthetic modification of MOFs on organic ligands and coordinatively unsaturated metal sites. Chemical Communications, 2012, 48: 6151-6153.

[35] Du J G, Tao M L, Zhang W Q. Fiber-supported acid-base bifunctional catalysts for efficient nucleophilic addition in water. ACS Sustainable Chemistry & Engineerin, 2016, 4: 4296-4304.

[36] Zhang F, Jiang H Y, Wu X T, Mao Z, Li H X. Organoamine-functionalized graphene oxide as a bifunctional carbocatalyst with remarkable acceleration in a one-pot multistep reaction. ACS Applied Materials & Interfaces, 2015, 7: 1669-1677.

[37] Sobhani S, Zarifi F. Pyridine-grafted graphene oxide: A reusable acid-base bifunctional catalyst for the one-pot synthesis of β-phosphonomalonates via a cascade Knoevenagel-phospha Michael addition reaction in water. RSC Advances, 2015, 5: 96532-96538.

[38] Alkordi M H, Liu Y, Larsen R W, Eubank J F, Eddaoudi M. Zeolite-like metal-organic frameworks as platforms for applications: On metalloporphyrin-based catalysts. Journal of the American Chemical Society, 2008, 130: 12639-12641.

[39] Dhakshinamoorthy A, Opanasenko M, Cejka J, Garcia H. Metal organic frameworks as heterogeneous catalysts for the production of fine chemicals. Catalysis Science & Technology, 2013, 3: 2509-2540.

[40] Wee L H, Bonino F, Lamberti C, Bordiga S, Martens J A. Cr-MIL-101 encapsulated Keggin phosphotungstic acid as active nanomaterial for catalysing the alcoholysis of styrene oxide. Green Chemistry, 2014, 16: 1351-1357.

[41] Garcia-Garcia P, Muller M, Corma A. MOF catalysis in relation to their homogeneous counterparts and conventional solid catalysts. Chemical Science, 2014, 5: 2979-3007.

[42] Fukuoka A, Dhepe P L. Catalytic conversion of cellulose into sugar alcohols. Angewandte Chemie International Edition, 2006, 45: 5161-5163.

[43] Dhepe P L, Fukuoka A. Cracking of cellulose over supported metal catalysts. Catalysis Surveys from Asia, 2007, 11: 186-191.

[44] Liu M, Deng W P, Zhang Q H, Wang Y L, Wang Y. Polyoxometalate-supported ruthenium nanoparticles as bifunctional heterogeneous catalysts for the conversions of cellobiose and cellulose into sorbitol under mild conditions. Chemical Communications, 2011, 47: 9717-9719.

[45] Onda A, Ochi T, Yanagisawa K. New direct production of gluconic acid from polysaccharides using a bifunctional catalyst in hot water. Catalysis Communications, 2011, 12: 421-425.

[46] Zhang J L, Liu X, Hedhili M N, Zhu Y H, Han Y. Highly selective and complete conversion of cellobiose to gluconic acid over $Au/Cs_2HPW_{12}O_{40}$ nanocomposite catalyst. ChemCatChem, 2011, 3: 1294-1298.

[47] An D L, Ye A H, Deng W P, Zhang Q H, Wang Y. Selective conversion of cellobiose and cellulose into gluconic acid in water in the presence of oxygen, catalyzed by polyoxometalate-supported gold nanoparticles. Chemistry-A European Journal, 2012, 18: 2938-2947.

[48] Chen J Z, Wang S P, Huang J, Chen L M, Ma L L, Huang X. Conversion of cellulose and cellobiose into sorbitol catalyzed by ruthenium supported on a polyoxometalate/metal-organic framework hybrid. ChemSusChem, 2013, 6: 1545-1555.

[49] Zhu F X, Wang W, Li H X. Water-medium and solvent-free organic reactions over a bifunctional catalyst with Au nanoparticles covalently bonded to HS/SO_3H functionalized periodic mesoporous organosilica. Journal of the American Chemical Society, 2011, 133: 11632-11640.

彩 图

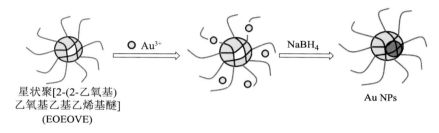

彩图 1　AuNPs@poly(EOEOVE)催化剂合成示意图

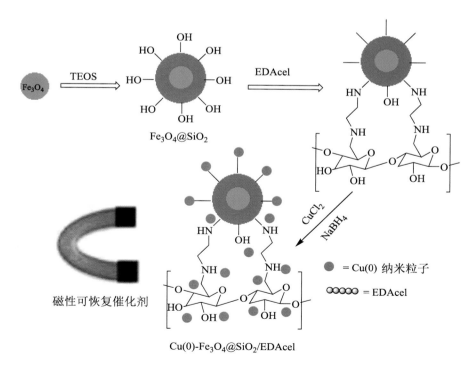

彩图 2　Cu(0)-Fe$_3$O$_4$@SiO$_2$/EDAcel 催化剂合成示意图

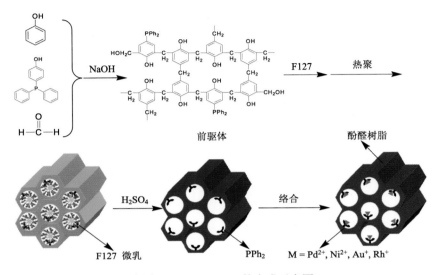

彩图 3　M-NCP-MPs 的合成示意图

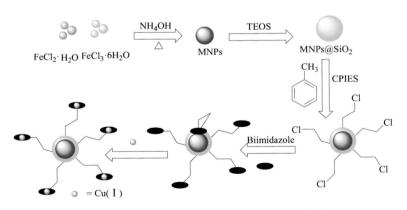

彩图 4　MNP@Biim Cu(Ⅰ)催化剂合成示意图

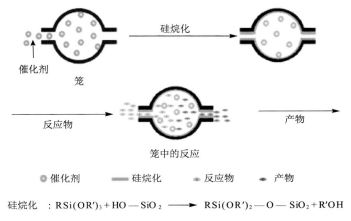

硅烷化：$RSi(OR')_3 + HO-SiO_2 \longrightarrow RSi(OR')_2-O-SiO_2 + R'OH$

彩图 5　TsDPEN-Ru/SBA-16 催化剂的制备示意图

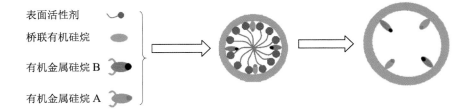

彩图 6　双有机金属多功能催化剂合成示意图

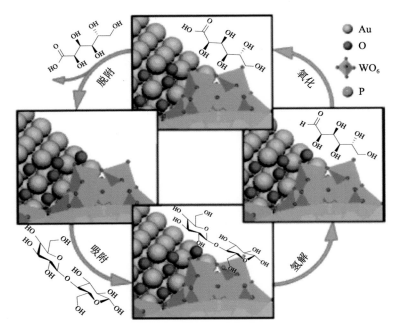

彩图 7 Au/Cs$_2$HPW$_{12}$O$_{40}$ 双功能催化剂催化纤维二糖水解氧化示意图

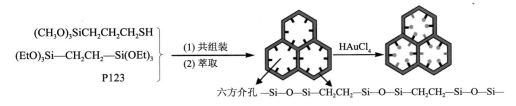

彩图 8 Au-SH/SO$_3$H-PMO(Et)催化剂的合成示意图（●=Au，▬=SO$_3$H，▬=SH）